FLUID MECHANICS

FLUID MECHANICS

David Pnueli and Chaim Gutfinger

Faculty of Mechanical Engineering
Technion – Israel Institute of Technology, Haifa, Israel

PUBLISHED BY THE PRESS SYNDICATE OF THE UNIVERSITY OF CAMBRIDGE
The Pitt Building, Trumpington Street, Cambridge CB2 1RP

CAMBRIDGE UNIVERSITY PRESS
The Edinburgh Building, Cambridge CB2 2RU, United Kingdom
40 West 20th Street, New York, NY 10011-4211, USA
10 Stamford Road, Oakleigh, Melbourne 3166, Australia

First published 1992
First paperback edition 1997

Library of Congress Cataloging-in-Publication Data is available.

A catalog record for this book is available from the British Library.

ISBN 0-521-41704-X hardback
ISBN 0-521-58797-2 paperback

Transferred to digital printing 2000

CONTENTS

PREFACE

This textbook is intended to be read by undergraduate students enrolled in engineering or engineering science curricula who wish to study fluid mechanics on an intermediate level. No previous knowledge of fluid mechanics is assumed, but the students who use this book are expected to have had two years of engineering education which include mathematics, physics, engineering mechanics and thermodynamics.

This book has been written with the intention to direct the readers to think in clear and correct terms of fluid mechanics, to make them understand the basic principles of the subject, to induce them to develop some intuitive grasp of flow phenomena and to show them some of the beauty of the subject.

Care has been taken in this book to strictly observe the chain of logic. When a concept evolves from a previous one, the connection is shown, and when a new start is made, the reason for it is clearly stated, e.g., as where the turbulent boundary layer equations are derived anew. As related later, there may be programs of study in which certain parts of this book may be skipped. However, from a pedagogical point of view it is important that these skipped parts are there and that the student feels that he or she has a complete presentation even though it is not required to read some particular section.

It is believed that an important objective of engineering education is to induce the students to think in clear and exact terms. In writing this book extreme effort has been made to present fluid mechanics in an exact manner, while at the same time to keep the mathematics under control such that the student's attention is not drawn away from the physics. In places where this delicate balance cannot be achieved, the physics prevails, and the student is told that some of that material has to be further considered in more advanced courses.

More effort has been made to reach a balance between exact and clear presentation, intuitive grasp of new ideas and creative application of these concepts. Some of this balance is achieved by following the new concepts, wherever possible, by examples of their applications, such that the new concept is reviewed from a less formal point of view.

Fluid mechanics has historically developed around the concept of an ideal, frictionless fluid. One reason for this is that the mathematics of frictionless flows is substantially simpler than that of viscous flow and that a fluid at rest really behaves like this. Also, *fluid statics* had been known long before all other branches of fluid mechanics. Another reason is that the development of *solid mechanics* has preceded that of the *mechanics of fluids*, and frictionless models in solid mechanics are in many cases fairly good approximations to reality. Moreover, in most cases these approximations may be later improved by the introduction of corrections and by modifications of coefficients.

The unqualified introduction of frictionless models in fluid mechanics may yield results which are totally wrong. The simplification resulting from the use of frictionless models should at best be applied by users who fully understand the limitations of the frictionless model. This can be achieved by first studying the mechanics of real viscous fluids and then considering simpler models which are special cases of the real fluid. This book starts with the real viscous fluid. It then shows that under certain particular conditions the ideal fluid models may yield acceptable approximations. By then, however, the student has a sufficient background and can place the ideal fluid model in its proper perspective.

The boundary layer concept is introduced as a necessary companion to ideal fluid flow. Slow flows are also treated as another extreme case of the general real fluid flow. One-dimensional compressible flow takes its place as both fast and compressible. An introduction to non-Newtonian flows is presented, with several examples solved.

Although the textbook has enough material for a two-semester course, it is understood that in most schools the realities brought about by the multitude of subjects in a modern engineering curriculum will limit the time allowed for fluid mechanics to a course of one semester. Hence, the instructor is faced with the decision of which parts of the book to exclude from the syllabus, while still providing a coherent structure on which the interested student may in the future fill in the gaps and expand his or her knowledge of the subject. To aid in this decision, some of the advanced topics that can be skipped in a basic exposition of the subject were marked with an asterisk. These topics are included in the book for the sake of completeness as well as for a comprehensive treatment of the subject in the framework of a more advanced course.

Students who read fluid mechanics for the first time may wonder why this subject seems sometimes more formal and less intuitive than solid mechanics. They are advised that most probably the encounter with fluids throughout their formal education is less extensive than the encounter with solids. A major objective of this book is to develop understanding and with it some intuitive grasp of fluid mechanics. Once this first obstacle is overcome, the pleasure of swimming in the sea of fluid mechanics compares favorably with that of walking in the field of solid mechanics.

1. INTRODUCTION

The Field of Fluid Mechanics

Fluid mechanics extends the ideas developed in mechanics and thermodynamics to the study of motion and equilibrium of fluids, namely of liquids and gases.

The beginner in the study of fluid mechanics may have some intuitive notion as to the nature of a fluid, a notion that centers around the idea of a fluid not having a fixed shape. This idea indicates at once that the field of fluid mechanics is more complex than that of solid mechanics. Fluid mechanics has to deal with the mechanics of bodies that continuously change their shape, or deform. Similarly, the ideas developed in classical equilibrium thermodynamics have to be extended to allow for the additional complexity of properties which vary continuously with space and time, normally encountered in fluid mechanics.

Fluid mechanics bases its description of a fluid on the concept of a continuum, with properties which have to be understood in a certain manner. So far, neither the idea of a continuum nor that of a fluid have been properly defined. We, therefore, begin with the explanation of what a continuum is and how local properties are defined. We then describe the various forces that act in a continuum leading to the definition of the concept of stress at a point. The concepts of stress and continuum are then used to define a fluid.

The Continuum

Consider a body of fluid, as shown in Fig. 1.1. Let A be a representative point in the fluid, and let S be a thermodynamic system which contains A. Let the volume of S be V and let the mass contained in it be m. The average density of the system S is defined as

$$\rho = \frac{m}{V}.$$

(1.1)

We now proceed to shrink V around A. If for a sufficiently small volume V_ε containing A there exists a limit such that

$$\lim_{V \to V_\varepsilon} \rho = \rho_A,$$

(1.2)

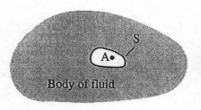

ρ_A is defined as the density at point A.

It should be noted that the existence of the limit in Eq. (1.2) is by no means assured. Furthermore, the limit ρ_A must be approached

Figure 1.1 Point A in a fluid.

for a finite V_ε. Indeed $V_\varepsilon \to 0$ is not considered at all, because for a very small V_ε the loss of single molecules while further decreasing V_ε may cause large fluctuations in ρ. The size of the "sufficiently small" V_ε must contain at least such a number of molecules that guarantees acceptable deviations from the equilibrium distributions of their velocities. These considerations may be repeated for another point B, resulting in ρ_B, and then for another point, and so on.

A fluid in a region which contains only points for which convergence occurs, i.e., in which the density is defined everywhere, is denoted *a continuum with respect to density*. A thermodynamic system defined inside such a region need not be in equilibrium, yet the density at each point in it is well defined. A fluid can also satisfy the requirement of being a continuum with respect to other fluid properties, e.g., *a continuum with respect to temperature*. A fluid which is a continuum with respect to all relevant properties, i.e., all properties of interest in the considered problem, shall be denoted a *continuum* with no further qualifications.

The requirements to be a continuum may be satisfied in one region but not in another or at one moment and not at another. Thus a fluid may be considered a continuum in parts of the region or part of the time only. The flow of a rarefied gas is a typical example of such a case. When the continuum model breaks down, some other models may become appropriate, e.g., those used in statistical mechanics or statistical thermodynamics. These are outside the scope of the book.

It is noted that the continuum model may have to be abandoned in cases where very fine details of the flow are required, i.e., in examples where regions smaller than V_ε are of interest.

Local Properties in a Continuum

The mental experiment which has led to the concepts of density and continuum with respect to density can be adjusted to yield some other properties, such as temperature, thermodynamic pressure, etc. The decision of what must be measured in these mental experiments is not always simple and is not necessarily unique. The emphasis, however, is on the realization that such an experiment can be performed and that the concept of, say, temperature can be defined as a result of a limiting process and with it the continuum with respect to temperature. The same, of course, holds for other measurable properties.

Properties which cannot be measured directly, but rather are calculated using other properties, are assumed to be here defined only after the measurable properties have been defined. With this understanding a continuum is now considered to have all its relevant properties defined everywhere. A somewhat picturesque description of what has been achieved up to this point is that of the fluid being divided into many small subregions, each of the order of V_ε in size and each having a list of its properties. This description is indeed correct. What follows is an elaboration, which does not increase accuracy but is more amenable to analysis. Hence, by changing somewhat the considered model now, fewer difficulties arise later in the analysis.

Denote the considered property, say, the density, by p. This p is known in any subregion of the continuum. Let a continuous function be defined such that for some reasonable subdivision of the region its values coincide with the values of p in the subregions. The definition of such a function does not add information but rather smoothes out the description. Let this continuous function also have continuous partial derivatives up to some order to be specified as needed. At a point where p is specified, this function coincides with its value. Between two adjacent points of specification in that particular division, the function interpolates between the two values and, therefore, does not change the approximation. By considering this function as p, instead of the previous discrete p values, no error is introduced. The continuous p function has, however, the advantage of having derivatives and being integrable, which proves invaluable in what follows.

A local property in a continuum is henceforth considered as the local value of this continuous function. What has been done so far is the establishment of a basis for translating physical reality into a form which is amenable to mathematical treatment. The part lost in this translation process is the molecular structure of matter. Hence, we consider regions of matter as continua of properties that change in a fairly smooth way. We may talk about density and temperature *fields* or velocity *profiles*. In describing the density of a continuum we disregard the fact that matter is composed of very dense protons and neutrons surrounded by regions of zero density. Similarly, while describing a velocity profile of a gas flowing in a pipe, we disregard the velocity of any given molecule and treat the velocity profile of the continuum as a continuous, well-behaved mathematical function.

Body and Surface Forces

Forces acting on a fluid divide naturally into body forces and surface forces. Body forces act at a distance and need not be associated with any transmission agent, or point of contact. In many cases they are the "effects" of external fields. Some typical examples of body forces are those of gravity, electrostatics and electromagnetism.

Surface forces are viewed as acting on any surface defined in the fluid, including its boundaries. Because each point in the fluid may belong to several surfaces, several such surface forces seem to coexist at the point. This is more complex than a single body force acting on a mass element, and therefore surface forces deserve some more elaborate consideration.

Stress at a Point

Let a surface passing through point A in a fluid be defined. Let a plane tangent to the surface pass also through A and a small disk be drawn on this plane, as shown in Fig. 1.2. This disk has two sides, one of which is assigned to the outer normal $\mathbf{n}$. Let the area of the disk be S and the force acting on its $\mathbf{n}$ surface be $\mathbf{F}$. The average stress acting on the surface is now defined as the force per unit surface area

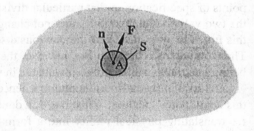

$$\mathbf{T} = \mathbf{F}\,/\,S. \qquad (1.3)$$

We now shrink S around A and look for a limit of the average stress. This limit is called the stress at point A. The considerations here are some-what similar to those leading to the definition of the concept of the local

Figure 1.2 A small disk tangent to a surface in the fluid at point A.

properties. The limit in Eq. (1.3) is a vector, which depends not only on the location of point A but also on the direction of the unit normal $\mathbf{n}$.

Henceforth, the term *stress* is used as a short form for *stress at a point*. The stress is a vector which depends on another vector $\mathbf{n}$. The component of the stress in the direction of this normal $\mathbf{n}$ is called *normal stress* and the component perpendicular to $\mathbf{n}$, i.e., which lies in the plane of the disk and is, therefore, tangent to the considered surface, is called *tangential stress* or *shear stress*.

Intuitively, the normal stress may be considered as that tending to pull the disk away from the surface while the shear stress is the one that tends to shear the disk off the surface while sliding on it tangentially.

Example 1.1

A body of water inside a container, as shown in Fig. 1.3, is moving to the right with a constant velocity of $U = 6$ m/s. The temperature in the water changes from $T = 20°\text{C}$ at the top to $T = 10°\text{C}$ at the bottom of the container.

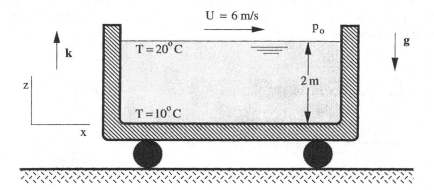

Figure 1.3 A body of water in a moving container.

Is the water a continuum with respect to density? With respect to temperature? What are the forces acting on the water? What are the stresses in the water?

Solution

The water is homogeneous and it is a continuum with respect to density. Its density, ρ, is constant, hence well defined at every point. Although the water temperature does change from top to bottom, still, we do not expect it to be discontinuous; hence the water is a continuum with respect to temperature too. The pressure in the water changes continuously from p_o at the top surface, increasing as we go down. The water is, therefore, a continuum with respect to pressure. In Chapter 3 we'll show that the *hydrostatic* pressure at any point in the body of water is given as $p = p_o + \rho g(2 - z)$.

The forces acting on the water are *body forces* and *surface forces*. The body force, $\mathbf{G}$, is due to gravity, $\mathbf{g}$; it is proportional to the mass of the water, and its direction is that of the vector $\mathbf{g}$, i.e., in the $(-\mathbf{k})$-direction,

$$dG = g\,dm = -kg\,\rho\,dV$$

The surface force **S** acting on any plane in the water is due to the normal stress, S_n, as explained in the next chapter. There are no shear stresses in this body of fluid.

Definition of a Fluid

We are now in a position to introduce the definition of a fluid. A fluid is defined as a *continuum which cannot support a shear stress while at rest*. The terms *continuum* and *shear stress* which appear here make it obvious why this definition had to wait until now.

top plate

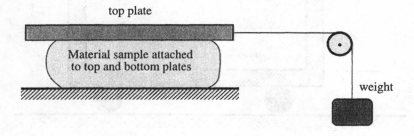

weight

Figure 1.4 Material subjected to shear stress.

Consider an experiment designed to use this definition to decide whether a given material is a fluid. Let the material be placed between two parallel plates and be subjected to an external shearing stress. An example of such an experiment is shown in Fig. 1.4, where the shear force is applied by means of a weight. A fluid should not be able to support *any* shear stress while at rest. The material will be considered a fluid if the top plate moves to the right as long as a shear stress is applied, no matter how small.

Example 1.2

A cylindrical rod is left on the surface of still water in a tank. It sinks to the bottom, Fig. 1.5. Neglecting forces at the bottom of the rod, which is thin and long, it is the shear stresses on its circumference which resist the sinking.

The experiment is repeated with tar instead of water. The same results are obtained, but the sinking takes many hours. Is water a fluid? Is tar?

Solution

The water is a fluid. It cannot support shear stresses while at rest. The decision concerning the tar depends on what the observer considers "at rest."

If the whole situation is of interest only for a few seconds, the tar is approximately at rest; it does support shear stresses, and a model of a solid fits better its behavior. However, for longer periods, e.g., to build a house on it, fluid mechanics considerations must be used and the tar must be considered a fluid. (Can a house be put safely on tar?)

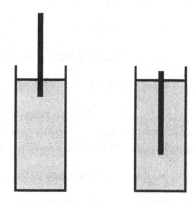

Figure 1.5 Cylindrical rod before and after sinking.

Units and Dimensions

Fluid mechanics, as many other branches of science and engineering, cannot be completely formulated by the use of pure numbers. Therefore, dimensions and basic units must be employed. The dimensions are considered modifiers for the pure numbers; the pure numbers describe quantities, i.e., "how much," while the dimensions tell us "of what." With this in mind unit vectors are also identified as modifiers, which tell us "in what direction." Thus a vector is given by its three components, which are pure numbers, modified by the system's three unit vectors, and with some possible additional modifiers – the physical dimensions of the vector, i.e., what it represents.

Transformations between different systems of units as well as between different systems of coordinates are possible. The point of view underlying these transformations is that all observers must see the *same physical phenomena.* The observers may choose their own means (units, coordinate systems) of reporting what they see, but they may not tamper with the phenomenon itself. Thus, if one "wouldn't touch it with a ten-foot pole" he or she would not touch it with a 3.048-meter pole either. Because no matter what units are used to report the length of the pole, it is still the same pole. Similarly, a vector viewed in different coordinate systems remains the same, no matter what systems the observer happens to favor.

We now return to the 10-foot pole and find what was involved in turning it

8   Fluid Mechanics

into a 3.048 meter stick. The original equation for the length l of the pole was

$$l = 10 \text{ ft.} \tag{1.4}$$

We also know that

$$1 \text{ ft} = 0.3048 \text{ m.} \tag{1.5}$$

Treating the dimension "ft" as an algebraic quantity, we now rewrite Eq. (1.5) as

$$1 = 0.3048 \text{ m/ft.} \tag{1.6}$$

As seen from the left-hand side of Eq. (1.6), the value of 0.3048 m/ft is unity (a pure number), and multiplication by 1 is always permissible. Performing this multiplication on Eq. (1.4) results in the length of the pole in meters:

$$l = 10 \text{ ft} \times 0.3048 \text{ m/ft} = 3.048 \text{ m,} \tag{1.7}$$

where again physical dimensions are treated as algebraic quantities.

Conversion factors similar to that of Eq. (1.6) are presented in Appendix A. The following example illustrates the use of these conversion factors.

Example 1.3

A small economy car runs 12 km per liter of gasoline. What is the car mileage in MPG (miles per gallon)?

Solution

Using the conversion factors in Appendix A,

$$12 \frac{\text{km}}{\text{L}} \times \frac{3.7854 \text{ L/gal}}{1.609 \text{ km/mi}} = 28.23 \frac{\text{mi}}{\text{gal}}.$$

The decision whether to multiply or divide by "one," i.e., by the conversion factor, is made such that the desired units are obtained after cancellation.

In principle the description of a vector in various coordinate systems, i.e., the conversion from one coordinate system to another, is analogous to the conversion of units. However, because vector algebra does not have the operation of division, one is restricted to multiplication only, and more details of this operation are necessary.

Dimensional Homogeneity

An equation is said to be *dimensionally homogeneous* if it does not depend on the system of units used. A dimensionally homogeneous equation has, therefore, the same dimensions for each term on both sides of the equation. In practice, this means that the numerical constants appearing in the equation are dimensionless. Thus, the equation

$$h = \tfrac{1}{2}gt^2 \tag{1.8}$$

is dimensionally homogeneous and the constant 1/2 is dimensionless. This homogeneous equation holds for any system of units.

A common practice in engineering is to write equations in a nonhomogeneous form. If, for example, we substitute in Eq. (1.8) the value $g = 32.174$ ft/s^2, we obtain

$$h = 16.09t^2. \tag{1.9}$$

This equation, although easier to use for repeated calculations, is nonhomogeneous and holds only for the case where t is given in seconds and h in feet. Here the constant 16.09 has dimensions of ft/s^2. In order to convert Eq. (1.9) to a different set of units, a new constant, c, has to be recalculated using the method outlined above. Thus, in SI units (i.e., International System of Units) this constant will be

$$c = 16.09 \text{ ft/s}^2 \times 0.3048 \text{ m/ft} = 4.90 \text{ m/s}^2, \tag{1.10}$$

resulting in a new dimensionally nonhomogeneous equation

$$h = 4.90t^2. \tag{1.11}$$

As demonstrated by this example, nonhomogeneous equations arise for practical reasons, such as to shorten repeated calculations or use more easily available data. It is rarely used in scientific presentations and, when used, must always be followed by a clear statement as to what units are employed.

Example 1.4

Atmospheric pressure at sea level is $p_a = 14.7$ psia, where the term "psia" stands for pounds force per square inch absolute. What is the pressure in N/m^2, i.e., in Pa (pascals)?

Solution

Using the conversion factors in Appendix A the pressure is

$$p_a = 14.7 \text{ lbf/in.}^2 = 14.7 \text{ lbf/in.}^2 \times 144 \text{ in.}^2 /\text{ft}^2 \times 10.764 \text{ ft}^2/\text{m}^2 \times 4.448 \text{ N/lbf}$$
$$= 101,300 \text{ N/m}^2 = 101,300 \text{ Pa.}$$

Fluid Properties

In this section some of the basic fluid properties used in this book are described.

Density, Specific Weight and Specific Gravity

The property density was used in the discussion of the concept of a continuum. It has been defined as

$$\rho = \lim_{V \to V_\varepsilon} \frac{m}{V}.$$

(1.12)

The units of density are mass per unit volume. Hence, in SI units density is measured in kilograms per cubic meter (kg/m^3). Other units frequently used for density are g/cm^3, lbm/ft^3 and $slug/ft^3$.

The density of water at room temperature and atmospheric pressure is

$$\rho = 1,000 \text{ kg/m}^3 = 1 \text{ g/cm}^3 = 62.4 \text{ lbm/ft}^3 = 1.94 \text{ slug/ft}^3.$$
(1.13)

The density of a fluid may change from point to point. However, if the fluid is of a single phase, the change in density will be almost everywhere continuous. The density may depend on both the pressure and the temperature. A fluid whose density may be assumed not to vary with pressure is called incompressible. Most liquids may be considered incompressible.

The reciprocal of the density is the specific volume, v:

$$v = \frac{1}{\rho}$$
(1.14)

A quantity related to the density is the specific weight, γ, defined as the weight per unit volume. Hence,

$\gamma = \rho g.$ \hfill (1.15)

The SI units of specific weight are N/m³. The specific weight of water at room temperature and atmospheric pressure is

$\gamma = 9806 \text{ N/m}^3 = 62.4 \text{ lbf/ft}^3.$ \hfill (1.16)

Occasionally one may encounter a term "specific gravity" which is defined as the ratio of the density of a given fluid to that of water. Specific gravity is therefore dimensionless, and its numerical value is equal to that of the density expressed in CGS (centimeter-gram-second) units, i.e., in g/cm³.

Viscosity

While deciding whether a given material is a fluid, we used an experiment in which the material was placed between two parallel plates, Fig. 1.4. One of the plates was held stationary while the other could move pulled by a force acting in a direction parallel to the plate. We have stated that if the plate moves, the material is considered a fluid. Now, the speed at which the plate moves depends on a property of the fluid called *viscosity*. The more viscous the fluid is, the slower the plate moves.

The force per unit plate area, i.e., the shear stress, T, exerted on the plate is proportional to the plate velocity V and inversely proportional to the gap between the plates, h:

$$T = \frac{F}{A} \propto \frac{V}{h}.$$ \hfill (1.17)

The proportionality constant that converts Eq. (1.17) into an equality is called *dynamic viscosity* and is denoted by μ. Thus, for that particular experiment

$$T = \mu \frac{V}{h}.$$ \hfill (1.18)

Equation (1.18) can be generalized for two adjacent fluid layers separated by a distance dy, both moving in the x-direction. If the difference in velocities between the layers is du, then Eq. (1.18) becomes

$$T = \mu \left(\frac{du}{dy} \right).$$ \hfill (1.19)

Equation (1.19) is sometimes referred to as Newton's law of viscosity. The term du/dy is called the rate of strain, or simply shear rate. The relationships between shear stress and the rate of strain are considered in detail in Chapter 5.

The dimensions of viscosity must be chosen such that Eqs. (1.19) and (1.18) are rendered homogeneous. Hence, as the shear stress, T, is given in N/m² and the shear rate in s⁻¹, the unit of viscosity becomes

$$1 \, [\mu] = 1 \, \frac{N \cdot s}{m^2} = 1 \, \frac{kg}{m \cdot s}. \tag{1.20}$$

In the British system the unit of viscosity is lbm/ft·s, while in the CGS system it is g/cm·s and is called the poise. Thus,

$$1 \, \text{poise} = 1 \frac{g}{cm \cdot s} = 0.1 \frac{kg}{m \cdot s}. \tag{1.21}$$

The viscosity of water at room temperature is roughly 0.01 poise, or 1 centipoise, which is written as 1 cp.

The viscosity of gases increases with temperature. This is due to the more vigorous molecular movement at higher temperatures. In liquids the viscosity depends mainly on the intermolecular cohesive forces. The forces between the molecules of a liquid decrease with an increase in temperature; so does the viscosity.

The ratio of the dynamic viscosity, μ, of a fluid to its density, ρ, is called *kinematic viscosity* and is denoted by ν,

$$\nu = \frac{\mu}{\rho}. \tag{1.22}$$

The unit of kinematic viscosity is cm²/s and is called the *stoke*. It is related to other units of kinematic viscosity as

$$1 \, m^2/s = 10^4 \, st = 10.764 \, ft^2/s. \tag{1.23}$$

Compressibility and Bulk Modulus of Elasticity

Fluid compressibility, κ, is defined as a relative change in fluid volume under the action of an external pressure, i.e.,

$$\kappa = -\frac{1}{V_o} \frac{\Delta V}{\Delta p}. \tag{1.24}$$

A more careful definition specifies under what conditions the compression occurs.* Hence, isothermal compressibility is defined as

$$\kappa_T = -\frac{1}{V} \left(\frac{\partial V}{\partial p} \right)_T. \tag{1.25}$$

while adiabatic compressibility performed reversibly (i.e., isentropically) is

---

* See, for example, A. Shavit and C. Gutfinger, *"Thermodynamics – From Concepts to Applications"*, Prentice-Hall International, 1995., pp. 291 – 292.

$$\kappa_s = -\frac{1}{V}\left(\frac{\partial V}{\partial p}\right)_s. \tag{1.26}$$

Equations (1.24) and (1.26) may be rewritten in terms of densities as

$$\kappa_T = \frac{1}{\rho}\left(\frac{\partial \rho}{\partial p}\right)_T \tag{1.27}$$

and

$$\kappa_s = \frac{1}{\rho}\left(\frac{\partial \rho}{\partial p}\right)_s. \tag{1.28}$$

The units of compressibility are reciprocal of those of pressure, e.g., atm^{-1}, psi^{-1}, m^2/N, etc. For example, the compressibility of water at room temperature and atmospheric pressure is 5×10^{-5} atm^{-1}. Hence, a pressure increase of 1 atm results in a relative volume reduction of 1/20,000.

The compressibility of a gas may be found from the perfect gas relationship

$$p = \rho RT, \tag{1.29}$$

which upon substitution into Eq. (1.27) yields, for isothermal compressibility,

$$\kappa_T = \frac{1}{p}. \tag{1.30}$$

Adiabatic compressibility is found by using the isentropic relationship

$$p\rho^{-k} = \text{const}, \tag{1.31}$$

where $k = c_p/c_v$. Equation (1.31) together with Eq. (1.28) result in the following expression for adiabatic compressibility of a perfect gas:

$$\kappa_s = \frac{1}{pk}. \tag{1.32}$$

The reciprocal of compressibility is known as the *volumetric* or *bulk modulus of elasticity*. The isothermal bulk modulus of elasticity is defined as

$$E_T = \frac{1}{\kappa_T} = \rho\left(\frac{\partial p}{\partial \rho}\right)_T. \tag{1.33}$$

Similarly, the isentropic bulk modulus is

$$E_s = \frac{1}{\kappa_s} = \rho\left(\frac{\partial p}{\partial \rho}\right)_s. \tag{1.34}$$

The bulk modulus of elasticity is expressed in units of pressure. The concepts of compressibility and bulk modulus of elasticity are used in the study of compressible flow. In calculations of flow phenomena liquids are usually considered incompressible. The assumption of incompressibility holds for gases at low speed only. In many flows, at speeds of up to 30% of the speed of sound, a gas may be treated as incompressible. The concept of compressibility is considered some more in connection with the equation of continuity, in Chapter 5, and in Chapter 13 on compressible flow.

Problems

1.1   In a certain city the amount of money people carry in their pockets is between $1 and $1,000. The average amount per person depends on the zone of the city. In trying to represent the city as a continuum with respect to solvency, i.e., amount of money per person, find the smallest number of people you have to include in V_ε such that the error caused by one person going in or out of V_ε is less than 1%.

1.2   A cylindrical container is filled with water of density $\rho = 1,000\,\text{kg/m}^3$ up to the height $h = 5\,\text{m}$, Fig. P1.2. The outside pressures $p_o = 10^5$ Pa.

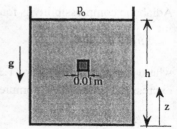

a. Is the water a continuum with respect to density? Is it with respect to pressure? What is the height Δz_ε of your V_ε if a deviation of 0.1% in p is negligible?

b. Is the water a simple thermodynamic system in equilibrium?

Figure P1.2 Water in a tank and small cube.

c. How are the answers modified for an outside pressure of $p_o = 6 \times 10^7$ Pa?

d. The top layer of the water is held for a long time at 310 K, while the bottom is held at 290 K. Is it a continuum with respect to temperature? Explain.

1.3   A body of water in the shape of a cube is selected inside the container of Fig. P1.2 such that the lower side of the cube coincides with $z = 2.0$ m. The side of the cube is 0.01 m.

a. What is the body force acting on the fluid inside the cube?

b. What are the surface forces acting on the six sides? What are the stresses?

c. Is the whole cube in mechanical equilibrium? Is it a stable equilibrium?
d. What are the forces and the stresses if the cylinder is put in space, i.e., for $\mathbf{g}=0$?
e. What are the forces if the cylinder falls freely? How would you keep the water together?

1.4   A boxlike block of wood has the dimensions of $1\,m \times 2\,m \times 3\,m$. The density of the wood is $\rho = 800\ kg/m^3$. The coefficient of dry friction between wood and concrete is 0.4, i.e., when the block is drawn on a concrete floor, it is pulled with a force of 0.4 N per each 1 N force pushing the block normal to the floor. Calculate and draw the average stress on that side of the block which touches the floor when this side is:
a. The $1\,m \times 2\,m$ side.
b. The $2\,m \times 3\,m$ side.
Note that stress is a vector.

1.5   A certain oil has the viscosity of 2 poise. Its density is 62 lbm/ft³. What is its kinematic viscosity in m²/s?

1.6   A certain slurry is filtered at constant pressure at the rate of

$$\dot{V} = \frac{52.5V + 6.2}{t},$$

where $\dot{V}$ [liters/s] is the volumetric flowrate, t [s] is the time and V [liters] is the filtered volume. Is the equation dimensionally homogeneous? Rewrite the equation with V in cubic feet. Can you rewrite the equation in a homogeneous form?

1.7   A falling body has its z-coordinate change in time as

$$z = z_o - 4.9t^2,$$

where t [s] is the time and z [m] is the height. Is the equation dimensionally homogeneous? Rewrite the equation with z [ft].
a. Can you rewrite the equation in a homogeneous form. Why is the answer here different from that in Problem 1.6?
b. Why do dimensionally homogeneous equations give more information? What is this information?

1.8   The following dimensionless numbers are defined:

Re = $dU\rho/\mu$, Reynolds number;
Pr = $\mu c_p/k$, Prandtl number;
Pe = Re·Pr, Peclet number;

where $U = 2$ m/s is the flow velocity in the pipe, $d = 2$ in. is the pipe diameter, $\rho = 1,000$ kg/m³ is the fluid density, $\mu = 3$ cp is its viscosity, $c_p = 0.5$ Btu/(lbm·ºF) is the specific heat of the fluid and $k = 0.65$ W/m·ºC is its thermal conductivity.
What are the numerical values of Re, Pr and Pe? In what system of units?

1.9 Using Appendix A, change into the S.I. system:

density	$\rho = 120$ lbm/ft³,
thermal conductivity	$k = 170$ Btu/(hr·ft·ºF),
thermal convection coefficient	$h = 211$ Btu/(hr·ft²· F),
specific heat	$c_p = 175$ Btu/(lbm·ºF),
viscosity	$\mu = 20$ cp,
viscosity	$\mu = 77$ lbf·s/ft²,
kinematic viscosity	$v = 3$ ft²/s,
Stefan–Boltzmann constant	$\alpha = 0.1713 \times 10^{-8}$ Btu/(ft²·hr·ºR⁴),
acceleration	$a = 12$ ft/s².

1.10 The following empirical equation gives the wall shear stress exerted on a fluid flowing in a concrete pipe:

$$\tau_w = 0.0021\rho V^2 r^{-\frac{1}{3}},$$

where τ_w is the shear stress in lbf/in.², ρ the fluid density in slug/ft³, V the average velocity of the fluid in ft/s and r the hydraulic radius of the pipe in feet. Rewrite the equation in terms of SI units.

1.11 The distance between the plates in an experimental system, as shown in Fig. 1.4, is $h = 1$ in. When the upper plate is pulled with the velocity $V = 40$ ft/min, the shear stress is $T = 12$ lbf/ft². Using Eq. (1.18) find the viscosity of the fluid, μ, in SI units.

1.12 A metal sphere of 1 ft in diameter is put on a scale. The scale shows a reading of 200 kg. Find the volume, the mass, the density, the specific volume and the specific weight of the sphere in:
a. SI units.
b. British units.

1.13   A volume of 30 liters of alcohol, subjected to a pressure of 500 atm at 25°C contracts to 28.8 liters.
   a. What is the modulus of elasticity of alcohol?
   b. What is its compressibility?

2. STRESS IN A FLUID

In this chapter we consider stresses in a fluid. We start by setting the forces resulting from these stresses in their proper perspective, i.e., in relation to body forces, together with which they raise accelerations. This results in a set of momentum equations, which are needed later.

We then consider the relations between the various stress components and proceed to inspect stress in fluids at rest and in moving fluids.

The Momentum Equations

In this section we establish relations between body forces, stresses and their corresponding surface forces and accelerations. Newton's second law of motion is used, and the results are the general momentum equations for fluid flow.

Consider a system consisting of a small cube of fluid, as shown in Fig. 2.1. A *system* is defined in classical thermodynamics as a given amount of matter with well-defined boundaries. The system always contains the same matter and none may flow through its boundaries.

As a rule a fluid system does not retain its shape, unless, of course, the fluid is at rest. This does not prevent the choice of a system with a certain particular shape, e.g., a cube. The choice means that imaginary surfaces are defined inside the fluid such that at the considered moment they enclose a system of fluid with a given shape. A moment later the system may have a different shape, because the shape is not a property of the fluid or of the location. Systems of different shapes may be chosen simultaneously at the same point, and the choice is made for convenience. Thus, two different shapes are chosen in this chapter, a cube and a tetrahedron, each leading conveniently to some particular conclusions.

Let a set of cartesian coordinates be selected and a system of fluid in the form of a small cube be defined, with its planes parallel to the coordinates directions as shown in Fig. 2.1.

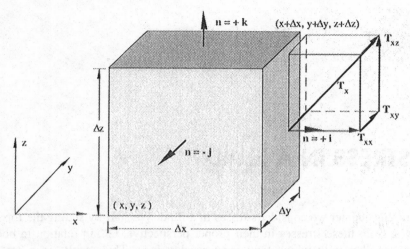

Figure 2.1 Small cube.

Now let all forces acting on this cube be considered. Using Newton's second law we may write for our system

$$m\mathbf{a} = \mathbf{G} + \mathbf{S},\qquad(2.1)$$

where $\mathbf{G}$ denotes the total body force acting on the cube and $\mathbf{S}$ is the surface force resulting from the stresses acting on the six sides of the cube.

Equation (2.1) is a vectorial equation and may be rewritten in component form. The x-component of Eq. (2.1) is

$$m a_x = G_x + S_x.\qquad(2.2)$$

Each surface of the cube is named after the direction of its outer normal $\mathbf{n}$. Thus, there is a positive x-surface located at $x + \Delta x$ in Fig. 2.1 and a negative x-surface at x. These two surfaces are denoted, respectively, as

$\mathbf{n} = \mathbf{i}$ for positive x-surface,
$\mathbf{n} = -\mathbf{i}$ for negative x-surface.

Similarly, the two y-surfaces located at $y + \Delta y$ and at y are $\mathbf{n} = \pm\mathbf{j}$, respectively; and the z-surfaces are $\mathbf{n} = \pm\mathbf{k}$. We denote the stress vector acting on a surface with the outer normal $\mathbf{n}$ by $\mathbf{T}_n$. Newton's third law then requires

$$\mathbf{T}_n = -\mathbf{T}_{-n}.\qquad(2.3)$$

We denote the stress vector acting on the positive x-surface of the cube by

$\mathbf{T}_x$, where the subscript x denotes the face of the cube. The force acting on this side is then

$$\mathbf{F}_x = \mathbf{T}_x \, \Delta y \, \Delta z \tag{2.4}$$

The term $\mathbf{F}_x$ can be resolved into components,

$$\mathbf{F}_x = \mathbf{i}(F_x)_x + \mathbf{j}(F_x)_y + \mathbf{k}(F_x)_z = \left[\mathbf{i}(T_x)_x + \mathbf{j}(T_x)_y + \mathbf{k}(T_x)_z \right] \Delta y \, \Delta z. \tag{2.5}$$

It is customary in fluid dynamics to drop the parentheses around the $\mathbf{F}_i$ and $\mathbf{T}_i$, with the understanding that the first subscript indicates the surface on which the force or the stress acts while the second subscript is used to show the direction of the component. Thus, T_{xx}, T_{xy} and T_{xz} are, respectively, the x-, y- and z-components of $\mathbf{T}_x$, which is the stress acting on the positive x-surface. Here T_{xx} is identified as the *normal* stress on the x-surface of the cube, while T_{xy} and T_{xz}, lying in the x-plane itself, are recognized as *shear stresses*.

The force on the positive x-surface, i.e., the one located at $x+\Delta x$, is therefore

$$F_x \big|_{x+\Delta x} = \left[\mathbf{i} T_{xx} + \mathbf{j} T_{xy} + \mathbf{k} T_{xz} \right] \Delta y \, \Delta z \tag{2.6}$$

while the force on the negative x-surface is

$$-F_x \big|_x = -\left[\mathbf{i} T_{xx} + \mathbf{j} T_{xy} + \mathbf{k} T_{xz} \right] \Delta y \, \Delta z \,. \tag{2.7}$$

Similarly, we have on the other four surfaces

$$
\begin{aligned}
F_y \big|_{y+\Delta y} &= \left[\mathbf{i} T_{yx} + \mathbf{j} T_{yy} + \mathbf{k} T_{yz} \right] \Delta x \, \Delta z, \\
-F_y \big|_y &= -\left[\mathbf{i} T_{yx} + \mathbf{j} T_{yy} + \mathbf{k} T_{yz} \right] \Delta x \, \Delta z, \\
F_z \big|_{z+\Delta z} &= \left[\mathbf{i} T_{zx} + \mathbf{j} T_{zy} + \mathbf{k} T_{zz} \right] \Delta x \, \Delta y, \\
-F_z \big|_z &= -\left[\mathbf{i} T_{zx} + \mathbf{j} T_{zy} + \mathbf{k} T_{zz} \right] \Delta x \, \Delta y.
\end{aligned}
\tag{2.8}
$$

We may now compute the x-component of the resultant surface force $\mathbf{S}$ acting on all the surfaces of the cube, Eq. (2.2). It is comprised of the normal stresses acting on the x-surfaces and the shear stresses acting on the y- and z-surfaces,

$$S_x = \left[T_{xx} \big|_{x+\Delta x} - T_{xx} \big|_x \right] \Delta y \, \Delta z + \left[T_{yx} \big|_{y+\Delta y} - T_{yx} \big|_y \right] \Delta x \, \Delta z + \left[T_{zx} \big|_{z+\Delta z} - T_{zx} \big|_z \right] \Delta x \, \Delta y. \tag{2.9}$$

Equation (2.9) may be rewritten as

$$S_x = \left[\frac{T_{xx} \big|_{x+\Delta x} - T_{xx} \big|_x}{\Delta x} + \frac{T_{yx} \big|_{y+\Delta y} - T_{yx} \big|_y}{\Delta y} + \frac{T_{zx} \big|_{z+\Delta z} - T_{zx} \big|_z}{\Delta z} \right] \Delta x \, \Delta y \, \Delta z. \tag{2.10}$$

Similar expressions may be formed for S_y and S_z.

Returning to Eq. (2.2), let a general body force per unit mass **g** be defined. The x-component of this body force acting on the cube is

$$G_x = g_x \rho \, \Delta x \, \Delta y \, \Delta z. \tag{2.11}$$

The left-hand side of Eq. (2.2) is written for the small cube as

$$ma_x = a_x \rho \, \Delta x \, \Delta y \, \Delta z. \tag{2.12}$$

Substitution of Eqs. (2.10), (2.11) and (2.12) into (2.2) and division by $\Delta x \, \Delta y \, \Delta z$ result in

$$\rho a_x = \rho g_x + \frac{T_{xx}|_{x+\Delta x} - T_{xx}|_x}{\Delta x} + \frac{T_{yx}|_{y+\Delta y} - T_{yx}|_y}{\Delta y} + \frac{T_{zx}|_{z+\Delta z} - T_{zx}|_z}{\Delta z}.$$

Shrinking the cube into a point, i.e., taking the limit as $\Delta x \to 0$, $\Delta y \to 0$ and $\Delta z \to 0$, leads to

$$\rho a_x = \rho g_x + \frac{\partial T_{xx}}{\partial x} + \frac{\partial T_{yx}}{\partial y} + \frac{\partial T_{zx}}{\partial z}. \tag{2.13}$$

Similarly, for the y- and z-directions,

$$\rho a_y = \rho g_y + \frac{\partial T_{xy}}{\partial x} + \frac{\partial T_{yy}}{\partial y} + \frac{\partial T_{zy}}{\partial z}, \tag{2.14}$$

$$\rho a_z = \rho g_z + \frac{\partial T_{xz}}{\partial x} + \frac{\partial T_{yz}}{\partial y} + \frac{\partial T_{zz}}{\partial z}. \tag{2.15}$$

Equations (2.13), (2.14) and (2.15) are, respectively, the x-, y- and z-components of the balance between forces and rates of change of momentum for a cube whose sides approach zero. They therefore hold at any point in the continuum and are called *momentum equations*.

Index Notation

Equations (2.13), (2.14) and (2.15) may be written in a more compact form by using index notation.

Let the subscript i have the *range* $i = 1,2,3$, x_i stand for $x_1 = x$, $x_2 = y$, $x_3 = z$ and g_i for g_x, g_y, g_z. Let also a repeated subscript indicate *summation* over that subscript, e.g.,

$$A_{ji}x_j = A_{1i}x_1 + A_{2i}x_2 + A_{3i}x_3, \qquad i = 1,2,3.$$
$$(2.16)$$

Making use of these *range* and *summation* conventions, the momentum equations (2.13) - (2.15) may be written as

$$\rho a_i = \rho g_i + \frac{\partial T_{ji}}{\partial x_j}.$$
$$(2.17)$$

The momentum equation, as given in Eq. (2.17), is entirely equivalent to Eqs. (2.13) - (2.15). It is yet another form of Newton's second law of motion as applied to fluids.

Equation (2.17) includes the acceleration **a**. From mechanics of solid bodies one expects this acceleration to be eventually expressed by the derivatives of the velocity of the system. But to take these derivatives, one must know where the thermodynamic system is at the time $t + \Delta t$, which is not known. Moreover, connections between the stress components, all nine of them, and the velocity must also be established, or Eq. (2.17) could not be solved. Both these problems, i.e., that of the acceleration and that of relating stresses to velocities, are treated in Chapter 5.

Because both i and j in $\mathbf{T}_{ij}$ have each the range 3, $\mathbf{T}_{ij}$ represents nine numbers, the so-called *stress components*, which may be arranged in a matrix form, the *stress matrix*:

$$\left[T_{ij}\right] = \begin{bmatrix} T_{11} & T_{12} & T_{13} \\ T_{21} & T_{22} & T_{23} \\ T_{31} & T_{32} & T_{33} \end{bmatrix} = \begin{bmatrix} T_{xx} & T_{xy} & T_{xz} \\ T_{yx} & T_{yy} & T_{yz} \\ T_{zx} & T_{zy} & T_{zz} \end{bmatrix}.$$

We now proceed to establish relations between the nine stress components at a point in one cartesian coordinate system and those at the same point in other cartesian systems, rotated with respect to the original one.

Moments on a Cube

In this section we show that the stress matrix is symmetrical, in the sense that $T_{yx} = T_{xy}$. This important feature of the stress matrix applies both for stationary and moving fluids.

Let the moments acting on the small cube be now computed. To obtain the x-component of the moments, we imagine a shaft piercing the centers of the $\mathbf{n} = \mathbf{i}$ and the $\mathbf{n} = -\mathbf{i}$ sides of the cube, Fig. 2.2, and compute the turning moments around this shaft.

The only contributions to the turning moment come from the shear stresses. For greater clarity these shear stresses are shown in detail in Fig. 2.2. We employ here the convention of taking the stresses represented by the arrows as positive when they act on positive surfaces, $\mathbf{n} = +\mathbf{j}$ and $\mathbf{n} = +\mathbf{k}$. Newton's third law, that of action and reaction, then implies that the stresses acting on the negative surface $\mathbf{n} = -\mathbf{j}$ and $\mathbf{n} = -\mathbf{k}$ are negative.

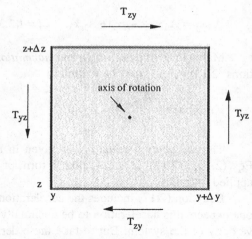

Figure 2.2 Shear stresses on a cube.

Note that the cube is small and in the limit it shrinks to a point. Computing moments we obtain

$$M_x = \left[T_{yz}\big|_{y+\Delta y}\, \Delta x\, \Delta z \right] \frac{\Delta y}{2} + \left[T_{yz}\big|_{y}\, \Delta x\, \Delta z \right] \frac{\Delta y}{2}$$
$$- \left[T_{zy}\big|_{z+\Delta z}\, \Delta x\, \Delta y \right] \frac{\Delta z}{2} - \left[T_{zy}\big|_{z}\, \Delta x\, \Delta y \right] \frac{\Delta z}{2} \qquad (2.18)$$
$$= \Delta x\, \Delta y\, \Delta z \left[\frac{1}{2}\left(T_{yz}\big|_{y+\Delta y} + T_{yz}\big|_{y} \right) - \frac{1}{2}\left(T_{zy}\big|_{z+\Delta z} + T_{zy}\big|_{z} \right) \right].$$

Newton's second law of motion states that the angular acceleration ω_x of the cube is

$$M_x = I_x \omega_x, \qquad (2.19)$$

where I_x is the polar moment of inertia of the cube around the piercing axis, i.e.,

$$I_x = \rho\, \Delta x \int_{-\frac{\Delta y}{2}}^{\frac{\Delta y}{2}} \int_{-\frac{\Delta z}{2}}^{\frac{\Delta z}{2}} \left(y^2 + z^2 \right) dz\, dy = \frac{\rho}{12} \Delta x\, \Delta y\, \Delta z \left[(\Delta y)^2 + (\Delta z)^2 \right]. \qquad (2.20)$$

Substitution of Eqs. (2.18) and (2.20) into Eq. (2.19) and division by the volume of the cube, $\Delta x\, \Delta y\, \Delta z$, yield

$$\frac{1}{2}\left(T_{yz}\big|_{y+\Delta y} + T_{yz}\big|_{y} \right) - \frac{1}{2}\left(T_{zy}\big|_{z+\Delta z} + T_{zy}\big|_{z} \right) = \left[(\Delta y)^2 + (\Delta z)^2 \right] \frac{\rho \omega_x}{12}. \qquad (2.21)$$

Now let the cube shrink to a point, i.e., let $\Delta x \to 0$, $\Delta y \to 0$ and $\Delta z \to 0$, resulting in

$$T_{yz} - T_{zy} = 0 \tag{2.22}$$

with T_{yz} and T_{zy} now acting at *the same point*. Repeating those considerations for M_y and M_z, two additional relations are obtained, i.e.,

$$T_{xy} = T_{yx}, \qquad T_{yz} = T_{zy}, \qquad T_{xz} = T_{zx}. \tag{2.23}$$

Equation (2.23) holds at a point. Furthermore, because there is nothing particular about the coordinate system used to derive it, it holds in any set of orthogonal coordinates.

We note that each of the three Eqs. (2.23) relates a stress on a plane to a stress on another plane perpendicular to the first plane. Thus the first equation states that the y-component of the stress acting on the x-plane equals the x-component of the stress acting on the y-plane. There is, therefore, a coupling between the stresses on the various planes passing through the same point. This coupling is investigated in more detail in the next section.

It is noted that normal stresses on the cube sides and body forces may also contribute to the moments. However, it can be shown that their contributions are of a higher differential order than those of the shear stresses, and when the cube shrinks to a point, i.e., in going from Eq. (2.21) to Eq. (2.22), their contributions vanish.

Forces at a Point on a Plane

In this section we find that once the stress matrix is known at a point in one cartesian system of coordinates, it is known at that point in all cartesian systems rotated with respect to the original one.

As stated earlier, surface forces acting at a point on a plane depend on the location of the point and on the plane orientation, i.e., on the direction of its outer normal. That the forces should depend on the coordinates of the considered point is quite clear. After all, the fluid properties may vary from point to point, and the velocity vector does, and it seems reasonable that the stresses and the forces should also change.

When *stress at a point* was defined, it seemed plausible that it should depend on the orientation of the plane's outer normal. The plane, or at least a small disk of it, was necessary to define the stress, and therefore the stress might depend on the plane. The previous section, however, indicates that there are fairly simple and quite clear relations between stresses acting at a point located on a particular plane and those acting at the same point on other planes, perpendicular to the original one.

We already know that there are nine stress components T_{ij}, which can be expressed in cartesian coordinates in the form of a 3×3 square matrix:

$$\begin{bmatrix} T_{11} & T_{12} & T_{13} \\ T_{21} & T_{22} & T_{23} \\ T_{31} & T_{32} & T_{33} \end{bmatrix}. \tag{2.24}$$

In this matrix the subscripts 1, 2, 3 stand for x, y, z, respectively. Equation (2.23), derived in the previous section, indicates that this matrix is symmetrical. It is quite important to realize that this symmetry establishes a relation between a stress acting on a particular plane and those acting on other planes perpendicular to it. But Eq. (2.23) holds for any coordinate system. Thus one may choose a new pair of y'- and z'-coordinates while x remains the same, and yet Eq. (2.23) connects $T_{xy'}$ with $T_{y'x}$. Therefore the components of the stress acting on one plane must be related to those acting on *any* other plane passing through the same point.

Consider now the stresses at a point on the x-plane, $\mathbf{n} = \mathbf{i}$, as shown in Fig. 2.3. The x-plane coincides with the printed page. The shear stress vector $\mathbf{T}_s$ in Fig. 2.3 may be expressed in terms of its components as

$$\mathbf{T}_s = \mathbf{j}T_{xy} + \mathbf{k}T_{xz}.$$

These components, in turn, uniquely determine T_{yx} and T_{zx} in the planes $\mathbf{n} = \mathbf{j}$ and $\mathbf{n} = \mathbf{k}$. Now let the coordinate system rotate around the x-axis. In the new (primed) system the shear stress vector remains the same and only its resolved components change:

$$\mathbf{T}_s = \mathbf{j}T_{xy} + \mathbf{k}T_{xz} = \mathbf{j}'T'_{xy} + \mathbf{k}'T'_{xz}. \tag{2.25}$$

These in turn uniquely determine T'_{yx} and T'_{zx}. Thus the shear stress vector seems to determine the shear stresses T_{yx} and T_{zx} for *all orientations* of the coordinate system. In other words, once the stress matrix (2.24) is known at a point for a particular set of coordinates, it is already uniquely determined at that point for all orientations. In the next section we develop the rules for this determination.

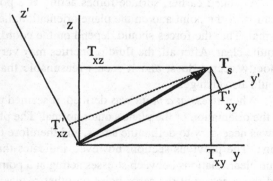

Figure 2.3 Shear stress at a point.

The Elementary Tetrahedron

We now proceed to obtain the rules which express the stress at a point on an arbitrary plane in terms of the components of the stress matrix T_{ij} in a given coordinate system.

A small system of fluid in the form of a tetrahedron is drawn in Fig. 2.4. The tetrahedron is generated by the intersection of four planes: three with their outer normals $-\mathbf{i}$, $-\mathbf{j}$, $-\mathbf{k}$, i.e., perpendicular to the coordinate axes, and the fourth with the outer normal $\mathbf{n} = \mathbf{i}n_1 + \mathbf{j}n_2 + \mathbf{k}n_3$, where n_1, n_2 and n_3 are the direction cosines of this normal.

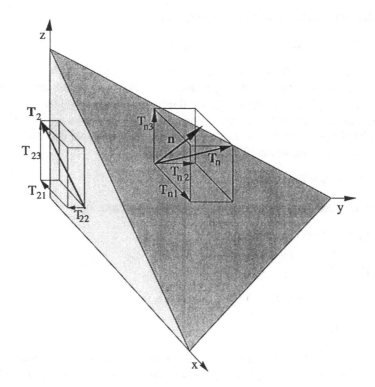

Figure 2.4 Elementary tetrahedron and stresses.

The tetrahedron thus obtained is convenient for what we have in mind. It has three sides on which the stress components T_{ij} are defined, as given in the stress matrix (2.24), and then a fourth side, with its normal $\mathbf{n}$. The $\mathbf{n}$ vector may have any orientation, and therefore the fourth side represents just any plane. Thus, the expression obtained for the stress on the $\mathbf{n}$ plane in terms of the stresses acting on the $-\mathbf{i}$, $-\mathbf{j}$, $-\mathbf{k}$ planes will be general and hold for any plane.

Figure 2.4 shows the stresses which act on the **n** plane and on one of the coordinate planes, i.e., the $-\mathbf{j}$ plane. Denoting the area of the **n** plane as A_n and those of the $-\mathbf{i}$, $-\mathbf{j}$ and $-\mathbf{k}$ planes as A_1, A_2 and A_3, respectively, we note that A_1, A_2 and A_3 are the x, y and z projections of A_n, or

$$A_1 = n_1 A_n, \qquad A_2 = n_2 A_n, \qquad A_3 = n_3 A_n.$$

Let the volume of the tetrahedron be V. Then the net force acting on the system is

$$-\mathbf{T}_1\left(n_1 A_n\right) - \mathbf{T}_2\left(n_2 A_n\right) - \mathbf{T}_3\left(n_3 A_n\right) + \mathbf{T}_n A_n + \rho V \mathbf{g} \tag{2.26}$$

and Newton's second law of motion requires

$$\rho V(\mathbf{a} - \mathbf{g}) = \left(\mathbf{T}_n - \mathbf{T}_1 n_1 - \mathbf{T}_2 n_2 - \mathbf{T}_3 n_3\right) A_n. \tag{2.27}$$

Equation (2.27) is now divided by A_n with $V/A_n = h_n/3$, where h_n is the "height" of the tetrahedron and A_n is its "base." We now let the tetrahedron shrink to a point, remaining geometrically similar to itself while shrinking. The shrinking is necessary because relations at a point are sought; its remaining similar to itself is essential if the **n** direction is to be preserved. As a result of this shrinking, $h_n \to 0$ and with it V/A_n. Equation (2.27) thus becomes

$$\mathbf{T}_n = \mathbf{T}_1 n_1 + \mathbf{T}_2 n_2 + \mathbf{T}_3 n_3. \tag{2.28}$$

Equation (2.28) may be also written in component form as

$$\begin{aligned}
T_{n1} &= T_{11} n_1 + T_{21} n_2 + T_{31} n_3, \\
T_{n2} &= T_{12} n_1 + T_{22} n_2 + T_{32} n_3, \\
T_{n3} &= T_{13} n_1 + T_{23} n_2 + T_{33} n_3.
\end{aligned} \tag{2.29}$$

Using index notation, Eq. (2.29) is recast in a more compact form,

$$T_{nj} = T_{ij} n_i. \tag{2.30}$$

The desired rule has thus been obtained. Indeed, for a given T_{ij} the stress vector $\mathbf{T}_n$ on any plane **n** can be computed, provided the direction cosines n_1, n_2 and n_3 of the normal to the plane are known.

Equations (2.28) - (2.30) indeed show coupling between the stresses acting on the various planes passing through the same point. Once the stress matrix is given in any one coordinate system, these equations show how the stresses may be obtained on *any* plane passing through the same point. All one has to do is specify the plane of interest by its outer normal, and the equations yield the stresses on that plane. Thus the stress matrix at a point in any one coordinate system contains all the information on stresses acting on other planes through the same point.

Example 2.1

The stress matrix at a point P is given by

$$\begin{bmatrix} T_{11} & T_{12} & T_{13} \\ T_{21} & T_{22} & T_{23} \\ T_{31} & T_{32} & T_{33} \end{bmatrix} = \begin{bmatrix} 2 & 1 & -3 \\ 1 & 1 & 2 \\ -3 & 2 & 1 \end{bmatrix}.$$

Find the stress vector on the plane passing through P and parallel to the plane whose unit normal is

$$\mathbf{n} = \tfrac{3}{7}\mathbf{i} + \tfrac{6}{7}\mathbf{j} + \tfrac{2}{7}\mathbf{k}.$$

Solution

We are essentially looking for $\mathbf{T}_1$, $\mathbf{T}_2$ and $\mathbf{T}_3$ in Eq. (2.28). Substitution of the components of the stress matrix into Eq. (2.29) yields

$$T_{n1} = T_{11}n_1 + T_{21}n_2 + T_{31}n_3 = \ 2n_1 + \ n_2 - 3n_3,$$
$$T_{n2} = T_{12}n_1 + T_{22}n_2 + T_{32}n_3 = \ \ n_1 + \ n_2 + 2n_3,$$
$$T_{n3} = T_{13}n_1 + T_{23}n_2 + T_{33}n_3 = -3n_1 + 2n_2 + \ n_3.$$

These equations may be written in matrix form as

$$\begin{bmatrix} T_{n1} & T_{n2} & T_{n3} \end{bmatrix} = \begin{bmatrix} n_1 & n_2 & n_3 \end{bmatrix} \cdot \begin{bmatrix} T_{11} & T_{12} & T_{13} \\ T_{21} & T_{22} & T_{23} \\ T_{31} & T_{32} & T_{33} \end{bmatrix}$$

$$= \begin{bmatrix} \tfrac{3}{7} & \tfrac{6}{7} & \tfrac{2}{7} \end{bmatrix} \cdot \begin{bmatrix} 2 & 1 & -3 \\ 1 & 1 & 2 \\ -3 & 2 & 1 \end{bmatrix} = \begin{bmatrix} \tfrac{6}{7} & \tfrac{13}{7} & \tfrac{5}{7} \end{bmatrix}.$$

Substitution into Eq. (2.28) yields the desired stress vector

$$\mathbf{T} = \tfrac{6}{7}\mathbf{i} + \tfrac{13}{7}\mathbf{j} + \tfrac{5}{7}\mathbf{k}.$$

Example 2.2

The stress matrix is given at the point P in the coordinate system x, y, z by

$$\begin{bmatrix} T_{11} & T_{12} & T_{13} \\ T_{21} & T_{22} & T_{23} \\ T_{31} & T_{32} & T_{33} \end{bmatrix} = \begin{bmatrix} 1 & 1 & -1 \\ 1 & 3 & 2 \\ 1 & 2 & 2 \end{bmatrix}.$$

a. Is such a matrix possible? If not, make a plausible correction.
b. Are the stresses known now on all the planes passing through P, e.g., on the planes whose normals are $\mathbf{i}$, $\mathbf{j}$, $\mathbf{k}$, $\mathbf{n} = \mathbf{i}n_1 + \mathbf{j}n_2 + \mathbf{k}n_3$?

Solution

a. No, this matrix is not symmetrical. We correct T_{31} from 1 to -1. Now the matrix may be right.
b. Yes, they are known. Inspecting the stress matrix, we write down the stress vectors on the x, y, z planes, respectively,

$$\mathbf{T}_x = \mathbf{i} + \mathbf{j} - \mathbf{k},$$
$$\mathbf{T}_y = \mathbf{i} + 3\mathbf{j} + 2\mathbf{k},$$
$$\mathbf{T}_z = -\mathbf{i} + 2\mathbf{j} + 2\mathbf{k}.$$

The stress vector acting on the $\mathbf{n}$ plane is

$$\mathbf{T}_n = \mathbf{i}(n_1 + n_2 - n_3) + \mathbf{j}(n_1 + 3n_2 + 2n_3) + \mathbf{k}(-n_1 + 2n_2 + 2n_3).$$

Stress in a Fluid at Rest

Here we show that in a stationary fluid the only remaining stress is the pressure and that this pressure stress is the same in all directions.

A fluid has been defined in Chapter 1 as a continuum which cannot support a shear stress while at rest. Thus a fluid at rest admits no shear stress and its stress matrix, e.g., Eq. (2.24), is a diagonal matrix in all systems of cartesian coordinates:

$$T_{ij} = 0 \quad \text{for } i \neq j. \tag{2.31}$$

Furthermore, Eq. (2.29) now reads

$$T_{n1} = T_{11}n_1,$$
$$T_{n2} = T_{22}n_2, \tag{2.32}$$
$$T_{n3} = T_{33}n_3.$$

We claim that because Eq.(2.32) must hold for *all* cartesian coordinate systems,

$$T_{11} = T_{22} = T_{33} = |T_n|, \tag{2.33}$$

a result known as Pascal's Law:

In a fluid at rest the pressure at a point is the same in all directions.

Example 2.3

Given Eq. (2.32) and the condition that it holds in all cartesian coordinates, prove Pascal's law.

Solution

Consider a plane with its outer normal **n** and the stress vector on it, viewed sideways, as shown in Fig. 2.5. Obviously there are no shear stresses and $\mathbf{T}_n$ is parallel to **n**. With Eq. (2.32) $\mathbf{T}_n$ becomes

$$\mathbf{T}_n = \mathbf{i}T_{n1} + \mathbf{j}T_{n2} + \mathbf{k}T_{n3}$$
$$= \mathbf{i}n_1 T_{11} + \mathbf{j}n_2 T_{22} + \mathbf{k}n_3 T_{33} = \mathbf{n}T_n,$$

where $\mathbf{T}_n = \mathbf{n}T_n$ is due to $\mathbf{T}_n$ being parallel to **n**.

We now form the scalar product of this expression with **i**, with **j** and with **k**. These products result in

$$T_{11} = T_n, \qquad T_{22} = T_n, \qquad T_{33} = T_n,$$

and therefore

$$T_{11} = T_{22} = T_{33} = T_n,$$

and the proof is complete.

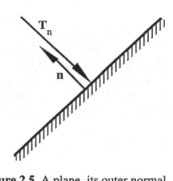

Figure 2.5 A plane, its outer normal and its stress vector in a fluid at rest.

Stress in a Moving Fluid

A moving fluid can support shear stress. Consider a fluid that fills the space between two parallel long plates, as shown in Fig. 2.6. The lower plate is stationary and the upper one moves at a uniform velocity V under the influence of a weight. The force pulling the upper plate thus must be counterbalanced by a force equal in magnitude and opposite in direction, or the upper plate and the weight must accelerate, which they do not. Consider now a rectangle of fluid Δx wide bounded by the two plates. We assume the pressure not to depend on x. Hence the shear stress on this fluid rectangle at the top, where it touches the upper plate, must be balanced by that at the bottom, and we see how this stress transfers to the bottom plate. Furthermore, the same considerations apply to a block which extends from the top plate to the A-A plane, and therefore the shear stress on the A-A plane is the same as on the top plate. As the A-A plane can be selected anywhere in the fluid, one must conclude that the shear stress is the same on all planes parallel to the plates.

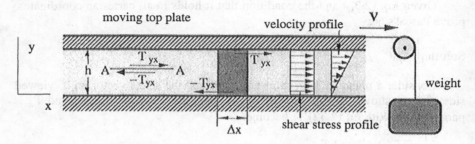

Figure 2.6 Fluid in shear flow.

It is known from experiments that the fluid adjacent to a solid surface adheres to it. Hence the fluid adjacent to the upper plate, at $y = h$, moves with the velocity V, while that at the lower plate, at $y = 0$, is at rest. The uniform stress, i.e., force per unit area, exerted on the viscous fluid by the upper plate, T_{yx}, can be expressed by Newton's law of viscosity, Eq. (1.19):

$$T_{yx} = \mu \frac{du}{dy}. \tag{1.19}$$

Equation (1.19) may be integrated with the boundary condition of $u = 0$ at $y = 0$ to yield

$$u = \frac{T_{yx}}{\mu} y. \tag{2.34}$$

Hence, once the shear stress is known, the velocity at any point in the fluid can be calculated by means of Eq. (2.34). The shear stress may be eliminated by substituting $u(h) = V$ into Eq. (2.34)

$$V = \frac{T_{yx}}{\mu} h, \tag{2.35}$$

resulting in

$$u = \frac{V}{h} y. \qquad (2.36)$$

The flow just obtained is known as Plane Couette Flow or Plane Shear Flow.

Example 2.4

Viscosity of printing ink is measured by first coating a metal plate with the ink and then passing the plate through a gap between two parallel stationary plates, Fig. 2.7. The velocity of the plate is measured by timing the descent of the weight between two marks on a yardstick.

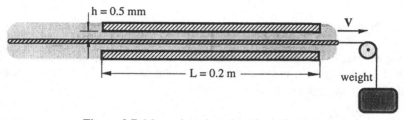

Figure 2.7 Measuring viscosity of printing ink.

What is the viscosity of the printing ink if the plate velocity is 1 cm/s and the length and width of the stationary plates are 20 cm and 10 cm, respectively; the gap on each side of the moving plate is 0.5 mm, and the weight is 5 N.

Solution

The shear stress applied by the weight on the fluid on both sides of the plate is

$$T_{yx} = \frac{F}{2A} = \frac{5}{2 \times 0.2 \times 0.1} = 125 \, N/m^2.$$

The viscosity of the ink is found by substitution into Eq. (2.35)

$$\mu = \frac{T_{yx}h}{V} = \frac{125 \times 0.5 \times 10^{-3}}{0.01} = 6.25 \, kg/m \cdot s = 6250 \, cp.$$

A Case where the Stress Is Not Constant

In the previous section we considered the case of constant shear in a moving fluid. In this section we examine a case where the stress in the moving fluid is not constant.

Consider the two-dimensional plane flow between two stationary plates, as shown in Fig. 2.8. The flow in this case is induced by a constant pressure drop, $-\Delta P/\Delta x$, along the x-axis.

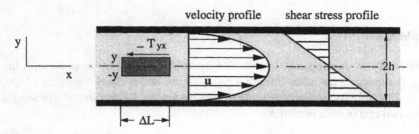

Figure 2.8 Flow between two stationary plates, plane Poiseuille flow.

In a way similar to the shear flow example we consider a rectangular fluid block of length ΔL and height $2y$, extending from $-y$ to $+y$, as shown in Fig. 2.8. The pressure difference between its two vertical sides is

$$\Delta P_L = \frac{\Delta P}{\Delta x} \, \Delta L,$$

and hence the force pushing it to the right is

$$2y \, \Delta P_L = 2y \left(-\frac{\Delta P}{\Delta x} \right) \Delta L$$

(note that $\Delta P/\Delta x < 0$). This force is balanced by the shear on the upper and lower sides:

$$-T_{yx} \cdot 2\Delta L.$$

Hence,

$$T_{yx} = -\left(-\frac{\Delta P}{\Delta x} \right) y, \tag{2.37}$$

and by Newton's law

$$\mu \frac{du}{dy} = -\left(-\frac{\Delta P}{\Delta x} \right) y.$$

Integration now yields

$$u = -\frac{1}{2\mu} \left(-\frac{\Delta P}{\Delta x} \right) y^2 + C,$$

which with the no-slip boundary condition $u(h) = 0$ yields the velocity profile

$$u = \frac{h^2}{2\mu} \left(-\frac{\Delta P}{\Delta x} \right) \left[1 - \left(\frac{y}{h} \right)^2 \right] \tag{2.38}$$

This flow is known as Plane Poiseuille Flow.

Inspection of Eq. (2.38) indicates that the highest velocity is at the midplane between the plates, i.e., at $y = 0$, and that its magnitude is

$$u_{max} = \left(-\frac{\Delta P}{\Delta L}\right)\frac{h^2}{2\mu}. \tag{2.39}$$

Hence, Eq. (2.38) may also be rewritten in terms of u_{max} as

$$u = u_{max}\left[1 - \left(\frac{y}{h}\right)^2\right]. \tag{2.40}$$

Example 2.5

The gap between the two plates, shown in Fig. 2.8, is $2h = 2 \times 0.005$ m. The pressure drop in the fluid between the plates is given by

$$-\frac{\Delta P}{\Delta L} = 20\,\frac{N/m^2}{m} = 20\,N/m^3.$$

The fluid between the plates is
a.   Water, $\mu = 0.001$ N·s/m².
b.   Glycerin, $\mu = 1.5$ N·s/m².

Find the highest velocity in the flow and the shear stress on the plates for each fluid.

Solution

The highest velocity in the flow is calculated from Eq.(2.39).

a.   For water: $u_{max} = 20 \times \dfrac{0.005^2}{2 \times 0.001} = 0.25$ m/s.

b.   For glycerin: $u_{max} = 20 \times \dfrac{0.005^2}{2 \times 1.5} = 1.667 \times 10^{-4}$ m/s $= 1$ cm/min.

The stress at the walls is conveniently obtained from Eq. (1.19) together with Eq. (2.38) set in the form

$$T_{yx}\big|_{y=h} = -\mu\frac{du}{dy}\bigg|_{y=h} = -\frac{\Delta P}{\Delta L}y\bigg|_{y=h} = \left(-\frac{\Delta P}{\Delta L}\right)h$$

Note, the stress on the wall is opposite in sign to the stress in the fluid.

The form just set does not include μ, and the result is the same for water and glycerin,

$$T_{yx} = 20 \times 0.005 = 0.1 \text{ N/m}^2.$$

That the result is the same for both fluids is not surprising and is brought by $-\Delta P/\Delta L$ being the same for both. Indeed, choosing the fluid block in Fig. 2.7 to have $y = h$ makes Eq. (2.38) the same as that used to compute T_{yx} on the walls here. In other words, the pressure-drop forces are eventually supported by the plates.

* The Stress Tensor

Equation (2.30) may be viewed as an operator equation which requires a vector, **n**, as an input for performing the operation; its output is also a vector, $\mathbf{T}_n$. The stress matrix $[T_{ij}]$ is the operator which is given at a point. For any unit vector **n** at that point a vector $\mathbf{T}_n$ is generated by the operation indicated in Eq. (2.30). Because this operation is linear, it is sometimes called a linear vector–vector dependence.

The operator T_{ij} is called a *tensor*, and because it relates to the stress vector $\mathbf{T}_n$, it is called the *stress tensor*. Its components T_{ij} are given by the stress matrix, Eq. (2.24).

Equation (2.30) holds quite generally. It describes a physical phenomenon and therefore does not depend on the coordinates used. Under coordinate transformation not only do the vectors **n** and $\mathbf{T}_n$ remain the same, but the form of the relation between them, Eq. (2.30), must also remain the same. The components of **n** and $\mathbf{T}_n$ are, however, coordinate dependent, and therefore the components T_{ij} must transform in such a way that the *form* of Eq. (2.30) is preserved. In other words, once the stress tensor components T_{ij} are known for a given coordinate system, the stress vector $\mathbf{T}_n$ acting on any plane with its outer normal **n** may be directly obtained by the use of Eq. (2.30).

The basic rules for vector and tensor transformations are now derived. The transformation of a vector is included as an introduction to that of the tensor.

* Transformation of Coordinates – Vectors

Consider the vector **S** given in Fig. 2.9 in terms of the $\mathbf{e}_i$ system:

$$\mathbf{S} = S_i \mathbf{e}_i,$$

where the summation convention holds. Let another system $\mathbf{e}'_k$ be chosen, in which the same vector assumes the form

$$\mathbf{S} = S'_k\, \mathbf{e}'_k.$$

Because $\mathbf{S} = \mathbf{S}'$,

$$S_i \mathbf{e}_i = S'_k\, \mathbf{e}'_k\ .$$

Scalar multiplication by $\mathbf{e}'_k$ on both sides results in

$$S'_k = S_i \left(\mathbf{e}_i \cdot \mathbf{e}'_k \right),$$

(2.41)

which is the transformation rule for vectors. Similarly

$$S_m = S'_n \left(\mathbf{e}'_n \cdot \mathbf{e}_m \right).$$

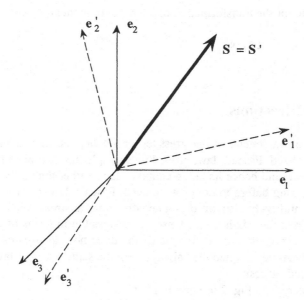

Figure 2.9 Components of a vector in two coordinate systems.

* Transformation of Coordinates – Tensors

The stress tensor T_{ij} yields the stress vector $\mathbf{S}$ for the surface with the outer normal $\mathbf{n}$:

$$\mathbf{S} = \mathbf{e}_i T_{ij} \left(\mathbf{e}_j \cdot \mathbf{n} \right) = \mathbf{e}_i T_{ij}\, n_j = \mathbf{e}_k S_k.$$

In the rotated coordinates $\mathbf{e}'_k$ the relation becomes

$$S' = e'_m T'_{mn} n'_n .$$

However, since S = S', substitution of

$$e'_m = e_i \left(e'_m \cdot e_i \right) \quad \text{and of} \quad n'_n = \left(e'_n \cdot e_j \right) n_j$$

into the expression for **S'** yields

$$S' = e_i \left(e'_m \cdot e_i \right) T'_{mn} \left(e'_n \cdot e_j \right) n_j$$

$$= e_i T'_{mn} \left(e'_m \cdot e_i \right) \left(e'_n \cdot e_j \right) n_j .$$

Comparison with the expression for **S** yields

$$T_{ij} = T'_{mn} \left(e'_m \cdot e_i \right) \left(e'_n \cdot e_j \right), \tag{2.42}$$

which is the rule for the transformation of a tensor component. Similarly

$$T'_{mn} = T_{ij} \left(e_i \cdot e'_m \right) \left(e_j \cdot e'_n \right).$$

* Principal Directions

Of the nine components of the stress tensor the three normal stresses are more intuitively visualized. Pascal's law, proved in Example 2.3, had been known as an experimental fact (and hence its being called a "law" rather than a theorem which can be proved) long before matrices were used. Pascal's law deals with the pressure, which is intuitively visualized as a negative normal stress. We already know that in fluid at rest the whole stress tensor is expressible in terms of the pressure only. We also suspect that in moving fluids the three normal stresses may not be the same, as otherwise, i.e., had they always been the same, Pascal's law would not be limited to fluids at rest.

We consider again Fig. 2.5. For a fluid at rest

$$T_n = nT_n = -np,$$

with p being the pressure. In a moving fluid the normal stress on the plane with its normal **n** is T_{nn}, i.e., the nth component of T_n.

Now there could be a direction **n** for which

$$T_n = nT_n .$$

In other words, we rotate **n** and with it, of course, the plane, until we find a direction for which the plane suffers no shear stress. When we find such a direction, we call it principal:

The Principal Directions of a stress tensor are defined as those in which all shear stresses vanish.

The condition for a principal direction may also be stated as the requirement that **n** and $\mathbf{T}_n$ lie on the same straight line, i.e.,

$$\mathbf{T}_n = \lambda \mathbf{n} \tag{2.43}$$

with λ being a proportionality factor which we already recognized as T_{nn}. Thus λ is the normal stress in the principal direction, called the principal stress. Substitution into Eq. (2.30) yields

$$\lambda n_j = T_{ij} n_i$$

or

$$T_{ij} n_i - \lambda n_j = 0, \tag{2.44}$$

where T_{ij} are the components of the stress tensor in the original coordinate system. We now rewrite Eq. (2.44) in component form as

$$
\begin{aligned}
(T_{11} - \lambda) n_1 + T_{21} n_2 + T_{31} n_3 &= 0, \\
T_{12} n_1 + (T_{22} - \lambda) n_2 + T_{32} n_3 &= 0, \\
T_{13} n_1 + T_{23} n_2 + (T_{33} - \lambda) n_3 &= 0.
\end{aligned}
\tag{2.45}
$$

This is a system of three linear homogeneous equations in three unknowns n_i. It has non-trivial solutions only when the determinant of the coefficients vanishes, i.e.,

$$
\begin{vmatrix}
(T_{11} - \lambda) & T_{12} & T_{13} \\
T_{21} & (T_{22} - \lambda) & T_{23} \\
T_{31} & T_{32} & (T_{33} - \lambda)
\end{vmatrix} = 0. \tag{2.46}
$$

It is noted that the stress tensor and the determinant of Eq. (2.46) are symmetrical. Expanding the determinant results in the following cubic equation in λ:

$$
\begin{aligned}
&\lambda^3 - [T_{11} + T_{22} + T_{33}] \lambda^2 + [T_{22} T_{33} - T_{23}^2 + T_{33} T_{11} - T_{31}^2 + T_{11} T_{22} - T_{12}^2] \lambda \\
&- [T_{11} T_{22} T_{33} - T_{12} T_{23} T_{31} + T_{13} T_{21} T_{32} - T_{13} T_{22} T_{31} - T_{11} T_{23} T_{32} - T_{12} T_{21} T_{33}] = 0
\end{aligned}
$$

Linear algebra then yields that this cubic equation has three real roots λ_i, $i = 1,2,3$, and therefore there are three principal directions. Using linear algebra, it can also be shown that these directions are orthogonal to one another. The rotation needed to reach one of these principal directions is obtained from

$$x T_{11} + y T_{12} + z T_{13} = \lambda_i x,$$
$$x T_{21} + y T_{22} + z T_{23} = \lambda_i y, \qquad\qquad (2.47)$$
$$x T_{31} + y T_{32} + z T_{33} = \lambda_i z,$$

which may be manipulated to express x/y, y/z and z/x, i.e., the tangents of the angles of rotation. The particular direction depends on the choice of i, $i = 1,2,3$.

Because the principal directions are orthogonal to one another, they may be used as a cartesian system of coordinates. In this system Eq. (2.47) attains the form

$$\begin{bmatrix} T_{11} & 0 & 0 \\ 0 & T_{22} & 0 \\ 0 & 0 & T_{33} \end{bmatrix} = \begin{bmatrix} \lambda_1 & 0 & 0 \\ 0 & \lambda_2 & 0 \\ 0 & 0 & \lambda_3 \end{bmatrix}. \qquad\qquad (2.48)$$

The nonvanishing terms in the stress tensor in its principal directions were called *principal stresses*. Thus the solutions λ_i, $i = 1,2,3$, have the direct physical interpretation of being the three normal stresses in the principal directions.

Example 2.6

Find the principal stresses and the principal directions for the stress tensor

$$[T_{ij}] = \begin{bmatrix} 3 & 1 & 1 \\ 1 & 0 & 2 \\ 1 & 2 & 0 \end{bmatrix}.$$

Solution

We first look for the three principal stresses λ_1, λ_2 and λ_3. The determinant equation (2.45) becomes

$$\begin{vmatrix} (T_{11}-\lambda) & T_{12} & T_{13} \\ T_{21} & (T_{22}-\lambda) & T_{23} \\ T_{31} & T_{32} & (T_{33}-\lambda) \end{vmatrix} = \begin{vmatrix} 3-\lambda & 1 & 1 \\ 1 & -\lambda & 2 \\ 1 & 2 & -\lambda \end{vmatrix} = 0.$$

Expanding, this yields the polynomial equation

$$(\lambda + 2)(\lambda - 1)(\lambda - 4) = 0,$$

whose roots give the principal stresses $\lambda_1 = -2$, $\lambda_2 = 1$, $\lambda_3 = 4$.

Hence, in terms of the principal stresses our stress tensor assumes the form

$$[T_{ij}] = \begin{bmatrix} \lambda_1 & 0 & 0 \\ 0 & \lambda_2 & 0 \\ 0 & 0 & \lambda_3 \end{bmatrix} = \begin{bmatrix} -2 & 0 & 0 \\ 0 & 1 & 0 \\ 0 & 0 & 4 \end{bmatrix}.$$

We now look for three unit vectors $\mathbf{n}_1$, $\mathbf{n}_2$ and $\mathbf{n}_3$ pointing in the principal directions.

Letting $\mathbf{n}_1$ be the direction associated with $\lambda_1 = -2$, we find upon substitution into Eq. (2.45)

$$5n_1 + n_2 + n_3 = 0,$$
$$n_1 + 2n_2 + 2n_3 = 0,$$
$$n_1 + 2n_2 + 2n_3 = 0.$$

The solution is $n_1 = 0$ and $n_2 = -n_3$.

Since $\mathbf{n}_1$ is a unit vector,

$$n_2^2 + n_3^2 = 1,$$

and we obtain

$$n_2 = 1/\sqrt{2}, \qquad n_3 = -1/\sqrt{2},$$

such that

$$\mathbf{n}_1 = 0\mathbf{i} + \frac{1}{\sqrt{2}}\mathbf{j} - \frac{1}{\sqrt{2}}\mathbf{k}.$$

where $\mathbf{i}$, $\mathbf{j}$, $\mathbf{k}$ are the unit vectors in the directions of the x, y, z coordinates.

Similarly, for λ_2, Eq. (2.45) yields

$$2n_1 + n_2 + n_3 = 0,$$
$$n_1 - n_2 + 2n_3 = 0,$$
$$n_1 + 2n_2 - n_3 = 0,$$

the solution of which leads to

$$\mathbf{n}_2 = \frac{1}{\sqrt{3}}\mathbf{i} - \frac{1}{\sqrt{3}}\mathbf{j} - \frac{1}{\sqrt{3}}\mathbf{k},$$

and for λ_3,

$$-n_1 + n_2 + n_3 = 0,$$
$$n_1 - 4n_2 + 2n_3 = 0,$$
$$n_1 + 2n_2 - 4n_3 = 0,$$

which yields

$$\mathbf{n}_3 = -\frac{2}{\sqrt{6}}\mathbf{i} - \frac{1}{\sqrt{6}}\mathbf{j} - \frac{1}{\sqrt{6}}\mathbf{k}.$$

Hence, the unit vectors $\mathbf{n}_1$, $\mathbf{n}_2$ and $\mathbf{n}_3$ in the principal directions have been obtained.

Example 2.7

The motion of the fluid between two parallel flat plates, with the lower one fixed and the upper one moving, is described in Figs. 2.6 and 2.10. The stress tensor at point B is given by

$$\begin{bmatrix} T_{xx} & T_{xy} \\ T_{yx} & T_{yy} \end{bmatrix} = \begin{bmatrix} -p & \mu V/h \\ \mu V/h & -p \end{bmatrix}.$$

Find the principal directions and the principal stresses, T_{ij}'', corresponding to these directions.

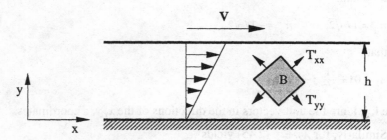

Figure 2.10 Velocity between a stationary and a moving plate.

Solution

Equation (2.46) yields for the present case

$$\begin{vmatrix} T_{xx} - \lambda & T_{xy} \\ T_{yx} & T_{yy} - \lambda \end{vmatrix} = 0.$$

Hence

$$(-p - \lambda)^2 - (\mu V / h)^2 = 0,$$

the solution of which yields

$$\lambda_1 = \frac{\mu V}{h} - p, \qquad T'_{xx} = \frac{\mu V}{h} - p$$

and

$$\lambda_2 = -\frac{\mu V}{h} - p, \qquad T'_{yy} = -\frac{\mu V}{h} - p.$$

Substitution of λ_1 and the expressions for the stress tensor into Eq. (2.45) yields

$$\left(\frac{\mu V}{h} - p\right)n_x = -pn_x + \frac{\mu V}{h}n_y,$$

$$\left(\frac{\mu V}{h} - p\right)n_y = \frac{\mu V}{h}n_x - pn_y,$$

with the solution

$$n_x = n_y.$$

Hence

$$\tan\alpha_1 = n_x / n_y; \quad \alpha_1 = 45^o.$$

Similarly for λ_2 one obtains $n_x = -n_y$ and $\alpha_2 = -45^o$.

Example 2.8

The velocity profile in a circular pipe is given by

$$q_r = 0, \qquad q_\theta = 0, \qquad q_z = 2U\left[1 - \left(\frac{r}{R}\right)^2\right],$$

where U is the average velocity and R is the pipe radius, Fig. 2.11. The x, y and r coordinates are related by $x^2 + y^2 = r^2$.

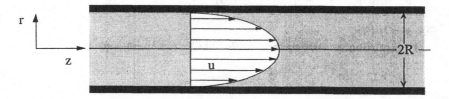

Figure 2.11 Velocity profile in a circular pipe.

Measurements and computations yield the following cartesian components for the stress tensor in this flow:

$$T_{xx} = T_{yy} = T_{zz} = -p, \qquad\qquad T_{xy} = T_{yx} = 0,$$

$$T_{xz} = T_{zx} = T_{yz} = T_{zy} = -\frac{4\mu Ur}{R^2},$$

all at $x = y$, i.e., at $\mathbf{r} = r(\mathbf{i} + \mathbf{j})/\sqrt{2}$, $\mathbf{r}$ is the radius vector in the x-y plane.

Write the stress tensor, and find the principal stresses and the principal directions for $r = 0$, $r = R/2$ and $r = R$.

Solution

For $r = 0$ the stress tensor is

$$T_{ij} = \begin{bmatrix} -p & 0 & 0 \\ 0 & -p & 0 \\ 0 & 0 & -p \end{bmatrix}.$$

All directions are principal, and all three principal stresses are $-p$.

For $r = R/2$:

$$T_{ij} = \begin{bmatrix} -p & 0 & -2\mu U/R \\ 0 & -p & -2\mu U/R \\ -2\mu U/R & -2\mu U/R & -p \end{bmatrix}.$$

The characteristic equation, Eq. (2.46), becomes

$$\begin{vmatrix} -p-\lambda & 0 & -2\mu U/R \\ 0 & -p-\lambda & -2\mu U/R \\ -2\mu U/R & -2\mu U/R & -p-\lambda \end{vmatrix} = 0$$

or

$$(-p-\lambda)\left[(-p-\lambda)^2 - a^2\right] - a\left[a(-p-\lambda)\right] = 0,$$

where $a = 2\mu U/R$.

The equation is satisfied by $\lambda_1 = -p$, and the remaining equation,

$$(-p-\lambda)^2 = 2a^2,$$

yields

$$\lambda_2 = -p + \sqrt{2}\,a,$$
$$\lambda_3 = -p - \sqrt{2}\,a.$$

For $\lambda_1 = -p$, Eq. (2.44) becomes

$$0 + 0 - an_n = 0,$$
$$0 + 0 - an_n = 0,$$
$$-an - an + 0 = 0,$$

with the solution $n_3 = 0$, $n_1 = -n_2$ and

$$\mathbf{n}_1 = \mathbf{i}\frac{1}{\sqrt{2}} - \mathbf{j}\frac{1}{\sqrt{2}},$$

where $\mathbf{n}_1$ is $\mathbf{n}$ for λ_1.

For $\lambda_2 = -p + \sqrt{2}a$, Eq. (2.32) becomes

$$-\sqrt{2}an_1 + 0 - an_3 = 0,$$
$$0 - \sqrt{2}an_1 - an_3 = 0,$$
$$an_1 + an_2 - \sqrt{2}an_3 = 0,$$

with the solution $n_1 = n_2$, $n_3 = -\sqrt{2}n_1$ and

$$\mathbf{n}_2 = \tfrac{1}{2}\left(\mathbf{i} + \mathbf{j} - \sqrt{2}\mathbf{k}\right),$$

where $\mathbf{n}_2$ is $\mathbf{n}$ for λ_2.

For $\lambda_3 = -p - \sqrt{2}a$, Eq. (2.32) becomes

$$\sqrt{2}an_1 + 0 - an_3 = 0,$$
$$0 + \sqrt{2}an_1 - an_3 = 0,$$
$$-an_1 - an_2 - \sqrt{2}an_3 = 0,$$

with the solution $n_1 = n_2$, $n_3 = \sqrt{2}n_1$ and

$$\mathbf{n}_3 = \tfrac{1}{2}\left(\mathbf{i} + \mathbf{j} + \sqrt{2}\mathbf{k}\right),$$

where $\mathbf{n}_3$ is $\mathbf{n}$ for λ_3.

The three principal directions have thus been obtained. To clarify the geometry of the results, it is noted that $\mathbf{i} + \mathbf{j} = \sqrt{2}\mathbf{r}$; hence $\mathbf{n}_2 = (\mathbf{r} - \mathbf{k})/\sqrt{2}$ and $\mathbf{n}_3 = (\mathbf{r} + \mathbf{k})/\sqrt{2}$. The principal directions are set at 45° to the direction of the flow and in the θ direction.

Note: $\mathbf{n}_1 \cdot \mathbf{n}_2 = \mathbf{n}_1 \cdot \mathbf{n}_3 = \mathbf{n}_2 \cdot \mathbf{n}_3 = 0$.

Note: $\lambda_1 + \lambda_2 + \lambda_3 = -3p = T_{11} + T_{22} + T_{33}$.

The sum of the normal stresses has been conserved.

For $r = R$: $a = 4\mu U/R$. The other results are the same as have been obtained for $r = R/2$.

* Identical Principal Stresses – Pascal's Law

There may be situations where all the principal stresses are identical. In that case all the directions are principal, i.e., no shear stresses are produced but the stress tensor is rotated. Furthermore, when no shear stresses appear in any direction, the normal shear stresses are identical.

A fluid has been defined as *a continuum which cannot support shear stress while at rest*. Thus, for a fluid at rest, the form of the stress tensor, in any coordinate system, is that of Eq. (2.48), i.e., with zeros in the locations reserved for the shear stresses. Therefore, for a fluid at rest, any coordinate system is principal. The equality of the normal stresses at a point is a direct conclusion of this property:

$$\lambda_1 = \lambda_2 = \lambda_3. \tag{2.49}$$

The normal stress in a fluid at rest is negative, i.e., compressive, and the term used for it is *pressure*. Thus, Eq. (2.49) becomes

$$p = -\lambda = \text{const.} \tag{2.50}$$

The result just obtained has been referred to as Pascal's law:

In a fluid at rest the pressure at a point is the same in all directions.

References

R. Aris, "Vectors, Tensors and the Basic Equations of Fluid Mechanics," Prentice-Hall, Englewood Cliffs, 1962.

D. Frederick and T.S. Chang, "Continuum Mechanics," Allyn and Bacon, Boston, 1965.

Y.C. Fung, "A First Course in Continuum Mechanics," 2nd ed., Prentice-Hall, Englewood Cliffs, 1977.

Problems

2.1   A unidirectional flow between three infinite flat plates is shown in Fig.
      P2.1. The two outer plates are stationary, while the midplate moves at a
      constant velocity U=1 m/s, as shown in the figure. The gap between the
      plates is $h = 0.005$ m. The fluid has a viscosity of $\mu = 2$ kg/(m·s) and is
      Newtonian, i.e., it obeys Newton's law of viscosity, Eq. (1.19).
      a. Consider Eqs. (2.34) - (2.36) and calculate the shear stress on the sur-
          faces of all the plates.
      b. Find the force, per unit area of plate, needed to maintain the steady
          motion of the midplate.

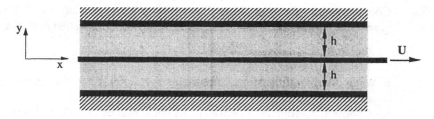

Figure P2.1

2.2   A two-dimensional flow past a semi-infinite stationary flat plate is shown in
      Fig. P2.2. The fluid far from the plate moves with a constant velocity **U** in a
      direction parallel to the plate. The fluid has the viscosity $\mu = 2$ kg/(m·s) and
      the density $\rho = 1,000$ kg/m³. The shear stress on the plate is measured as

$$\tau_{xy} = 0.332 \frac{\rho U^2}{\sqrt{Re_x}} \qquad \text{where} \qquad Re_x = \frac{Ux\rho}{\mu}$$

and x is the distance from the leading edge of the plate. This shear integrates
into a force pushing the plate in the flow direction. Find the force, per unit
width of plate, acting on the regions $0<x<1$ m, $0<x<2$ m, $0<x<10$ m.

Figure P2.2

2.3   Consider again the flow between the plates in Problem 2.1. Measurements show that on all the wetted surfaces,

$$T_{xx} = T_{yy} = -p = \text{const.}$$

a. Write the two-dimensional stress tensor at a point on the upper surface of the lower plate.
b. Choose a coordinate system x'-y', which is rotated by $\pi/4$ with respect to the x-y system. Write the two-dimensional stress tensor at a point on the upper surface of the lower plate in the x'-y' coordinate system.
c. Write again those two expressions for the tensors at a point in the fluid midway between the lower stationary plate and the moving plate.

2.4   Consider again the two-dimensional flow past a semi-infinite flat plate given in Problem 2.2. Measurements show that on both wetted surfaces,

$$T_{xx} = T_{yy} = -p = \text{const.}$$

a. Write the two-dimensional stress tensor at a point $x = 1.5\,\text{m}$ on the upper surface of the plate.
b. Choose a coordinate system x'-y', which is rotated by $\pi/4$ with respect to the x-y system, and write the two-dimensional stress tensor at the same point in the x'-y' coordinate system.
c. Write again those two expressions for the tensors at the point $x = 5\,\text{m}$ on the upper surface of the plate.

2.5   The stress tensor at a point is given as

$$\begin{bmatrix} -1 & 4 \\ 4 & -3 \end{bmatrix}.$$

a. Find the principal directions.
b. Find the direction in which the shear stress is the largest.
c. Can the continuum in which this tensor exists be a fluid?
    Hint: a fluid cannot withstand a positive normal stress.

2.6   Three coordinate systems are given as

$$(x_1,\ y_1,\ z_1); \qquad (x_2,\ y_2,\ z_2); \qquad (x_3,\ y_3,\ z_3).$$

System 2 is obtained by turning system 1 by $\pi/4$ around the z_1-axis.
System 3 is obtained by turning system 2 by $\pi/4$ around the x_2-axis.
Express the vectors

$$\mathbf{R} = 5\mathbf{i}_1 + 6\mathbf{j}_1 + 7\mathbf{k}_1 \qquad \text{and} \qquad \mathbf{S} = \mathbf{i}_2 + 2\mathbf{j}_2 + 3\mathbf{k}_2$$

in all three systems.

2.7   Four coordinate systems are given as

$$(x_1,\ y_1,\ z_1); \qquad (x_2,\ y_2,\ z_2); \qquad (x_3,\ y_3,\ z_3); \qquad (x_4,\ y_4,\ z_4).$$

System 2 is obtained by turning system 1 by $\pi/4$ around the z_1-axis.
System 3 is obtained by turning system 2 by $\pi/4$ around the x_2-axis.
System 4 is obtained by turning system 3 by $(-\pi/4)$ around the z_3-axis.
Express the vectors $\mathbf{i}=\mathbf{i}_1, \mathbf{j}=\mathbf{j}_1, \mathbf{k}=\mathbf{k}_1$ in systems 2, 3, 4.

2.8   a. A solid rod of cross-sectional area A is subjected to a compressive force
       $\mathbf{F}$, Fig. P2.8. Assuming that the stress is uniform across the section, find
       the shear stress in the plane B-B.
       b. A fluid enclosed in a pipe is compressed by two pistons, as shown in the
       same scheme as for part a. Find the shear stress on the plane B-B. Why
       is the result different from that in part a?

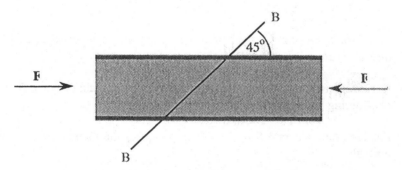

Figure P2.8

2.9   The stress at a point P is given by the matrix

$$\begin{bmatrix} 7 & 0 & -2 \\ 0 & 5 & 0 \\ -2 & 0 & 4 \end{bmatrix}.$$

Find the principal directions, i.e., the directions in which no shear stress
exists, and the stresses in these directions.

2.10   The state of stress at a point (x_o, y_o, z_o) is given by

$$\begin{bmatrix} 100 & 0 & 0 \\ 0 & 50 & 0 \\ 0 & 0 & -100 \end{bmatrix}.$$

Find the stress vector and the magnitudes of the normal and shearing stresses acting on the plane whose normal is

$$n = \frac{1}{\sqrt{14}}(i + 2j + 3k).$$

Is this material liquid or solid?

2.11   a. Show that in a two-dimensional flow field where the stress tensor is reduced to

$$T_{ij} = \begin{vmatrix} T_{xx} & T_{xy} \\ T_{yx} & T_{yy} \end{vmatrix} = \begin{vmatrix} A & B \\ C & D \end{vmatrix},$$

   the two principal directions are perpendicular to one another.
   b. Write the tensor in its principal directions and show that it contains no shear components.

2.12   Show for a two-dimensional field that if the fluid has no shear at all, all directions are principal.

2.13   Extend the results of Problems 2.11 and 2.12 to three-dimensional flows by considering pairs from i, j, k.

2.14   For the plane Couette flow find the stresses on the planes with the outer normals

$$n_1 = \frac{i + j}{\sqrt{2}}, \qquad n_2 = \frac{-i + j}{\sqrt{2}}.$$

2.15   For the Poiseuille flow between two plates find the stresses on the planes with the outer normals

$$n_1 = \frac{i + j}{\sqrt{2}}, \qquad n_2 = \frac{-i + j}{\sqrt{2}}$$

   at the locations $y = 0$, $y = h/2$ and $y = h$.

2.16   A flat plate of 1 m² is put inside a wind tunnel, as shown in Fig. P2.16. When the tunnel is run, the plate is held in place by a force of 50 N, which balances the drag force exerted by the air on the plate. The intake side of the tunnel is open to the atmosphere.

a. Find the stress on the two sides of the plate before the tunnel starts running.

b. Find the approximate stress on the surfaces of the plate while the tunnel is running.

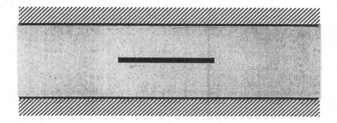

Figure P2.16

2.17   A model of a certain airplane wing, which is essentially a flat plate of 1 m², is put inside a wind tunnel such that its angle of attack is α, as shown in Fig. P2.17. The angle of attack is defined as the angle by which the plate is tilted up relative to the incoming flow direction. When the tunnel is run, the dynamometer balancing the plate shows a force of 50 N in the direction of the air velocity (drag force) and a force of 500 N in the upward direction (lift force). The intake side of the tunnel is open to the atmosphere.

a. Find the stress on the two sides of the plate before the tunnel starts running.

b. Find the approximate stress on the surfaces of the plate while the tunnel is running.

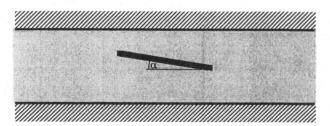

Figure P2.17

3. FLUID STATICS

The Equation of Hydrostatics

Hydrostatics is a branch of fluid mechanics which considers static fluids, i.e., fluids at rest.

A fluid was defined in Chapter 1 as a continuum which cannot support shear stresses when at rest. At that stage this definition was quite adequate. Now, however, one may inquire further: "at rest" - relative to what? The definition of a fluid is therefore extended now to "at rest relative to *any* coordinate system." The coordinate system itself may have any velocity or accelerations, and as long as a whole body of fluid is at rest relative to that coordinate system, it may support no shear stresses. It is quite important to realize that this coordinate system is not necessarily inertial. Review of Chapter 2 shows that all accelerations and body forces drop out of the considerations leading to the relations between the stress components. Therefore, all these relations, including Pascal's law, are valid for a fluid static in noninertial coordinates too. The momentum equations, (2.13), (2.14) and (2.15), however, do contain acceleration terms and may, therefore, require some further considerations on this account.

We start with the simpler case, where the system in which the fluid rests is inertial. The momentum equations, (2.13), (2.14) and (2.15), contain no shear terms for the case of the static fluid and become

$$\rho a_x = \frac{\partial T_{xx}}{\partial x} + \rho g_x,$$

$$\rho a_y = \frac{\partial T_{yy}}{\partial y} + \rho g_y,$$

$$\rho a_z = \frac{\partial T_{zz}}{\partial z} + \rho g_z.$$

(3.1)

Pascal's law states

$$T_{xx} = T_{yy} = T_{zz} = -p,$$ (3.2)

and we obtain

$$\rho a_x = -\frac{\partial p}{\partial x} + \rho g_x,$$

$$\rho a_y = -\frac{\partial p}{\partial y} + \rho g_y,$$ (3.3)

$$\rho a_z = -\frac{\partial p}{\partial z} + \rho g_z,$$

or, using vector notation,

$$\rho \mathbf{a} = -\nabla p + \rho \mathbf{g}.$$ (3.4)

For the special case of a fluid at rest with respect to a static or non-accelerating frame of reference the acceleration vector is zero. Hence

$$0 = -\nabla p + \rho \mathbf{g}.$$ (3.5)

Equation (3.5) is known as the *equation of hydrostatics*. It may be further simplified for the special case where $\mathbf{g}$ is the body force of gravity only and the coordinate system is oriented such that the direction of $\mathbf{g}$ coincides with the negative z-direction, i.e., $\mathbf{g} = -\mathbf{k}g$. With these simplifications it becomes

$$\partial p / \partial z = -\rho g,$$

$$\partial p / \partial y = 0,$$ (3.6)

$$\partial p / \partial x = 0.$$

The last two expressions in Eq. (3.6) imply that the pressure p is not a function of either x or y; hence

$$dp / dz = -\rho g,$$ (3.7)

which may be integrated for any ρg given as a function of z. The simplest case is, of course, that of an incompressible fluid for which the specific weight $\gamma = \rho g$ is constant. For this case integration of Eq. (3.7) between some reference position z_o and z yields

$$p = p_o + \gamma(z_o - z).$$ (3.8)

Here p depends on z only. An experiment illustrating this fact is shown in Fig. 3.1. A corollary of Eq. (3.8) states that the fluid in all branches of the apparatus shown in the figure rises to the same level.

Denoting the elevation difference between z_o and z by h, Eq. (3.8) becomes

$$p = p_o + \gamma h.$$ (3.9)

When the reference surface is the free surface, as in Fig. 3.1, p_o is the atmospheric pressure. Sometimes one is interested in the pressure excess above that of the atmosphere. This pressure is called the *gage pressure*, as distinguished from the *absolute pressure p*, and is expressible in terms of the liquid column height:

$$p_{\text{gage}} = p - p_o = \gamma h. \tag{3.10}$$

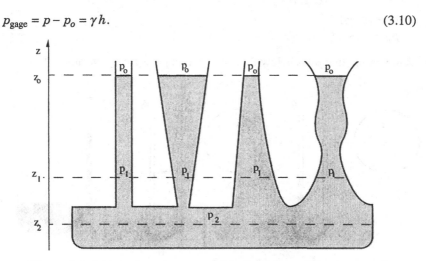

Figure 3.1 Pressure in an incompressible fluid at rest depends on z only.

Manometers

Manometers are instruments which measure fluid pressure. In the context of this chapter, however, manometers will denote arrangements which indicate fluid pressures by the measurement of the heights of fluid columns in them and the application of Eq. (3.8). We now survey several such arrangements, all shown in Fig. 3.2.

(a)   The Simple (U-Tube) Manometer, Fig. 3.2a

In order to find the pressure p_A, we write expressions for the pressure p_B in terms of the liquid in the left and right legs of the manometer, respectively,

$$p_B = p_A + \gamma(z_A - z_B)$$

and

$$p_B = p_o + \gamma(z_o - z_B).$$

Equating both equations for p_B, one obtains

$$p_A = p_o + \gamma(z_o - z_A) = p_o + \gamma h. \tag{3.11}$$

This result could be inferred directly by noting that the pressure at $z = z_A$ is the same in both legs. Similarly, for Fig. 3.2b one may write

$$p_o = p_A + \gamma h, \tag{3.12}$$
$$p_A = p_o - \gamma h, \tag{3.13}$$

thus indicating that p_A is below the atmospheric pressure by the amount γh.

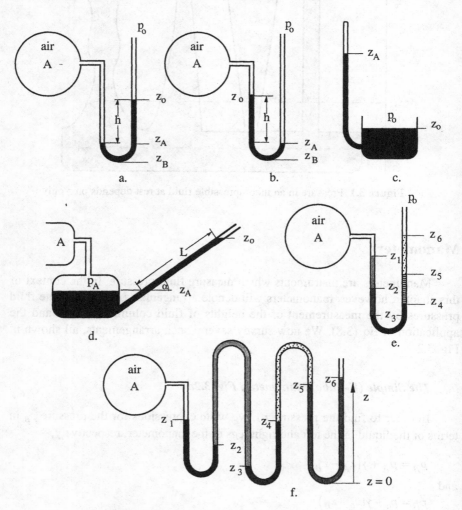

Figure 3.2 Manometers.

(b) The Barometer, Fig. 3.2c

$$p_A = p_0 - \gamma(z_A - z_0). \qquad (3.14)$$

This instrument is first completely filled with a fluid, usually mercury because of its high density and low vapor pressure. It is then inverted to its shown position. Thus, p_A indicates the vapor pressure of the manometric fluid. If one assumes that p_A is negligible, then the measurement of $z_A - z_0$ directly yields the atmospheric pressure

$$p_0 = \gamma(z_A - z_0). \qquad (3.15)$$

(c) The Inclined Manometer, Fig. 3.2d

Because of the large cross-sectional area of one side of this manometer, z_A is practically constant. Instead of measuring z_0, L is measured, which because of the inclination results in higher precision.

(d) The Multifluid Manometer, the Differential Manometer, Fig. 3.2e

Computing the pressure for each leg, one obtains

$$p = p_A + \gamma_{23}(z_2 - z_3) + \gamma_{12}(z_1 - z_2),$$
$$p = p_0 + \gamma_{34}(z_4 - z_3) + \gamma_{45}(z_5 - z_4) + \gamma_{56}(z_6 - z_5).$$

Elimination of p yields p_A:

$$p_A = p_0 + \gamma_{56}(z_6 - z_5) + \gamma_{45}(z_5 - z_4) + \gamma_{34}(z_4 - z_3) + \gamma_{23}(z_3 - z_2) + \gamma_{12}(z_2 - z_1)$$

or

$$p_A = p_0 + \sum_{i=1}^{n} \gamma_{i(i+1)}(z_{i+1} - z_i). \qquad (3.16)$$

Equation (3.16) also holds for the case shown in Fig. 3.2f. This is so because the hydrostatic pressure depends only on Δz of each fluid and not on the number of wiggles of the manometer tube. Moreover, if one of the fluids is a gas, the density of which might be negligible, its contribution to the pressure difference might also be neglected. The gas column does, however, transmit the pressure without modifying it.

Example 3.1

A liquid of density ρ flows in a tube in which an orifice plate is installed. The pressure drop across the orifice plate $(p_A - p_B)$ may be measured by the open stand

pipes or by the U-tube manometer, as shown in Fig. 3.3. The density of the manometric fluid is ρ_m.

a.   Find $p_A - p_B$ in terms of the manometer reading Δh_m.
b.   Express Δh in terms of Δh_m.

Solution

a.   Equating pressures in both legs of the manometer,

$$p_A + \gamma(h_2 - \Delta h_m) = p_B + \gamma_m \Delta h_m + \gamma h_2.$$

Hence

$$p_A - p_B = \Delta h_m (\gamma_m - \gamma).$$

b.   For each stand pipe

$$p_A = p_o + \gamma h_A,$$
$$p_B = p_o + \gamma h_B,$$
$$p_A - p_B = \gamma(h_A - h_B) = \gamma \Delta h.$$

Comparing this with part a,

$$\gamma \Delta h = \Delta h_m (\gamma_m - \gamma),$$

$$\Delta h = \Delta h_m \frac{\gamma_m - \gamma}{\gamma}$$

or

$$\Delta h = \Delta h_m \frac{\rho_m - \rho}{\rho}.$$

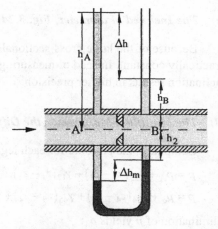

Figure 3.3 Pressure measurement.

Thus, for a large pressure drop, i.e., large Δh, one must select a heavy manometric
fluid, say mercury, in order to keep the manometer reading Δh_m within reason. For
small Δh one can amplify Δh_m by selecting a manometric fluid with a density ρ_m
only slightly higher than ρ.

Example 3.2

A mercury barometer consists of a long glass tube of 4 mm diameter, closed
at the top and immersed in a mercury reservoir of 40 mm diameter, Fig. 3.4. The
barometer was calibrated and was found to give an exact reading at 760 mm
mercury. Now it reads 750 mm. What is the atmospheric pressure?

Solution

The atmospheric pressure is approximately

$$p = \gamma h = 13,600 \times 9.81 \times 0.750$$

$$= 100,062 \text{ N / m}^2.$$

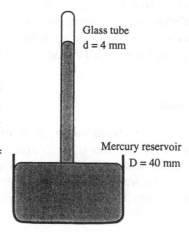

Glass tube
d = 4 mm

Mercury reservoir
D = 40 mm

To obtain the exact pressure, it is noted that when the column of mercury went down in the tube, it went up in the reservoir. It went down by Δh=10 mm and therefore had to rise in the reservoir by $\Delta H = \Delta h (d/D)^2 = 10(4/40)^2 = 0.1$ mm. The exact pressure is therefore

$$p = 13,600 \times 9.81 \times (760100.1) \times 10^{-3}$$

$$= 100,048.7 \text{ N/m}^2.$$

Note: The outer diameter of the glass tube is 6 mm. Is the "exact" pressure above correct? Answer: No. The exact rise in the reservoir is

Figure 3.4 Mercury barometer.

$$\Delta h \frac{d^2}{D^2 - d_o^2} = 10\frac{4^2}{40^2 - 6^2} = 0.102 \text{ mm}$$

and therefore,

$$p = 13,600 \times 9.81 \times [760 - 10 - 0.102] \times 10^{-3} = 100,048.4 \text{ N/m}^2.$$

Equation of Hydrostatics in Accelerating Frames of Reference

Hydrostatics is now extended to accelerating coordinate systems. Consider the case where a body of fluid moves in such a way that there is no relative motion between the fluid particles. This is sometimes referred to as a fluid in "rigid body motion." A coordinate system may be chosen such that this "rigid body" is fixed to it, i.e., is at rest relative to it. However, this coordinate system is not inertial. As already stated, Pascal's law holds for this case and no shear stress exists in the fluid. The only terms left are pressure, and the momentum equations which describe this case take the form of Eqs. (3.3) or (3.4), i.e.,

$$\rho\mathbf{a} = -\nabla p + \rho\mathbf{g}$$

or

$$0 = -\nabla p + \rho(\mathbf{g} - \mathbf{a}). \tag{3.17}$$

Equation (3.17) is identical to the equation of hydrostatics, Eq. (3.5), except for the body force term which is now replaced by $\mathbf{g} - \mathbf{a}$. It applies to a fluid at rest relative to a coordinate system having an acceleration $\mathbf{a}$. The equation of hydrostatics in a nonaccelerating frame, Eq. (3.4), can thus be considered a particular case of Eq. (3.17). The following example illustrates the use of Eq. (3.17).

Example 3.3

A cylindrical bucket, originally filled with water to a level h, rotates about its axis of symmetry with an angular velocity ω, as shown in Fig. 3.5. After some time the water rotates like a rigid body. Find the pressure in the fluid and the shape of the free surface, $z_o = f(r)$.

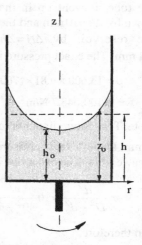

Figure 3.5 Rotating bucket.

Solution

The gravity and acceleration vectors for this case are, respectively,

$$\mathbf{g} = -\mathbf{e}_z g,$$
$$\mathbf{a} = -\mathbf{e}_r \omega^2 r,$$

and the three components of Eq. (3.17) are

$$0 = -\frac{\partial p}{\partial r} + \rho \omega^2 r,$$

$$0 = -\frac{1}{r}\frac{\partial p}{\partial \theta},$$

$$0 = -\frac{\partial p}{\partial z} - \rho g,$$

where $\partial p / \partial \theta = 0$ implies that p is not a function of θ. The other two equations are integrated to

$$p = \tfrac{1}{2}\rho \omega^2 r^2 + f_1(z),$$
$$p = -\rho g z + f_2(r).$$

The same p must satisfy both expressions; hence

$$p(r,z) = p_o + \tfrac{1}{2}\rho \omega^2 r^2 - \rho g z + c.$$

The constant c is eliminated by the use of the boundary condition $p(0, h_o) = p_o$; hence

$$p(r,z) = p_o + \tfrac{1}{2}\rho\omega^2 r^2 - \rho g(z - h_o).$$

The equation of the free surface $z_o(r)$ is found from the condition $p(r, z_o) = p_o$, and thus

$$z_o = h_o + \frac{\omega^2 r^2}{2g}.$$

The magnitude of h_o depends, of course, on the original volume of water in the bucket:

$$V = \pi R^2 h = \int_0^R z_o\, 2\pi r\, dr = \int_0^R \left(h_o + \frac{\omega^2 r^2}{2g} \right) 2\pi r\, dr.$$

Integration yields

$$\pi R^2 h \;=\; \pi R^2 \left(h_o + \frac{\omega^2 R^2}{4g} \right)$$

and finally

$$h_o = h - \frac{\omega^2 R^2}{4g}.$$

Forces Acting on Submerged Surfaces

The only stress present in a static fluid is the normal one, i.e., the pressure, given by Eq. (3.5). For any given body force **g**, Eq. (3.5) is considered integrable. In many cases gravity is the only body force present; hence

$$\mathbf{g} = -\,\mathbf{k}g.$$

For this case Eq. (3.5) assumes the form of Eq. (3.7), and integration results in Eq. (3.8) or (3.9). Surfaces in contact with a fluid are called *submerged surfaces* and must satisfy Eqs. (3.8) and (3.9). The force acting on an element of a submerged surface dS is then

$$d\mathbf{F} = -p\mathbf{n}\,dS, \tag{3.18}$$

n being the element's outer normal. The total force acting on the surface S is

$$\mathbf{F} = -\int_S p\mathbf{n}\,dS = -\int_S (p_o + \gamma h)\mathbf{n}\,dS, \tag{3.19}$$

where h is the depth of the fluid below the $p = p_o$ level. Equation (3.19) is literally used to compute the details of the forces acting on submerged surfaces. Such detailed information may be necessary in the design of individual plates in the sides of boats or in flood gates of dams.

Equations (3.18) and (3.19) are vectorial and may be resolved into their components. Thus the x-component of the force in Eq. (3.18) is

$$dF_x = \mathbf{i} \cdot d\mathbf{F} = -p(\mathbf{i} \cdot \mathbf{n})dS = -p\,dS_x.$$

Here dS_x is the projection of dS on the x-plane. Similarly from Eq. (3.19)

$$F_x = -\int_S p\,dS_x = -\int_S (p_o + \gamma h)dS_x, \qquad (3.20)$$

$$F_y = -\int_S p\,dS_y = -\int_S (p_o + \gamma h)dS_y, \qquad (3.21)$$

$$F_z = -\int_S p\,dS_z = -\int_S (p_o + \gamma h)dS_z. \qquad (3.22)$$

We first consider a special case of Eq. (3.19), i.e., that of the hydrostatic force on a plane surface. For this case one may choose a coordinate system such that two of its axes lie in that plane. In this system the direction of the force coincides with the axis perpendicular to that plane and just one of the Eqs. (3.20) - (3.22) need be used.

Force on a Submerged Plane Surface

Plane surfaces, as the one shown in Fig. 3.6, are represented by sides of liquid containers, flood gates of dams, etc. One side of such a surface is usually submerged in the liquid while the other side is exposed to the atmosphere of p_o. Thus the net force exerted on such a surface is due to the excess pressure above atmospheric.

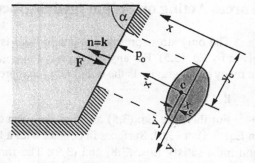

Figure 3.6 Force on a plane surface. The centroid of the surface is at C.

Let the coordinate system for this case be chosen such that the surface coincides with the z-plane, i.e., $\mathbf{n}\,dS_z = \mathbf{k}\,dS$ and the origin is at a pressure $p = p_o$, Fig. 3.6. Equation (3.19) now simplifies to Eq. (3.22),

$$F_z = -\int_S p\,dS = -\int_S \gamma h\,dS. \qquad (3.23)$$

Substitution of $h = y \sin \alpha$ results in

$$F_z = -\gamma \sin \alpha \int_S y\,dS = -\gamma y_c\, S \sin \alpha, \qquad (3.24)$$

$$F_z = -p_c S, \tag{3.25}$$

where y_c is the y-coordinate of the centroid of the area S, which is defined by

$$y_c = \frac{1}{S} \int_S y \, dS, \tag{3.26}$$

and $p_c = \gamma y_c \sin \alpha$ is the hydrostatic pressure at a depth corresponding to y_c.

Equation (3.25) is particularly useful for surfaces with known centroids. Thus, for example, the resultant force acting on a circular plate is $\pi R^2 p_c$ and its direction is normal to the plate surface. A resultant single force may be substituted for the distributed pressure forces acting on the total surface. To represent the distributed forces correctly, this resultant must

i. equal these forces in magnitude and direction;
ii. have its point of application such that its moment about any axis parallel to the x-coordinate axis equals the total moment of the distributed forces about the same axis; and
iii. have its point of application such that its moment about any axis parallel to the y-coordinate axis equals the total moment of the distributed forces about the same axis.

Requirement (i) is already satisfied by Eqs. (3.25) and (3.26). With the point of application denoted (x_F, y_F), requirement (ii) determines y_F, and requirement (iii) determines x_F. It is noted that for surfaces symmetrical with respect to the y-axis, x_F comes out to be zero. The same, however, does not apply to surfaces symmetrical with respect to the x-axis, because the pressure field is not symmetrical about the x-axis.

To find this point of application (x_F, y_F), also known as the center of pressure, requirement (ii) implies that moments be taken about the x-axis:

$$F_z y_F = -\int_S y \, dF = -\gamma \sin \alpha \int_S y^2 \, dS. \tag{3.27}$$

The surface integral in Eq. (3.27) is identified as the second moment of the area, or the moment of inertia of the area with respect to the x-axis, i.e., I_{xx}. This moment of inertia about the x-axis can be related to the moment of inertia about a parallel x-axis, passing through the centroid of the surface, Fig. 3.6, by the use of a geometrical relation known as Steiner's Theorem.

Let a coordinate system parallel to the x, y system but with its origin at the centroid c be x', y', Fig. 3.6. Steiner's theorem then states

$$I_{xx} = \int_S y^2 \, dS = I_{x'x'} + S y_c^2. \tag{3.28}$$

Equation (3.27) together with Eqs. (3.28) and (3.24) yield the y-coordinate of the center of pressure,

$$y_F = \frac{1}{y_c S} \int_S y^2 \, dS = \frac{I_{x'x}}{y_c S} + y_c \, . \qquad (3.29)$$

Some moments of inertia $I_{x'x'}$ are tabulated in standard handbooks and may be used instead of the integration in Eq. (3.29).

For a plane surface not symmetrical with respect to the y-axis x_F may be found in a similar way from

$$F_z x_F = \int_S x \, dF = -\gamma \sin \alpha \int_S xy \, dS = -\gamma \sin \alpha \, I_{xy} \, . \qquad (3.30)$$

Steiner's theorem is used again in the form

$$I_{xy} = I_{x'y'} + S x_c y_c \, . \qquad (3.31)$$

This yields

$$x_F = \frac{1}{y_c S} \int_S xy \, dS = \frac{I_{x'y'}}{y_c S} + x_c \, , \qquad (3.32)$$

where the x-coordinate of the area centroid is defined by

$$x_c = \frac{1}{S} \int_S x \, dS. \qquad (3.33)$$

Example 3.4

a. A vertical rectangular plate $AA'B'B$ is set under water ($\rho = 1{,}000 \text{ kg/m}^3$), Fig. 3.7a. Find the resultant force, its direction and its point of application, y_F.

b. A larger plate is now shown in Fig. 3.7b. Find the resultant and its point of application.

Solution

a. The centroid of the plate is at C, which is $L/2$ deep. The force is obtained from Eq. (3.24) as

$$\mathbf{F} = -\mathbf{k}\gamma \frac{L}{2}(D \cdot L) = -\mathbf{k} \, 500 g D L^2 \, .$$

The moment of inertia of the rectangular plate about its centroid is

$$I_{x'x'} = 2\int_0^{L/2} D y^2 \, dy = \frac{2}{3} D \left(\frac{L}{2}\right)^3 = \frac{1}{12} D L^3 \, .$$

Equation (3.29) then yields the y-coordinate of the center of pressure

$$y_F = \frac{I_{x'x'}}{y_c S} + y_c = \frac{\frac{1}{12}DL^3}{\frac{L}{2}LD} + \frac{L}{2} = \frac{2}{3}L.$$

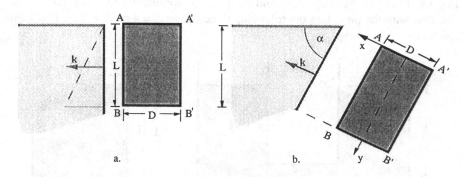

Figure 3.7 Rectangular plate under water.

Note: A diagram of the pressure as a function of depth, drawn along AB (dashed line in Fig. 3.7a), is triangular. The center of gravity of this triangle is $2/3\ L$ deep. The point of application of the resultant is also $2/3\ L$ deep. Is there a connection?

b.   Equation (3.24) yields the force exerted on the plate,

$$\mathbf{F} = -\mathbf{k}\gamma\frac{L}{2}\left(D\cdot\frac{L}{\sin\alpha}\right) = -\mathbf{k}\frac{500gDL^2}{\sin\alpha}.$$

Substitution of the moment of inertia and the y-coordinate of the centroid of the area, respectively,

$$I_{x'x'} = \frac{\frac{1}{12}DL^3}{\sin^3\alpha}, \qquad y_c = \frac{L}{2\sin\alpha},$$

into Eq. (3.29) yields the y-coordinate of the center of pressure,

$$y_F = \frac{\frac{\frac{1}{12}DL^3}{\sin^3\alpha}}{\frac{L}{2\sin\alpha}\left(\frac{DL}{\sin\alpha}\right)} + \frac{L}{2\sin\alpha} = \left(\frac{1}{6}+\frac{1}{2}\right)\frac{L}{\sin\alpha} = \frac{2L}{3\sin\alpha}.$$

The point of application is again at two-thirds of the plate and at two-thirds of the maximal depth.

Example 3.5

Two reservoirs, A and B, are filled with water, and connected by a pipe, Fig. 3.8. Find the resultant force and the point of application of this force on the partition with the pipe. Use the following dimensions: $L = 4$ m, $H = 3$ m, $D = 1$ m, $b = 3$ m.

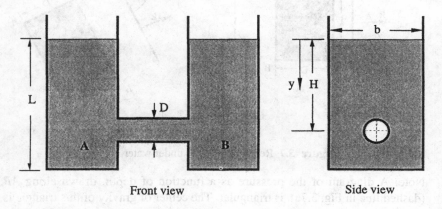

Figure 3.8 Two reservoirs connected by a pipe.

Solution

To find the force and its point of application on a plate with a circular hole, we perform the following steps:

Find the force and point of application for the plate without the hole.
Find the force and point of application for a circular plate having the same dimensions and location as the hole.

The sought force is the difference between that of the rectangular plate and that of the circle. The combined point of application is obtained by the requirement that the moment of the resultant force must equal the difference between the moments of the plate and of the circle.

Let the subscripts 1 and 2 refer to the plate and to the circle, respectively. For the rectangular plate without the hole

$$-F_1 = L \times b \times \gamma \frac{L}{2} = \frac{\gamma L^2 b}{2} = \frac{9{,}810 \times 4^2 \times 3}{2} = 235{,}440 \text{ N.}$$

In Example 3.4 it was shown that the point of application is at $2/3\,L$. Hence

$$y_{F1} = \frac{2L}{3} = 2.667 \text{ m}.$$

The force on a circular plate the size of the hole is

$$-F_2 = \gamma H \frac{\pi D^2}{4} = 9,810 \times 3 \frac{\pi \times 1^2}{4} = 23,114 \text{ N}.$$

For a circular plate

$$I_{x'x'} = \tfrac{1}{2} I_r = \tfrac{1}{2} \int_0^R r^2 \, 2\pi r \, dr = \frac{\pi R^4}{4} = \frac{\pi D^4}{64}$$

and Eq. (3.29) yields

$$y_{F2} = \frac{\pi D^4 / 64}{\pi D^2 H / 4} + H = \frac{D^2}{16H} + 3 = 3.02 \text{ m}.$$

The resultant force on the plate with the hole is

$$-F_{12} = -(F_1 - F_2) = 235,440 - 23,114 = 212,326 \text{ N}.$$

The point of application is found from

$$y_F F_{12} = y_{F1} F_1 - y_{F2} F_2.$$

Hence

$$y_F = \frac{y_{F1} F_1 - y_{F2} F_2}{F_{12}} = \frac{2.667 \times 235,440 - 3.02 \times 23,114}{212,326} = 2.63 \text{ m}.$$

Example 3.6

a. The inclined triangular wall in Fig. 3.9a with $\alpha = 60°$, $L = 4\,\text{m}$, $H = 6\,\text{m}$ is immersed in water (γ=9,810 N/m³). Find the resultant force representing correctly the effects of the hydrostatic pressure.
b. Repeat part a for the triangle in Fig. 3.9b.

Solution

a. The centroid of a triangle is at the point of intersection of its medians, which is at one-third of its height. This can be shown analytically by

$$S y_c = \int_{y=0}^{H} y \, dS = \int_{y=0}^{H} y \int_{x=0}^{b} dx \, dy = \int_{y=0}^{H} by \, dy,$$

$$S y_c = \int\limits_{y=0}^{H} L\left(1-\frac{y}{H}\right) y\, dy = LH^2\left(\tfrac{1}{2}-\tfrac{1}{3}\right) = \tfrac{1}{6}LH^2 = \frac{H}{3}\times\frac{LH}{2} = \frac{H}{3}\times S.$$

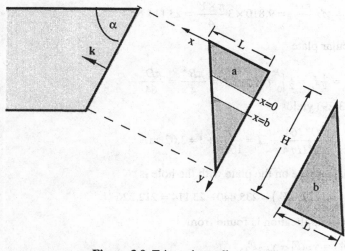

Figure 3.9 Triangular walls.

Hence, by Eq. (3.24),

$$-F_z = \gamma S y_c \sin\alpha = \gamma\left(\tfrac{1}{6}LH^2\right)\sin\alpha = 9{,}810\times\left(\tfrac{1}{6}\times 4\times 6^2\right)\sin 60^\circ = 203{,}897 \text{ N}.$$

In this case it is easier to obtain I_{xx} rather than $I_{x'x'}$; hence

$$I_{xx} = \int\limits_{y=0}^{H} L\left(1-\frac{y}{H}\right) y^2\, dy = LH^3\left(\tfrac{1}{3}-\tfrac{1}{4}\right) = \frac{LH^3}{12} = 72\,\text{m}^4,$$

and by Eq. (3.27),

$$y_F = \frac{I_{xx}\gamma\sin\alpha}{-F_z} = \frac{9{,}810\times\sin 60^\circ\times 72}{203{,}897} = 3 \text{ m}.$$

Similarly,

$$I_{xy} = \int\limits_{y=0}^{H} y \int\limits_{x=0}^{b} x\, dx\, dy = \int\limits_{y=0}^{H} \frac{L^2}{2}\left(1-\frac{y}{H}\right)^2 y\, dy = \frac{L^2}{2}\int\limits_{y=0}^{H}\left(1-\frac{2y}{H}+\frac{y^2}{H^2}\right) y\, dy$$

$$= \frac{L^2 H^2}{2}\left(\frac{1}{2}-\frac{2}{3}+\frac{1}{4}\right) = \frac{L^2 H^2}{24} = 24\,\text{m}^4,$$

and by Eq. (3.30),

$$x_F = \frac{I_{xy}\gamma\sin\alpha}{-F_z} = \frac{9,810\times\sin 60°\times 24}{203,897} = 1 \text{ m}.$$

b.   Here we have $y_c = 4$ m. Hence

$$-F_z = \gamma S y_c \sin\alpha = \gamma\left(\tfrac{1}{3}LH^2\right)\sin\alpha = 9,810\times\left(\tfrac{1}{3}\times 4\times 6^2\right)\sin 60° = 407,794 \text{ N}.$$

From part a above

$$I_{xx_a} = 72 = I_{x'x'} + y_{c_a}^2 \times S = I_{x'x'} + 4\times 12 = I_{x'x'} + 48,$$
$$I_{x'x'} = 24 \text{ m}^4.$$

Equation (3.29) now yields

$$y_F = \frac{24}{4\times 12} + 4 = 4.5 \text{ m}.$$

Similarly, from a above

$$I_{xy_a} = 24 = I_{x'y'} + x_{c_a}y_{c_a}\times S = I_{x'y'} + 32,$$
$$I_{x'y'_a} = -8 \text{ m}^4,$$

$$I_{xy} = \int_{y=0}^{H} y\int_{x=-b}^{0} x\,dx\,dy = -\int_{y=0}^{H}\frac{L^2 y^2}{2H^2}y\,dy = -\frac{L^2 H^2}{2}\times\frac{1}{4} = -\frac{L^2 H^2}{8} - 72\,\text{m}^4.$$

Finally, by Eq. (3.30)

$$x_F = \frac{I_{xy}\gamma\sin\alpha}{-F_z} = \frac{9,810\times\sin 60°\times(-72)}{407,794} = -1.5 \text{ m}.$$

Note: y_F for case a is at one-half the depth, and y_F for case b is at three-fourths of the depth. Compare this to the rectangular case, Example 3.4, where y_F came out to be at two-thirds of the depth.

Forces on Submerged Surfaces – The General Case

As shown in Fig. 3.10, submerged surfaces divide into three distinct groups, namely those belonging to

a.   completely submerged bodies (S_1, S_2);
b.   floating bodies (S_3, S_4); and
c.   submerged boundaries (S_5, S_6).

Of these the easiest to analyze is the completely submerged body, because it is bounded by surfaces of one kind only, i.e., submerged surfaces. We therefore consider a body of this type first.

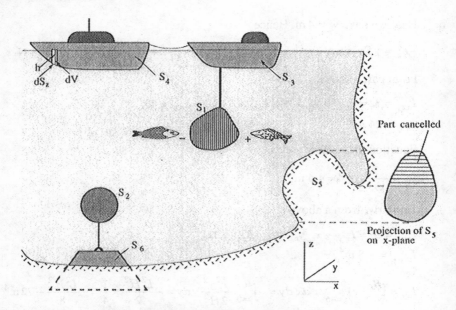

Figure 3.10 Submerged bodies.

a. Completely Submerged Bodies (S_1, S_2, Fig. 3.10)

The total force acting on a completely submerged surface, such as S_1, is given by Eq. (3.19), which, upon substitution of p from Eq. (3.8), reads

$$F = -\int_S p\mathbf{n}\, dS = -\int_S (p_o + \gamma z_o - \gamma z)\mathbf{n}\, dS. \tag{3.34}$$

The horizontal components of $\mathbf{F}$, F_x and F_y are given by Eqs. (3.20) and (3.21), respectively,

$$F_x = -\int_S p\, dS_x = -\int_S (p_o + \gamma z_o - \gamma z)\, dS_x, \tag{3.35}$$

$$F_y = -\int_S p\, dS_y = -\int_S (p_o + \gamma z_o - \gamma z)\, dS_y. \tag{3.36}$$

We now proceed to find F_x by integrating Eq. (3.35). It is convenient to carry out the integration in two stages: first by performing it at constant z on horizontal strips of thickness dz and then by integrating over all strips, i.e., on all z. Equation (3.35) thus becomes

$$F_x = - \int_{\substack{\text{all strips}}} (p_o + \gamma z_o - \gamma z) \int_{\substack{\text{one strip}}} dS_x .$$

(3.37)

The integral $\int dS_x$ over a single strip yields twice the projection of the strip on the x-plane, once with a positive sign as viewed by the happy fish in Fig. 3.10 and once with a negative sign as viewed by the unhappy fish. The integral's total contribution is therefore nil, and F_x vanishes. The same result is obtained for F_y with the conclusion that *completely submerged bodies suffer no net horizontal force.*

The argument which led to vanishing horizontal resultants does not apply to the vertical component. The pressure, p, depends on z and cannot be set outside the integral over the single strip, as done in Eq. (3.37). For S_1 this vertical force is easily obtained by the application of Gauss' theorem to Eq. (3.19):

$$\mathbf{F} = -\int_S p\mathbf{n}\, dS = -\int_V \nabla p\, dV = -\mathbf{k} \int_V \frac{dp}{dz}\, dV = \mathbf{k} \int_V \gamma\, dV = \mathbf{k}\gamma V ,$$

(3.38)

where γV is the weight of the fluid displaced by the submerged body. Denoting by γ_B the average specific weight of the submerged body, the force applied to the boat by the hanging rock S_1 is $-\mathbf{k}V_1(\gamma_{B1} - \gamma)$, while that applied to the concrete block S_6 by the float S_2 is $\mathbf{k}V_2(\gamma - \gamma_{B2})$. The vertical force given by Eq. (3.38) is called *buoyancy force*

The *center of buoyancy* of a submerged body is defined as the point of application of the buoyancy force, i.e., the point with respect to which the moment of the forces exerted on the submerged body by the fluid vanishes. Specifically, for an arbitrary choice of the origin of the coordinate system the x-coordinate of the center of buoyancy, x_B, is found by

$$x_B \gamma V = \int_V x\gamma\, dV .$$

(3.39)

Hence

$$x_B = \frac{1}{V} \int_V x\, dV ,$$

(3.40)

and similarly

$$y_B = \frac{1}{V} \int_V y\, dV .$$

(3.41)

Let z_b be defined as $z_B = \frac{1}{V} \int_V z\, dV$. Then the point (x_b, y_b, z_b) comes out to be the center of gravity of the displaced fluid. The center of buoyancy thus coincides with the center of gravity of the displaced fluid. Considering this point as that at which the buoyancy force acts results in the buoyancy force correctly representing the fluid forces as far as forces and moments are concerned.

Example 3.7

A quick and fairly accurate way to measure the density ρ of a solid sample is to weigh the sample in air and then again in a known liquid, e.g., in water, Fig. 3.11. Find ρ if the ratio of the weight in water to that in air is $G_w / G_a = 0.8883$.

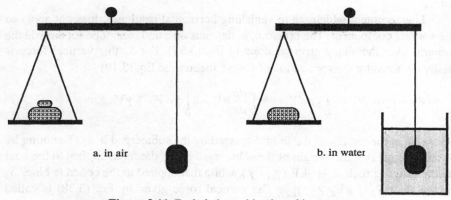

a. in air   b. in water

Figure 3.11 Body balanced in air and in water.

Solution

Let the sample weight in air be G_a and that in water be G_w. Then

$$G_a = V\rho g,$$
$$G_w = V\rho g - V\rho_w g = Vg(\rho - \rho_w),$$

where $V\rho_w g$ is the buoyancy force. Hence

$$\frac{G_w}{G_a} = \frac{\rho - \rho_w}{\rho} = 1 - \frac{\rho_w}{\rho}$$

or

$$\frac{\rho_w}{\rho} = 1 - \frac{G_w}{G_a} = 1 - 0.8883 = 0.1117.$$

Hence,

$$\rho = \rho_w / 0.1117 = 1,000 / 0.1117 = 8,953 \text{ kg/m}^3,$$

and the sample could be made of nickel.

Question: How do you perform this procedure for samples lighter than water?
Answer: Tie a piece of metal to the sample.

Example 3.8

In preparing astronauts for moving on the surface of the moon, they are put with their space outfits at the bottom of a pool filled with water ($\rho_w = 1000$ kg/m³). The volume of a 75-kg astronaut wearing a space outfit is 0.085 m³. Gravity on the surface of the moon is about $g/6$. The mass of the suit is about 50 kg. Should astronauts be equipped with weights or with floats to have them feel as if under moon gravity?

Solution

The force acting on the astronaut on the moon is

$$F_m = (75 + 50)\frac{9.81}{6} = 204.4 \text{ N.}$$

The force under water is

$$F_w = g\left[(75 + 50) - 0.085\rho_w\right] = 9.81 \times [75 + 50 - 85] = 392.4 \text{ N.}$$

The astronaut needs floats of about 0.02 m³ (neglecting the floats' own weight) in order to reduce the force by $392.4 - 204.4 = 188$ N.

b.   Floating Bodies (S_3, S_4, Fig. 3.10)

Floating bodies differ from completely submerged bodies by not being exclusively enclosed by surfaces in contact with the fluid. Rather, they stick out of the fluid. Floating bodies are always associated with a free surface of the fluid. Let this free surface have the coordinate z_o. The submerged part of the floating body is now defined as that part of the body bounded by its wetted parts and by the extension of the z_o plane into the floating body. That part of a ship which is below the water line is an example of the submerged part of a floating body.

Inspection of the analysis of the completely submerged bodies reveals that the same analysis may be repeated here for the computation of the horizontal components of the force, with the conclusion that no horizontal resultant exists for the floating body.

The vertical force F_z may be easily calculated using Eq. (3.22):

$$F_z = -\int_S \left(p_o + \gamma h \right) dS_z = -\int_S p_o\, dS_z - \gamma \int_S h\, dS_z = \gamma V.$$

The contribution of p_o on the upward acting force F_z cancels out by the downward acting p_o on the exposed part of the body. Thus the total force exerted on the floating body by the fluid is still $\mathbf{F} = \mathbf{k}\gamma V$, as in Eq.(3.38), but with V denoting the volume of the submerged part only. The buoyancy force obtained is again equal to the weight of the displaced fluid. This last result for both the completely submerged body and for the floating one is also known as *Archimedes' Law*.

The center of buoyancy of the floating body is again located at the center of gravity of the displaced fluid, as can be easily verified by repeating the considerations used for the submerged body.

Example 3.9

Ice at -10°C has the density $\rho_i = 998.15$ kg/m^3. A 1,000-ton (10^6-kg) spherical iceberg floats at sea, Fig. 3.12. The salty water has a density of $\rho_S = 1,025$ kg/m^3. By how much does "the tip of the iceberg" stick out of the water?

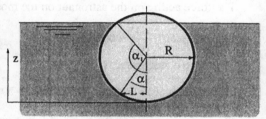

Figure 3.12 Spherical iceberg.

Solution

The volume of the iceberg is

$$V = \frac{G_i}{\rho_i} = \frac{10^6}{998.15} = 1001.9 \text{ m}^3,$$

which corresponds to a sphere of radius $R = 6.207$ m. The volume of sea water it must displace, by Archimedes' law, is

$$V_w = 10^6 / 1{,}025 = 975.6 \text{ m}^3.$$

From Fig. 3.12,

$$z = R(1 - \cos\alpha), \qquad dz = R\sin\alpha\, d\alpha,$$
$$L = R\sin\alpha, \qquad dV = \pi L^2 dz = \pi R^3 \sin^3\alpha\, d\alpha,$$

$$V = \int_0^{\alpha_1} \pi R^3 \sin^3\alpha \, d\alpha = \pi R^3 \int_0^{\alpha_1}\left(1 - \cos^2\alpha\right)\sin\alpha \, d\alpha$$

$$= \pi R^3\left(-\cos\alpha_1 + \tfrac{1}{3}\cos^3\alpha_1\right) + \tfrac{2}{3}\pi R^3 = 975.6 \text{ m}^3.$$

Hence, for $R = 6.207$ m we obtain

$$\alpha_1 = 144^\circ, \qquad z_1 = 11.229 \text{ m}.$$

The height of the part sticking out of the water is

$$2R - z_1 = 1.185 \text{ m}.$$

For an iceberg that is cubic, the side of the cube is $1,001.9^{1/3} = 10.006$ m, and the height of the part sticking out is

$$10.006 - \frac{975.6}{10.006^2} = 0.262 \text{ m}.$$

c. Submerged Boundaries (S5, S6, Fig. 3.10)

The plane surface has already been treated in detail in a previous section. The surface S_5 in Fig. 3.10 is a more general example of such a boundary. The upper part of S_5 is completely girdled by horizontal strips, and therefore this part has its projections on the x- and y-planes canceled and does not contribute to the horizontal force. However, the lower part of S_5 cannot be thus girdled. There are therefore net horizontal forces acting on S_5. Each case of such a general submerged surface must be calculated separately by the use of Eqs. (3.34), (3.35) and (3.36). The only somewhat simplifying statements which can be made for these surfaces are:

For horizontal components: Equations (3.35) and (3.36) do not change if the surface S is replaced by its projections, i.e., dS_x, dS_y, respectively. Different surfaces which have the same projections in a certain horizontal direction thus suffer the same horizontal component of the force in that direction.

For vertical components: Equation (3.34) becomes

$$F_z = -\int_S p \, ds_z.$$

A *vertical ray* entering S at a particular dS_z may sometimes pierce S again. When this happens, the two pierced dS_z may be counted together in the integral to yield $dF_z = \gamma \,(\Delta z)\, dS_z$, where Δz is the distance along the ray between its entry and exit points.

Thus dF_z becomes just $\gamma \, dV$, with dV the volume of the cylinder formed by the base dS_z and the height Δz. When a part of the submerged surface can be completely separated from the rest of it by a surface formed entirely by such piercing rays, the resultant vertical force acting on this separated section is the buoyancy force corresponding to its volume.

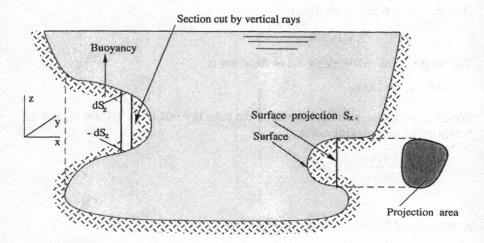

Figure 3.13 Submerged surfaces.

Examples for the use of these statements are illustrated in Fig. 3.13.

Example 3.10

Figure 3.14 shows three types of water gates. Assume each gate is 1 m wide and that the depth of the water is $H = 4$ m. What is the resultant force of the water pressure exerted on each gate, and what is its line of application, such that the correct moment results.

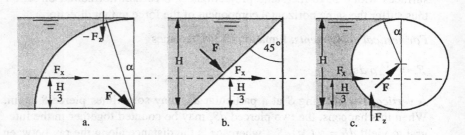

Figure 3.14 Shapes of gates.

Solution

For gate a

$$p = \rho g h = \gamma H (1 - \cos \alpha).$$

Using Eq. (3.18), we have

$$d\mathbf{F} = -p \mathbf{n} dS = -p \mathbf{n} H \, d\alpha,$$

where $\mathbf{n} = \mathbf{r}$; hence

$$dF_x = -pH \sin \alpha \, d\alpha = \gamma H^2 (1 - \cos \alpha) \sin \alpha \, d\alpha,$$

$$F_x = \int_0^{\pi/2} \gamma H^2 (1 - \cos \alpha) \sin \alpha \, d\alpha = \gamma H^2 \left[1 - \tfrac{1}{2} \right] = \tfrac{1}{2} \gamma H^2,$$

$$z_c F_x = \int_0^{\pi/2} z \gamma H^2 (1 - \cos \alpha) \sin \alpha \, d\alpha = H^3 \gamma \int_0^{\pi/2} \left(\cos \alpha - \cos^2 \alpha \right) \sin \alpha \, d\alpha$$

$$= H^3 \gamma \left[\tfrac{1}{2} - \tfrac{1}{3} \right] = \tfrac{1}{6} H^3 \gamma, \qquad z_c = H / 3,$$

$$-dF_z = pH \cos \alpha \, d\alpha = \gamma H^2 (1 - \cos \alpha) \cos \alpha \, d\alpha,$$

$$-F_z = \int_0^{\pi/2} \gamma H^2 \left(\cos \alpha - \cos^2 \alpha \right) d\alpha = \gamma H^2 \left[1 - \frac{\pi}{4} \right] = 0.2146 \gamma H^2,$$

$$(-F_z)(-x_c) = \int_0^{\pi/2} \gamma H^3 \left(\cos \alpha - \cos^2 \alpha \right) \sin \alpha \, d\alpha = \gamma H^3 \left[\frac{1}{2} - \frac{1}{3} \right] = \frac{\gamma H^3}{6},$$

$$(-x_c) = 0.7766 \, H,$$

$$|\mathbf{F}| = \left| \sqrt{F_x^2 + F_z^2} \right| = \left| \sqrt{\left(\tfrac{1}{2} \gamma H^2 \right)^2 + \left(0.2146 \gamma H^2 \right)^2} \right|$$

$$= 0.5441 \gamma H^2 = 0.5441 \times 9,810 \times 4^2 = 85,403 \, \text{N}$$

and

$$\alpha_F = \arctan \left| \frac{F_x}{F_z} \right| = \arctan \left| \frac{\tfrac{1}{2}}{0.2146} \right| = 66.7^\circ.$$

For gate b, we have from Example 3.4

$$z_c = \frac{H}{3},$$

$$F = H \sqrt{2} \times \gamma \times \tfrac{1}{2} H = 0.7071 \gamma H^2 = 110,986 \, \text{N}.$$

The point of application of that force is at $H/3$, and its direction is perpendicular to the inclined plane.

For gate c

$$p = \tfrac{1}{2}\gamma H(1-\cos\alpha),$$

$$dF_x = p\tfrac{H}{2}\sin\alpha\, d\alpha = \tfrac{1}{4}\gamma H^2(1-\cos\alpha)\sin\alpha\, d\alpha,$$

$$F_x = \int_0^\pi \tfrac{1}{4}\gamma H^2(1-\cos\alpha)\sin\alpha\, d\alpha = \tfrac{1}{4}\gamma H^2[2-0] = \tfrac{1}{2}\gamma H^2,$$

$$z_c F_x = \int_0^\pi \tfrac{1}{4}z\gamma H^2(1-\cos\alpha)\sin\alpha\, d\alpha = \tfrac{1}{8}\gamma H^3\int_0^\pi(1+\cos\alpha)(1-\cos\alpha)\sin\alpha\, d\alpha$$

$$= \tfrac{1}{8}\gamma H^3\int_0^\pi(1-\cos^2\alpha)\sin\alpha\, d\alpha = \tfrac{1}{8}\gamma H^3[2-\tfrac{2}{3}] = \tfrac{1}{6}\gamma H^3.$$

Hence

$$z_x = \frac{z_x F_x}{F_x} = \frac{H}{3}.$$

The force acting upward, F_z, is found by Archimedes' law as the buoyancy force acting on half the cylinder:

$$F_z = \gamma V = \gamma\left(\frac{\pi H^2 \times 1}{4\times 2}\right) = 0.3927\gamma H^2.$$

We also know from solid mechanics that the center of gravity of half a circle is at

$$-x_c = \frac{4}{3\pi}r = \frac{2}{3\pi}H = 0.2122\,H,$$

$$|\mathbf{F}| = \left|\sqrt{F_x^2 + F_z^2}\right| = \left|\sqrt{\left(\tfrac{1}{2}\gamma H^2\right)^2 + \left(0.3927\gamma H^2\right)^2}\right|$$

$$= 0.6358\gamma H^2 = 0.6358\times 9{,}810\times 4^2 = 99{,}792 \text{ N}$$

and finally

$$\alpha_F = 360 - \arctan\left|\frac{F_x}{F_z}\right| = 308°.$$

Note that in both cases a and c there is no moment about the center of the cylinder.

Example 3.11

A float valve is schematically shown in Fig. 3.15. In its closed position the sphere is pressed by the water pressure onto the valve seat. To open the valve, force F must be applied to overcome the water pressure and the valve weight G.

Once open, the valve must stay open until the flush tank is emptied. It is kept open by the buoyancy forces overcoming its weight G. Given the radius of the ball R, find bounds for permissible G and required F.

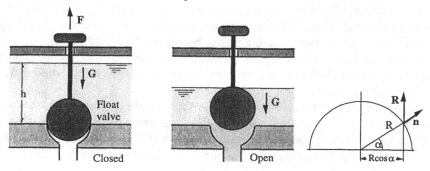

Figure 3.15 Float valve.

Solution

The weight of the valve G should be less than that of the water it displaces, i.e.,

$$G < \tfrac{4}{3}\pi R^3 \gamma_w .$$

The applied force F should be

$$F \geq G + \gamma \int_o^{\pi/2} (h - R\sin\alpha)2\pi R\cos\alpha\, R\sin\alpha\, d\alpha$$

$$= G + \gamma 2\pi R^2 \int_o^{\pi/2} \left(\sin\alpha\cos\alpha - R\sin^2\alpha\,\cos\alpha\right)d\alpha$$

or

$$F \geq G + \gamma\pi R^2\left(h - \tfrac{2}{3}R\right).$$

For $R = 3$ cm, $h = 30$ cm,

$$G < \tfrac{4}{3}\pi R^3 \gamma_w = \tfrac{4}{3}\pi \times 0.03^3 \times 9{,}810 = 1.11\,\text{N}$$

and

$$F \geq G + \gamma\pi R^2\left(h - \tfrac{2}{3}R\right) = 1.11 + 9{,}810\pi \times 0.03^2\left(0.3 - \tfrac{2}{3}\times 0.03\right) = 8.88\ \text{N} .$$

Hydrostatic Stability

A completely submerged body or a floating one has its buoyancy force acting through its center of buoyancy in the positive z-direction. It also has its weight acting through its center of gravity in the negative z-direction.

The body can remain static, i.e., not move up or down only when these two forces are equal in magnitude. Furthermore, it must have its center of buoyancy and its center of gravity on the same vertical line or it cannot be in equilibrium, i.e., it will roll. When this equilibrium exists, however, it can still be stable, unstable or neutral. Static stability is important both for structures which are apparently static, such as dams, floating cranes, buoys, etc., and for those which move at noticeable speeds, such as ships or submarines. It serves as a necessary condition in the design of the structures and also as an approximation for their dynamic stability near equilibrium.

At static equilibrium the moment of the couple of forces, buoyancy and weight is zero, because they act along the same vertical line, e.g., Fig. 3.16a,c. When the considered body rolls with the angle α, the relative positions of the center of gravity, G, and the center of buoyancy, B, change, while the forces themselves remain vertical.

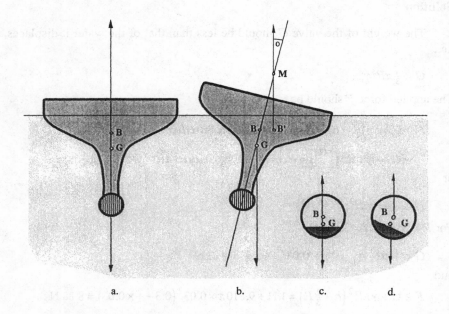

Figure 3.16 A submarine and a sail boat with center of gravity below center of buoyancy.

A moment may appear, which tends to further increase the roll angle α, in which case the situation is called unstable; or the moment may tend to decrease α and diminish the roll, in which case the situation is denoted stable. When no moment appears, a neutral equilibrium is inferred. The submarine with its cylindrical cross section in Fig. 3.16c,d is in stable equilibrium. It is seen at once that a submarine, while underwater, can be in stable equilibrium only if its center of gravity is below its center of buoyancy. Indeed, such a submarine would correct its position even from a roll of $\alpha = 180°$, assuming that nothing has been shifted inside it while rolling.

This manifestation of absolute stability seems very attractive and some sail boats are designed this way (Fig. 3.16a,b), mainly boats that must withstand very large moments caused by high masts and large sails. Unfortunately, when a wave strikes from the side a boat so designed, the boat rolls through large angles, because the wave's momentum is transferred to it quite above its center of gravity. For this reason most boats and all ships have their center of gravity in the vicinity of the water line, where waves usually strike. This results, however in the center of gravity being above the center of buoyancy. Such a boat and a stable one is shown in Fig. 3.17.

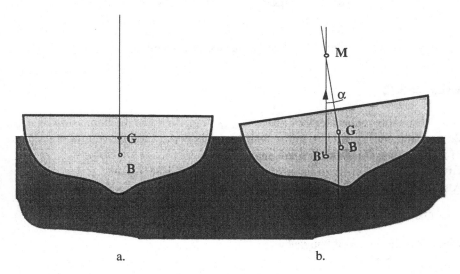

Figure 3.17 A stable boat.

The point M in Fig. 3.17 is obtained by the intersection of the line of action of the instantaneous buoyancy force with the straight line passing through the center of gravity and the center of buoyancy of the structure at equilibrium. The length GM is called the *metacentric height,* and one notes at once that as long as $GM > 0$ (i.e., M is above G), the α range of stable equilibrium has not been exceeded. Neglecting dynamical effects, this means that the ship will not capsize.

Historically the line BG was provided by the mast, and a ship master who knew his ship could estimate the metacentric height at different roll angles. This criterion, however, is as good as any other and is still used by naval architects.

Some more examples of stable and unstable equilibria are shown in Fig. 3.18.

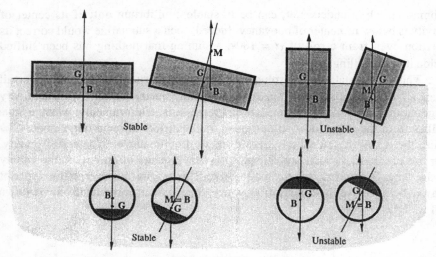

Figure 3.18 Stable and unstable equilibria.

Example 3.12

The rectangular block of balsa wood in Fig. 3.19 has the dimensions $a \times b \times 2c$ and the density ρ_b. It is put in water of density ρ_w. Find the returning moment of the block as it rolls with small angles.
Given: $a = b = 4$ m, $2c = 2$ m, $\rho_b = 500$ kg/m³, $\rho_w = 1000$ kg/m³.
Also given from trigonometric considerations,

$$\overline{QB} = \tfrac{5}{6}\sin\alpha + \tfrac{2}{3}\sin\alpha\tan^2\alpha \approx \tfrac{5}{6}\alpha,$$

where B is the center of buoyancy.

Solution

The weight force of the block is

$$\mathbf{W} = \rho_b g V = \rho_b g \times a \times b \times (2c),$$

which is counterbalanced by the buoyancy force

$$\mathbf{F} = \rho_W g \times a \times b \times s,$$

where s is its submerged part. Because $\rho_W = 2\rho_b$, $s = c$, and the block sinks by $c = 1$ m into the water. The buoyancy force is therefore

$$\mathbf{F} = \rho_W g\, a \times b \times c = 1{,}000 \times 9.806 \times 4 \times 4 \times 1 = 156{,}900\,\mathrm{N}.$$

When the block tilts at an angle α, the returning moment is

$$\mathbf{M}_\alpha = \mathbf{F} \times \overline{QB} = 156,900 \times \tfrac{5}{6}\alpha = 130,750\,\alpha \text{ Nm.}$$

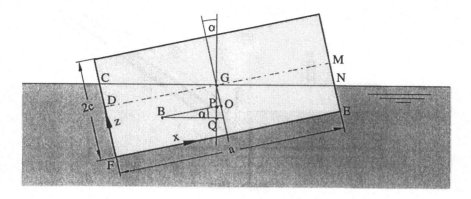

Figure 3.19 Rolling block.

Problems

3.1   What is the pressure at point A in Fig. P3.1?

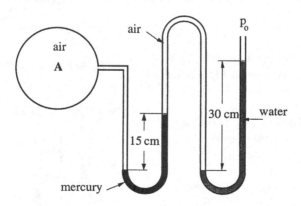

Figure P3.1 Multifluid manometer.

3.2   The inclined manometer shown in Fig. P3.2 is filled with water which reaches mark B on the inclined leg when the pressure tap is open to the atmosphere. A pressure p is applied and the water level rises by $L = 100$ mm. If $d = 2$ mm and $D = 100$ mm, what is the pressure p? What is the error introduced by not measuring the water level drop in the wide leg?

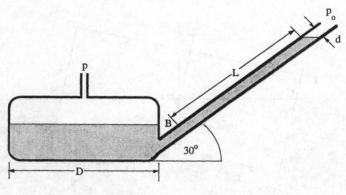

Figure P3.2

3.3   The pressure difference between points 1 and 2 in the water pipe shown in Fig. P3.3 is measured by the manometer shown, where $H = 3$ m and $h = 0.2$ m. The pressure at point 2 is 1.5 bars. What is the pressure at point 1? Which way does the pump pump?

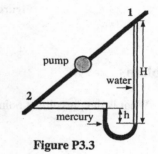

Figure P3.3

3.4   A kerosene line is fitted with a differential manometer, Fig. P3.4. Find the absolute pressure at points A and B, given
$p_o = 100$ kPa, $\rho_{water} = 1000$ kg/m³,
$\rho_{kerosene} = 900$ kg/m³,
$\rho_{mercury} = 13600$ kg/m³.

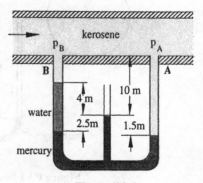

Figure P3.4

3.5   All known materials behave like fluids at pressure higher than 10,000 - 20,000 bars. Assuming the upper solid layers of the earth to have a density of 2.5 g/cm³, estimate the thickness of the earth's crust that still behaves like solid.

Estimate its thickness below the bottom of the sea, assuming sea depth of 10 km.

3.6   A mercury barometer gave a true reading of H_1 mm Hg. A quantity of fluid B was then injected into the bottom of the vertical barometer tube. The B fluid rose on top of the mercury, and the situation became that of Fig. P3.6. What is the vapor pressure of the B fluid?

Data:

$H_1 = 760$ mm, $H = 660$ mm, $h = 100$ mm,
$\rho_{Hg} = 13,600$ kg/m³, $\rho_B = 850$ kg/m³.

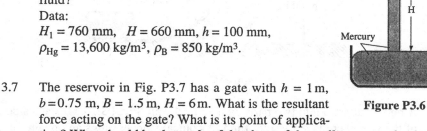

Figure P3.6

3.7   The reservoir in Fig. P3.7 has a gate with $h = 1$ m, $b = 0.75$ m, $B = 1.5$ m, $H = 6$ m. What is the resultant force acting on the gate? What is its point of application? What should be the angle of the slope of the walls, α, to make the line of action of the resultant force on the wall, R, pass through C?

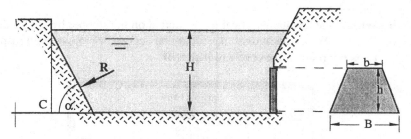

Figure P3.7

3.8   What is the resultant of the forces of the fluid acting on the gate AB in Fig. P3.8? What is the moment acting at point A? (Give the answer per unit width.)

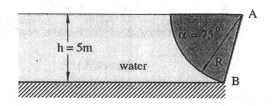

Figure P3.8

3.9   A rectangular water channel is 4 m wide and has a circular gate as shown
      in Fig. P3.9. The gate is kept closed by a weight at point E. Neglect the
      weight of the gate itself and find what weight just keeps the gate closed.

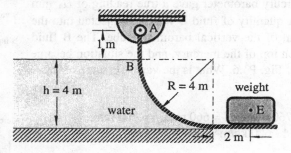

Figure P3.9

3.10 A tank is divided into two parts by a vertical wall, Fig. P3.10. On each side
      of the wall there is a different fluid. A conical stopper plugs an opening in
      the wall.
      a. What is the resultant horizontal force exerted on the stopper by the
          fluids?
      b. What is the resultant vertical force exerted on the stopper by the fluids?
      c. What is the moment of all the forces that act on the stopper with respect
          to its midpoint at the center of the wall?

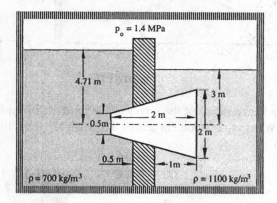

Figure P3.10

3.11 Assume air to satisfy $pv = RT, R = 0.287$ kJ/kg·K, all through the atmo-
      sphere, and the atmosphere to be isothermal at 300°C. Find the height
      below which lies 99% of the mass of the atmosphere.

3.12   A spherical balloon with a diameter of 5 m floats in the air at a height of 200 m. The atmospheric pressure at sea level is 100 kPa. Assume the atmosphere to be an ideal gas satisfying $p = \rho RT$, $R = 287$ J/kg K at the constant temperature $T = 300$ K. Find the total mass of the balloon.

3.13   The balloon in Fig. P3.13 has a diameter of 10 m and is filled with hydrogen at 0.1 MPa and 300 K. Neglect the weight of the balloon itself and find what force it applies to the basket?
What is its force when the balloon is filled with helium?
Air, hydrogen and helium satisfy the perfect gas equation, $pv = RT$.

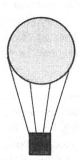

Figure P3.13

3.14   The float shown in Fig. P3.14 is used to control the specific density of brine with $\rho = 1100$ kg/m³. It is desired that variations of 1% in the specific density manifest themselves by variations of 1 cm in the height of the dry part of the float. For $d = 4$ mm, what is the weight of the float?

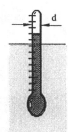

Figure P3.14

3.15   A cylinder with a bottom section made of lead ($\rho_{lead} = 11{,}370$ kg/m³) is filled with water, as shown in Fig. P3.15. The total volume of the cylinder is 15.7 liters and its weight together with the water is 16.5 kg. The cylinder is placed in a solar pond. The pond density varies as $\rho = 1+0.01y$ g/cm³ (y is in m). Find the vertical location of the cylinder.

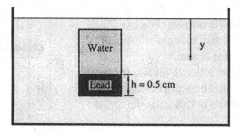

Figure P3.15

3.16 A wooden cylinder ($\gamma = 700$ kg/m³) is shown in Fig. P3.16. Calculate the forces and moments that act on the cylinder (per unit length). Will the cylinder rotate? Would it if it were an elliptical cylinder? Assume that the fluid does not flow from the tank through the seal.

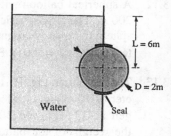

Figure P3.16

3.17 An open tank shaped as a truncated cone is filled with oil to its top. The tank is suspended from a rope and is partly immersed in water, as shown in Fig. P3.17. A

piece of wood floats in the oil, as shown. Calculate the force **F** exerted by the rope on the tank assuming the weight of the empty tank is negligible.

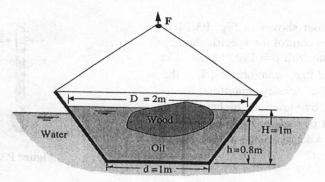

Figure P3.17

3.18 The diving bell in Fig. P3.18 is lowered into the water. Its weight in air is $W = 4,800$ kg, and $L = 2$ m, $D = 10$ m, $d = 2$ m. The air inside the bell is compressed isothermally, i.e., $pv = $ const. The atmospheric pressure is $p_0 = 10^5$ N/m². Neglect the volume of the metal of which the bell is made.

How high will the water level rise inside the bell? Will the bell sink by itself? Once sunk, will the bell float by itself?

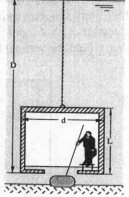

Figure P3.18

3.19   A steel bell (ρ_s = 8,000 kg/m³) is shown
       in Fig. P3.19. The weight of the bell in
       air is W = 8,000 kg. Its inside height is
       3 m, and it contains 6 m³ of air. The bell
       is lowered into the sea until its bottom is
       10 m below water level. The air inside
       the bell is assumed to be compressed
       isothermally.

       a. Find the tension in the cable which
          holds the bell.

       b. A window of a 60° circular arc sector
          is fitted in the bell as shown in the
          figure. Find what force acts on the
          window.

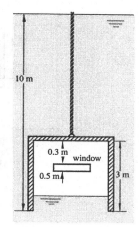

Figure P3.19

3.20   A man named Archimedes had a 100-g sphere which looked like gold. He
       balanced the sphere on a scale using a cube of gold, as shown in Fig. P3.20,
       and then took a bath. Being absent-minded, he took the scale with him into
       the water. Now he suspects the sphere to be gold-plated silver. How much
       silver should he add beside the sphere to have the scale balance in water?
       ρ_{gold} = 19,300 kg/m³, ρ_{silver} = 10,500 kg/m³.

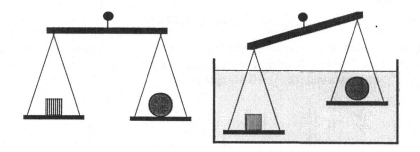

Figure P3.20

3.21   A boxlike barge is shown in Fig. P3.21. When loaded, D = 1 m, B = 5 m and
       C = 3.5 m. The barge rolls to the angle α = 15°. Assuming the load to be
       well secured, what is its righting moment? Assuming the load to be water-
       like, with a free surface, as shown by b in the figure, what is the righting
       moment? How is the last result modified if the barge had a vertical partition
       dividing it into two equal halves?

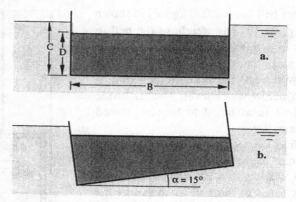

Figure P3.21

3.22   What is the restoring torque of the body that is shown in Fig. P3.22? Sketch
the torque as a function of θ, $M = M(\theta)$. Assume all other information that
you need.

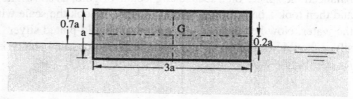

Figure P3.22

3.23   A safety gate is constructed as shown in Fig. P3.23. The width of the gate is
1 m, $b = 1$ m, $h = 4$ m. Find the moment the water exerts on the gate. At
what water height h_1 will the gate open?

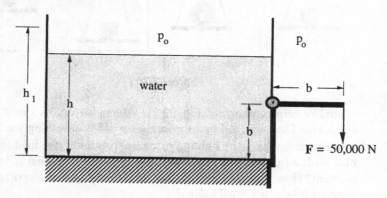

Figure P3.23 A safety gate.

3.24   A rectangular tank containing oil accelerates at **a** = 0.2 **g**, as shown in Fig.
       P3.24. Once the oil is again at rest (with respect to what?), what is α?
       (Ignore the manometer inside the oil.)

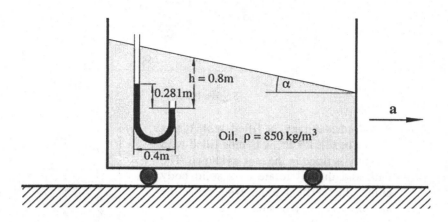

Figure P3.24

3.25   The tank in Fig. P3.24 accelerates at constant **a**. From the reading of the
       mercury manometer, what is α? What is **a**? What will the manometer show
       when the experiment is repeated without the oil?

3.26   A pipe inclined at an angle $\alpha = 30°$ is
       closed at the bottom, as shown in
       Fig. P3.26. It is filled with water and
       then rotated with $\omega = 8$ rad/s.
       What is the pressure, p_B, at the bottom
       of the pipe?
       What is the pressure, p_o, at the axis of
       rotation?

3.27   A cart on wheels accelerates freely,
       because of gravity, on a slope,
       Fig. P3.27. Neglecting friction, find at
       what angle β would a fluid rest,
       relative to the cart, for $\alpha = 30°$,
       $\alpha = 50°$.

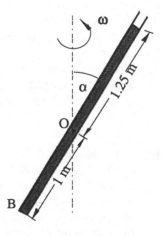

Figure P3.26

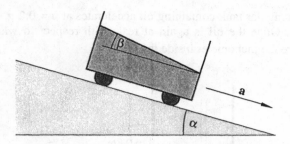

Figure P3.27

3.28 A cylindrical tank is filled with oil, as shown in Fig. P3.28. A U-tube filled with mercury is fitted in the tank as shown. The tank and the oil rotate as a solid body about the vertical axis. Determine the angular speed of the tank.

Given: ρ_{oil} = 900 kg/m³,

ρ_M = 13,600 kg/m³.

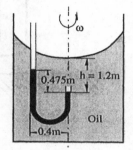

Figure P3.28

3.29 A circular hollow top, Fig. P3.29, is filled with water. A small hole is made at point A. The top spins about its vertical axis with an angular velocity of 10 rad/s; $H=1$ m, $R=0.75$ m.

Find the pressure on the internal surfaces, 1, 2 and 3.

Find the forces acting on the seams connecting the flat plate 1 with the cylindrical shell 2 and that with the conical shell 3.

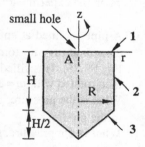

Figure P3.29

3.30 A cylindrical bell, Fig. P3.30, has a mass of 1,500 kg and an inner volume of 6 m³ while just touching the water surface. The bell inner height is $H = 3$ m. The bell is pushed downward with the initial speed **q** m/s, and the air inside is assumed to be compressed isothermally, i.e., with pV = const. Neglect the volume of the metal from which the bell is made, and consider the hydrodynamic resistance of the water to the motion of the bell an additional margin of safety. Find the maximal value of **q** which still permits the bell to float back by its own buoyancy.

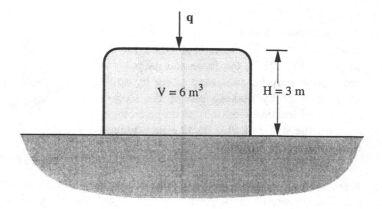

Figure P3.30

3.31 Figure P3.31 shows a chamber designed to test underwater windows. The windows can be set in three orientations as shown: horizontal (A), vertical (B) and at an angle α to the vertical (C). The windows themselves are plane and have the shapes shown in Fig. P3.31a-e. Find the total force exerted on the windows, its direction and its point of application. Solve the problem for all the windows and for each of the three orientations.

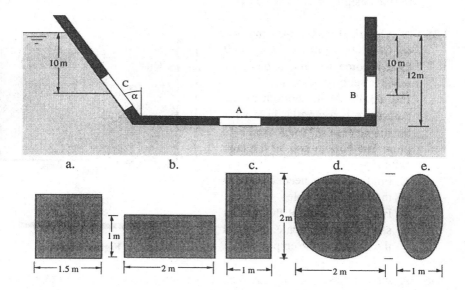

Figure P3.31 A chamber for testing windows.

3.32   Repeat Problem 3.31 for the case of the window given in Fig. P3.32. Note that while for the shapes in Fig. P3.31 the force and its point of application may be computed either by direct integration or by making use of the shape's moment of inertia together with Steiner's theorem, for the shape of Fig. P3.32 direct integration is extremely difficult.

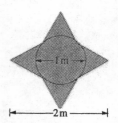

Figure P3.32

3.33   A research minisubmarine is designed to operate at a depth of 500 m in the sea. The density of sea water is $\rho = 1050\,\text{kg/m}$. The glass windows of the submarine are round, with a diameter of 0.3 m, and flat, to prevent optical distortion. The submarine operates at all orientations, i.e., the windows may look downward, sideways or in any other direction. Find for what force and for what moment must the windows be designed.

3.34   Considerations of strength led engineers to consider spherical shells made of glass for the windows of the submarine in Problem 3.33. The wet part of the windows has the shape of a section of a sphere - all variations are considered, from a flat plane to half a sphere. Find the forces and the moments for the designs of these windows.

3.35   A cylindrical tank containing water ($\rho = 1000\,\text{kg/m}$), oil ($\rho = 850\,\text{kg/m}$), some ice and a piece of wood is set on a scale as shown in Fig. P3.35. At the bottom of the tank there is a circular hole with the diameter of 0.3 m blocked by a plug. The bottom area of the tank is 2 m².

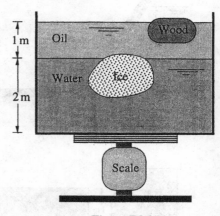

a.   Find the force needed to keep the plug in the hole.

b.   Find the weight of the tank as shown by the scale. Neglect the weight of the empty tank.

Figure P3.35

3.36   A wharf is protected against being rammed by a ship by a bumper connected to a float as shown in Fig. P3.36. When no ship touches the bumper, the force in the chain is 20 kN.

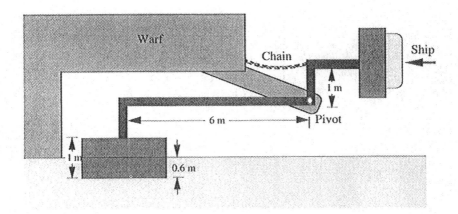

Figure P3.36

a. Find the maximal resistance force the bumper can offer.
b. Find the amount of energy absorbed by the bumper–float system before this maximal force appears.
c. Find the amount of energy taken from the moving ship as it pushes in the bumper by 0.4 m.

4. FLUIDS IN MOTION – INTEGRAL ANALYSIS

Moving fluids are subject to the same laws of physics as are moving rigid bodies or fluids at rest. The problem of identification of a fluid particle in a moving fluid is, however, much more difficult than the identification of a solid body or of a fluid particle in a static fluid. When the fluid properties, e.g., its velocity, also vary from point to point in the field, the analysis of these properties may become quite elaborate.

There are cases where such an identification of a particle is made, and then this particle is followed and the change of its properties is investigated. This is known as the Lagrangian approach. In most cases one tries to avoid the need for such an identification and presents the phenomena in some *field equations*, i.e., equations that describe what takes place at each point at all times. This differential analysis, which is called the Eulerian approach, is the subject of the next chapter.

There exist, however, quite a few cases where there are sufficient restrictions on the flow field, such that while the individual fluid particles still elude identification, whole chunks of fluid can be identified. When this is the case, some useful engineering results may be obtained by integrating the relevant properties over the appropriate chunk. This approach is known as integral analysis.

In most cases where integral representation is possible, there are walls impermeable to the fluid. The moving fluid which cannot pass through these walls must flow along them, and while doing so its location is better kept under control. We call the volume enclosed by these walls a *control volume*.

We begin this chapter with the definition of a control volume and with relations between thermodynamic systems and control volumes. We then present the Reynolds transport theorem and apply it to obtain integral relations of conservation of mass and Newton's second law of motion for a control volume.

Thermodynamic Systems and Control Volumes

A very useful concept in integral analysis is that of the *control volume*, which consists of an *enclosure* with well-defined *impermeable walls* and well-defined *openings*. The walls may deform and the openings may change dimensions, but as long as they remain well defined, the enclosure may still serve as a control volume. The control volume may move in space and may even accelerate. It is the choice of a proper control volume which determines whether a particular problem can be treated successfully by integral methods.

The three physical laws most important in fluid mechanics are *conservation of mass, Newton's second law of motion,* and *the first law of thermodynamics.* All three laws are formulated for a *thermodynamic system* which is classically defined as a specified amount of matter with well-defined boundaries. In a moving fluid such a thermodynamic system will, in general, continuously change its location and shape and will be hard to follow.

The object of the general considerations presented here is to reformulate two of the three laws mentioned above, i.e., for mass and momentum, in terms convenient to apply for control volumes. This control volume formulation is done by the application of the Reynolds transport theorem, which states what takes place when a thermodynamic system moves through a control volume.

Reynolds Transport Theorem

Let a thermodynamic system be chosen inside the fluid, Fig. 4.1. At the time t it occupies the volume V. Let B be some general extensive property of the thermodynamic system inside the volume V and let b be its specific value, that is, the value of B per unit mass; then

$$B = \int_V \rho b \, dV. \tag{4.1}$$

The theorem deals with the rate of change of B with time. We are interested in this because for conservation of mass B is identified with the total mass of the thermodynamic system, and therefore its rate of change in time vanishes, and for B identified with, say, the x-component of the total momentum of the system, Newton's second law demands that the rate of change in time of this component equals the x-component of the resultant force acting on the system.

We note that B belongs to the system and that the system may change its location and its shape in time. Had our system been a rigid body, say a steel ball, the notation dB/dt might have sufficed because it would be clear that we talked about the steel ball. However, for a fluid system it is useful to emphasize that the rate of change of B is sought while following the system. This is done by writing DB/Dt, which is called the material derivative of B.

Thus the rate of change of the property B is

$$\frac{DB}{Dt} = \frac{D}{Dt}\int_V \rho b \, dV. \tag{4.2}$$

For example, conservation of mass states that a given system cannot change its mass. This may be expressed by substituting $B = m$ and $b = 1$ in Eq. (4.2), leading to

$$\frac{Dm}{Dt} = \frac{D}{Dt}\int_V \rho \, dV = 0. \tag{4.3}$$

In Eq. (4.2) the domain of the integration, V, depends on time because the considered thermodynamic system changes its location and its shape. These changes are not arbitrary. They are determined by the flow field, and the velocity vector is expected to appear in the expressions for the rate of change of B.

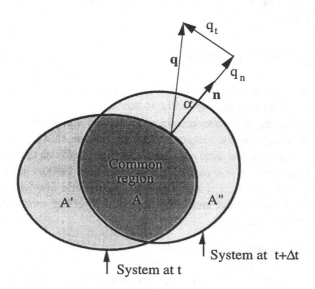

Figure 4.1 Thermodynamic system moving in velocity field.

Consider the system of fluid inside the volume $V(t)$ shown in Fig. 4.1, which has the extensive property B as defined by Eq. (4.1). The property B may depend on time and location. We look now for the rate of change of B, as expressed by Eq.(4.2), while the system moves with the flow field. This rate of change is DB/Dt.

In Fig. 4.1 we see the system at the time t and at the time $t + \Delta t$. Since our considerations are for the limit of vanishing Δt, there is always a region common to both locations of the system, as designated by A in Fig. 4.1. As we see, at the time t the system occupies the volume $A + A'$, while at $t + \Delta t$ it extends over $A + A''$. The common part, A, is, of course, the same at t and at $t + \Delta t$, but the specific property b in it may be different because b depends also on time. Thus the rate of change of B becomes

$$\frac{DB}{Dt} = \lim_{\Delta t \to 0} \frac{B(t + \Delta t) - B(t)}{\Delta t} = \lim_{\Delta t \to 0} \frac{B_A(t + \Delta t) + B_{A''}(t + \Delta t) - B_A(t) - B_{A'}(t)}{\Delta t}$$

or

$$\frac{DB}{Dt} = \lim_{\Delta t \to 0} \frac{B_A(t + \Delta t) - B_A(t)}{\Delta t} + \lim_{\Delta t \to 0} \frac{B_{A''}(t + \Delta t) - B_{A'}(t)}{\Delta t} = \frac{\partial B}{\partial t} + \dot{B}. \quad (4.4)$$

The first term on the right-hand side of Eq. (4.4) gives the change of B at constant position. This result is what would be expected had the fluid been at rest. This term is just

$$\frac{\partial B}{\partial t} = \frac{\partial}{\partial t} \int_V \rho b \, dV = \int_V \frac{\partial}{\partial t} (\rho b) \, dV, \quad (4.5)$$

where the order of the differentiation and the integration may be interchanged because V is fixed in space during both operations.

The second term, $\dot{B}$, expresses the contribution due to the flow of the fluid. Thus the total change in the property B of the system that originally occupied regions $A + A'$ is due to the time change of B inside A plus the flux of part of the system with its property B into region A'' minus the flux that has left region A'.

Referring to Fig. 4.1, it is noted that the volumetric flow rate of fluid, dQ, passing through a differential area dS is the product of the area and the velocity component normal to the area:

$$dQ = q_n \, dS = \mathbf{q} \cdot |dS| \cdot \cos \alpha = \mathbf{q} \cdot \mathbf{n} \, dS. \quad (4.6)$$

A positive $\mathbf{q} \cdot \mathbf{n}$ denotes outflow, i.e., a flow from A to A'', while a negative $\mathbf{q} \cdot \mathbf{n}$ denotes inflow, i.e., flow from A' to A. The flux through dS is

$$d\dot{B} = \rho b \, dQ = \rho b \mathbf{q} \cdot \mathbf{n} \, dS$$

and the total contribution

$$\dot{B} = \lim_{\Delta t \to 0} \frac{B_{A''}(t + \Delta t) - B_{A'}(t)}{\Delta t} = \int_S \rho b \mathbf{q} \cdot \mathbf{n} dS. \tag{4.7}$$

Thus the total rate of change of B for a system, DB/Dt, is

$$\frac{DB}{Dt} = \frac{\partial B}{\partial t} + \int_S \rho b \mathbf{q} \cdot \mathbf{n} dS, \tag{4.8}$$

$$\frac{DB}{Dt} = \frac{\partial}{\partial t} \int_V \rho b \, dV + \int_S \rho b \mathbf{q} \cdot \mathbf{n} dS \tag{4.9}$$

and by Eqs. (4.1) and (4.5)

$$\frac{D}{Dt} \int_V (\rho b) dV = \int_V \frac{\partial}{\partial t} (\rho b) dV + \int_S \rho b \mathbf{q} \cdot \mathbf{n} dS. \tag{4.10}$$

Equations (4.8) - (4.10) state the fact that the rate of change of B for a given material is due to the change of B inside a volume $V(t)$ plus the net outflow of B through the boundaries of the volume. Equations (4.9) - (4.10) are known as the *Reynolds transport theorem.*

Suppose we have a control volume with the volume V. Equation (4.9) now states that the rate of change of the extensive property B of the thermodynamic system, which just fills the control volume, is obtained by the rate of change of B inside the control volume plus the flux of the property B from the control volume. This flux can take place, of course, only through the openings of the control volume.

Control Volume Analysis of Conservation of Mass

In many practical cases, the details of the flow field are not known. Such cases arise either because these details cannot be obtained or because they are not really required. Still, an overall inventory of the mass for a given volume is desired. This information can be obtained by an integral mass balance performed for the control volume using the Reynolds transport theorem.

Let B in Eq. (4.9) be the mass of the system located inside the control volume at the instant of consideration. The mass per unit mass, to be substituted for b, is just 1, and Eq. (4.9) becomes

$$\frac{Dm}{Dt} = \frac{\partial}{\partial t} \int_V \rho \, dV + \int_S \rho \mathbf{q} \cdot \mathbf{n} \, dS. \tag{4.11}$$

Conservation of mass requires that the system has the same mass, m, at all times, thus

$$\frac{Dm}{Dt} = 0,$$

which upon substitution into Eq. (4.11) yields

$$0 = \frac{\partial}{\partial t}\int_V \rho \, dV + \int_S \rho \mathbf{q} \cdot \mathbf{n} \, dS. \tag{4.12}$$

Let the surface S which encloses the thermodynamic system be split into

$$S = S' + S_o,$$

where S' is where the system actually touches the walls of the control volume, through which there is no flow, and S_o denotes the openings of the control volume. Obviously

$$\int_{S'} \rho \mathbf{q} \cdot \mathbf{n} \, dS = 0$$

and Eq. (4.12) becomes

$$-\frac{\partial}{\partial t}\int_V \rho \, dV = \int_{S_o} \rho \mathbf{q} \cdot \mathbf{n} \, dS. \tag{4.13}$$

Equation (4.13) is general and holds for deformable, moving and accelerating control volumes, provided that the velocity $\mathbf{q}$ is always *taken relative to the differential surface* $\mathbf{n} \, dS$. Equation (4.13) may be rewritten in a slightly different form by noting that

$$\int_V \rho \, dV = m_{cv},$$

where m_{cv} is the total mass inside the control volume. Hence

$$-\frac{\partial m_{cv}}{\partial t} = \int_{S_o} \rho \mathbf{q} \cdot \mathbf{n} \, dS. \tag{4.14}$$

The surface integral in Eq. (4.14) stands for the total mass flowrate leaving through the control surface. Thus, Eq. (4.14) indicates that the rate of mass reduction inside the control volume equals the total rate of mass leaving through the control surface.

Both Eqs. (4.13) and (4.14) require the details of the velocity field to enable integration over the control surface. However, in many cases there is just a finite number of well-defined openings in the control volume, e.g., several pipes feeding and emptying a water tank, and the integration over the surface excluding these openings contributes nothing. When the flow through the cross section of each of these openings is uniform, i.e., $(\rho q_n)_i = $const, Eqs. (4.13) and (4.14) simplify to

$$-\frac{\partial}{\partial t}\int_V \rho \, dV = \sum_{i=1}(\rho \mathbf{q} \cdot \mathbf{n} A)_i \tag{4.15}$$

or

$$-\frac{\partial m_{cv}}{\partial t} = \sum_{i=1} (\rho q_n A)_i, \tag{4.16}$$

where the area S_o now consists of the sum of A_i, and where $\mathbf{q} \cdot \mathbf{n} = q_n$ is the component of the velocity vector normal to the corresponding area A_i.

Example 4.1

A solid-fuel rocket has a nozzle with a critical cross-sectional area of $A_{2n} = 0.01 \text{ m}^2$, Fig. 4.12 The average velocity of the gas flowing through the critical section is 880 m/s, relative to the rocket, and its density is 1.4 kg/m³. As the rocket starts its vertical motion up at sea level, its mass is 300 kg. The solid fuel is used up after 20 s. Find the mass of the rocket after 20 s.

Solution

The given gas velocity is relative to the critical cross section. Therefore, Eq. (4.16) yields

$$\frac{\partial m_{cv}}{\partial t} = -\rho q_n A_c = -1.4 \times 880 \times 0.01 = -12.32 \text{ kg / s}$$

or

$$dm_{cv} = -\rho q_n A_c \, dt = -12.32 \, dt.$$

Integration yields

$$m_{cv_{20}} = m_{cv_0} + \int_0^{20} (-12.32) \, dt = 300 + 20 \times (-12.32) = 53.6 \text{ kg}.$$

Conservation of Mass under Steady State Conditions

An important special case of Eqs. (4.15) and (4.16) is that of steady state, i.e., time-independent flow. At steady state there is no change with time of any property at any point inside the control volume, that is, $\partial / \partial t = 0$. Thus, for example, the pressure, p, at a given point remains constant at steady state, although there may be a pressure drop, Δp, between two given points. This pressure drop, again, does not change with time at steady state. For steady state Eq. (4.16) simplifies to

$$\sum_{i=1} \rho_i V_i A_i = 0 \tag{4.17}$$

and for incompressible flows, where $\rho = \text{const}$,

$$\sum_{i=1} V_i A_i = 0. \tag{4.18}$$

In Eqs. (4.17) and (4.18), A_i is the cross-sectional area normal to V_i. The terms in the sums of Eqs. (4.17) and (4.18) should be taken algebraically with their proper signs, *positive* for *outgoing* flows and *negative* for *incoming* flows.

There are many examples of control volumes to which Eqs. (4.17) and (4.18) apply and which have only two openings. Such examples are pipe sections, nozzles, pumps, compressors, turbines, etc. Here the mass flowrate, $\dot{m}$, through the control volume is constant and Eq. (4.17) becomes

$$-\rho_1 V_1 A_1 = \rho_2 V_2 A_2 = \dot{m}. \tag{4.19}$$

Equations (4.17), (4.18) and (4.19) can be modified to apply to cases where $(\rho q)_i$ is not uniform provided that some average mass velocity $\overline{G_i} = \overline{(\rho q)}_i$ can be found. For the average values

$$\overline{G_i} = \overline{(\rho q)}_i = \frac{1}{A_i} \int_{A_i} \rho \mathbf{q} \cdot \mathbf{n} \, dA. \tag{4.20}$$

Equation (4.17) becomes

$$\sum_{i=1} \overline{(\rho q_n)}_i A_i = 0. \tag{4.21}$$

For constant density flows, ρ can be canceled out, and the average velocity, V_i, used in Eq. (4.18) is defined in a way similar to Eq. (4.20),

$$V_i = \frac{1}{A_i} \int_{A_i} \mathbf{q} \cdot \mathbf{n} \, dA. \tag{4.22}$$

It is noted that in many practical situations it is easier to evaluate or to measure these mean velocities than to obtain velocity distributions over the openings. In many cases the easiest quantity to measure is the mass flux, $\dot{m} = \rho V A$.

Example 4.2

A cylindrical water tank, shown in Fig. 4.2, has a cross section of 5 m². Water flows out of the tank through a pipe of 0.2 m diameter with an average velocity of $q_1 = 6$ m/s. Water flows into the tank through a 0.1-m-diameter pipe with an average velocity $q_2 = 12$ m/s. Define a control volume and find at what rate the water level rises in the tank.

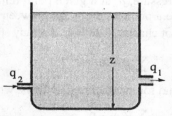

Figure 4.2 Water tank and piping.

Solution

A convenient control volume is chosen as bounded by the entrance and the exit to the pipes, by the wetted part of the tank walls and by the free water surface. From Eq. (4.16)

$$\left(\frac{\partial m}{\partial t}\right)_{cv} = -\rho \sum A_i q_{ni}$$

or

$$\left(\frac{\partial V}{\partial t}\right)_{cv} = -\sum A_i q_{ni} = -\left(A_1 q_{n1} + A_2 q_{n2}\right)$$

$$= -\frac{\pi}{4}\left(0.2^2 \times 6 - 0.1^2 \times 12\right) = -0.094 \ \text{m}^3/\text{s}.$$

Here $q_{n2} = -12\,\text{m/s}$ is negative because the velocity at the inflow points in a direction opposite to that of the outer normal to the surface. The rate of change in the water level, $\partial z/\partial t$, is found from

$$\frac{\partial V}{\partial t} = \frac{\partial (Az)}{\partial t} = A\frac{\partial z}{\partial t} = 5\frac{\partial z}{\partial t} = -0.094 \ \text{m}^3 / \text{s}$$

and finally

$$\frac{\partial z}{\partial t} = -\frac{0.094}{5} = -0.0188\,\text{m}/\text{s} = -1.88\ \text{cm}/\text{s}.$$

Hence, the water level is *reduced* at a rate of 1.88 cm/s.

Example 4.3

Water flows in a pipe of diameter $d_1 = 0.05$ m with an average velocity $V_1 = 0.2\,\text{m/s}$ and enters a sprinkler, Fig. 4.3. When the flow starts, the sprinkler accelerates from $\omega = 0$ to $\omega = 4\,\text{rad/s}$. The sprinkler pipe ends with two nozzles, each of diameter $d = 0.005$ m. Find the exit speed of the water relative to the nozzle, V_2.

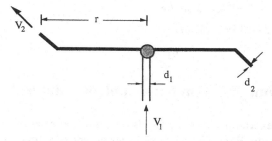

Figure 4.3 Water sprinkler.

Solution

The acceleration is not relevant at all, because the velocity sought is relative to the nozzle. Assuming that half of the inflow leaves through each nozzle, Eq. (4.18) yields

$$V_2 = \left(\frac{V_1}{2}\right)\left(\frac{d_1}{d_2}\right)^2 = 0.1\left(\frac{0.05}{0.005}\right)^2 = 10 \text{ m / s}.$$

Example 4.4

A two-dimensional water flow divider is shown in Fig. 4.4. The flow in through opening A varies linearly from 3 m/s to 6 m/s and is inclined with an angle $\alpha = 45^\circ$ to the opening itself. The area of the A opening is 0.4 m²/m, that of B is 0.2 m²/m, and that of C is 0.15 m²/m. The flow at B, normal to the opening varies linearly from 3 m/s to 5 m/s, and the flow at C, normal to the opening, is approximately constant. Find q_c.

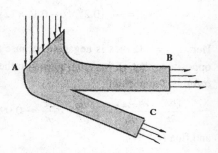

Figure 4.4 Flow divider.

Solution

The flow is incompressible, and from Eq. (4.14)

$$\int_A \mathbf{q} \cdot \mathbf{n} \, dA = -\bar{q}_A A_A \cos 45^\circ = -\frac{(3+6)}{2} \times 0.4 \times 0.7071 = -1.273 \text{ m}^3/\text{m} \cdot \text{s},$$

$$\int_B \mathbf{q} \cdot \mathbf{n} \, dA = \bar{q}_B A_B = \frac{(3+5)}{2} \times 0.2 = 0.8 \text{ m}^3/\text{m} \cdot \text{s}.$$

Hence,

$$q_C A_C = 1.273 - 0.8 = 0.473 \text{ m}^3/\text{m} \cdot \text{s},$$

$$q_C = 0.473 / 0.15 = 3.153 \text{ m / s}.$$

The Momentum Theorem for a Control Volume

The forces exerted by moving fluids on rigid bodies are important in many fields of engineering. Examples of such forces are those acting on airplane wings, hulls of ships, turbine blades, pipe bends and weather vanes. These forces can be computed either by integrating the stresses at the boundaries or by an integral approach utilizing directly Newton's second law of motion. In many cases the integral approach, which relies on control volume analysis, is easier to use.

Newton's second law of motion applied to a thermodynamic system of fluid states that the sum of all the forces acting on the system, i.e., body and surface forces, equals the rate of change of the momentum of this system:

$$\mathbf{F}_B + \mathbf{F}_S = \frac{D}{Dt}\int_V \rho \mathbf{q}\ dV. \tag{4.23}$$

Substitution of the body force, $\mathbf{F}_B$, and the surface force, $\mathbf{F}_S$, in Eq. (4.23) results in a momentum equation of the form

$$\int_V \rho \mathbf{g}\ dV + \int_S \mathbf{T}\ dS = \frac{D}{Dt}\int_V \rho \mathbf{q}\ dV, \tag{4.24}$$

where $\mathbf{g}$ is the general body force per unit mass and $\mathbf{T}$ is the stress at the system boundary.

Consider now the x-component of this momentum equation, still written for the thermodynamic system. For this case Eq. (4.24) becomes

$$\int_V \rho g_x\ dV + \int_S T_x\ dS = \frac{D}{Dt}\int_V \rho u\ dV \tag{4.25}$$

in which u is the x-component of the velocity vector, $\mathbf{q} = \mathbf{i}u + \mathbf{j}v + \mathbf{k}w$.

We would like to apply the Reynolds transport theorem to the integral on the right-hand side of Eq. (4.25). Before doing this, it is noted that Newton's second law of motion requires that the coordinate system in which Eqs. (4.23), (4.24) and (4.25) hold be an inertial system, i.e., the vector $\mathbf{q} = \mathbf{i}u + \mathbf{j}v + \mathbf{k}w$ is given in an inertial coordinate system. It is also noted that $q_n = \mathbf{q} \cdot \mathbf{n}$ which appears in the Reynolds transport theorem, e.g., in Eqs. (4.9) and (4.13), is the velocity relative to the control volume. Finally, we want a momentum theorem that applies to accelerating control volumes as well as to ones which may be considered inertial systems. Therefore, to remove any possible confusion, the notation $\mathbf{q}_r \cdot \mathbf{n}$ will be used to remind one that this velocity is relative to the control volume, while $\mathbf{q}$ elsewhere is absolute, given in an inertial coordinate system.

The Reynolds transport theorem, Eq. (4.9), is now applied to the property $B = mu$ and $b = u$, in Eq. (4.25), resulting in

$$\frac{D}{Dt}\int_V \rho u\ dV = \frac{\partial}{\partial t}\int_V \rho u\ dV + \int_S \rho u(\mathbf{q}_r \cdot \mathbf{n})\ dS. \tag{4.26}$$

Comparison of Eqs. (4.25) and (4.26) yields another form for the x-component of the momentum equation,

$$\int_V \rho g_x \, dV + \int_S T_x \, dS = \frac{\partial}{\partial t} \int_V \rho u \, dV + \int_S \rho u (\mathbf{q}_r \cdot \mathbf{n}) \, dS. \qquad (4.27)$$

On the right-hand side of Eq. (4.27) the first term stands for the instantaneous rate of change of the total momentum enclosed inside the volume V, as it is now (hence the partial $\partial/\partial t$), while the second term indicates the instantaneous flux of momentum through the surface S, which encloses V. These two operations are control volume operations. Thus Eq. (4.27) makes control volume analysis applicable for Newton's second law of motion.

The y- and z-components of the momentum equation are obtained in a manner similar to that of the x-component and have the forms

$$\int_V \rho g_y \, dV + \int_S T_y \, dS = \frac{\partial}{\partial t} \int_V \rho v \, dV + \int_S \rho v (\mathbf{q}_r \cdot \mathbf{n}) \, dS, \qquad (4.28)$$

$$\int_V \rho g_z \, dV + \int_S T_z \, dS = \frac{\partial}{\partial t} \int_V \rho w \, dV + \int_S \rho w (\mathbf{q}_r \cdot \mathbf{n}) \, dS. \qquad (4.29)$$

The three components of the integral momentum equation, Eqs. (4.27), (4.28) and (4.29), may now be combined back into a vectorial expression,

$$\int_V \rho \mathbf{g} \, dV + \int_S \mathbf{T} \, dS = \frac{\partial}{\partial t} \int_V \rho \mathbf{q} \, dV + \int_S \rho \mathbf{q} (\mathbf{q}_r \cdot \mathbf{n}) \, dS. \qquad (4.30)$$

Equation (4.30) is known as the Integral Momentum Theorem for a control volume. This theorem states that the rate of change of momentum in the control volume and the outflow of momentum through the control surfaces are balanced by the action of the body force and by the surface stress on the fluid.

Equation (4.30) holds for any control volume, whether stationary, moving or deforming. Again it is emphasized that the velocity vector $\mathbf{q}_r$ in the expression for the differential volumetric flowrate $(\mathbf{q}_r \cdot \mathbf{n}) \, dS$ should be taken relative to the surface dS, and for a moving or deforming control volume it must be taken relative to the moving surface.

The surface forces acting on the fluid in the control volume consist of those exerted by adjacent fluid layers and those acting on the fluid by the solid boundaries.

As before, the surface S is split into

$$S = S' + S_o,$$

where S' is the contact surface between the fluid inside the control volume and the walls of the control volume and S_o is the area of the openings. At the walls $\mathbf{q}_r \cdot \mathbf{n} = 0$, and therefore the momentum flux term applies to the openings only, i.e.,

$$\int_S \rho \mathbf{q}(\mathbf{q}_r \cdot \mathbf{n}) dS = \int_{S_o} \rho \mathbf{q}(\mathbf{q}_r \cdot \mathbf{n}) dS. \tag{4.31}$$

The surface forces in Eq. (4.30) can also be split into those acting on the solid walls and those felt over the openings

$$\int_S \mathbf{T} dS = \int_{S'} \mathbf{T} dS + \int_{S_o} \mathbf{T} dS. \tag{4.32}$$

The integral over the solid walls equals the total force applied to the fluid system by these walls of the control volume; let this integral be denoted $-\mathbf{R}$, i.e.,

$$\int_{S'} \mathbf{T} dS = -\mathbf{R}. \tag{4.33}$$

Then, by Newton's third law, $+\mathbf{R}$ is the force applied by the fluid to the surface S' of the control volume.

Combining Eqs. (4.30) and (4.33), an expression for $\mathbf{R}$ is obtained,

$$\mathbf{R} = \int_{S_o} \mathbf{T} dS + \int_V \rho \mathbf{g} dV - \frac{\partial}{\partial t} \int_V \rho \mathbf{q} dV - \int_{S_o} \rho \mathbf{q}(\mathbf{q}_r \cdot \mathbf{n}) dS. \tag{4.34}$$

It is customary to resolve the stress $\mathbf{T}$ on the surfaces of the openings into normal stress, i.e., pressure, and into shear stress:

$$\int_{S_o} \mathbf{T} dS = -\int_{S_o} p\mathbf{n} \, dS + \int_{S_o} \tau dS. \tag{4.35}$$

In many cases the last integral which represents shear force on the openings is very small and may be neglected. In such cases Eq. (4.34) simplifies to

$$\mathbf{R} = -\int_{S_o} p\mathbf{n} \, dS + \int_V \rho \mathbf{g} \, dV - \frac{\partial}{\partial t} \int_V \rho \mathbf{q} \, dV - \int_{S_o} \rho \mathbf{q}(\mathbf{q}_r \cdot \mathbf{n}) \, dS. \tag{4.36}$$

Equation (4.36) is used to calculate $\mathbf{R}$, which is the resultant of all the forces exerted by the fluid passing through the control volume on the inner walls of the control volume. This equation is known as the momentum theorem for a control volume.

There are situations in which Eq. (4.36) can be further simplified. One such situation is where the last integral on the right, i.e., momentum flux through the openings, may be evaluated by the use of mean values for the velocities. In such a case we substitute

$$-\int_{S_o} \rho \mathbf{q}(\mathbf{q}_r \cdot \mathbf{n}) \, dS = -\sum_{i=1}^k \left[\rho \mathbf{q}(\mathbf{q}_r \cdot \mathbf{n}) A \right]_i, \tag{4.37}$$

where the velocities on the right-hand side are mean velocities, A_i are the opening areas and the summation is over all openings, say k. Indeed, when no details of the flow are known, the average velocities are more readily available because their measurements involve total flow rates only.

Another rather common simplifying situation is where the engineering device may be classified as a single-input–single-output (SISO) device. In such a case there are only two terms in the summation on the right-hand side of Eq. (4.37). The two last simplifications, when applicable, change Eq. (4.36) into

$$\mathbf{R} = -p_1(\mathbf{n}A)_1 - p_2(\mathbf{n}A)_2 + m\mathbf{g} - \frac{\partial}{\partial t}(m\mathbf{q}) - \rho_1 q_1(\mathbf{q}_{1r} \cdot \mathbf{n})A_1 - \rho_2 q_2(\mathbf{q}_{2r} \cdot \mathbf{n})A_2, \quad (4.38)$$

where average pressures have also been assumed.

Finally, some cases may be one dimensional. We assign the coordinate x to the relevant direction and rewrite Eqs. (4.36) and (4.38) in their x-component terms, i.e., as scalar equations. Hence, Eq.(4.36) becomes

$$R = -\int_{S_o} p \, dS_n + \int_V \rho g_x \, dV - \frac{\partial}{\partial t}\int_V \rho u \, dV - \int_{S_o} \rho u(u_r) \, dS_n, \quad (4.39)$$

where dS_n is positive or negative according to the sign of $(\mathbf{i} \cdot \mathbf{n})$, and u is the velocity in the x-direction. Equation (4.38) simplifies for a one-dimensional SISO device to

$$R = -p_1 A_{1n} - p_2 A_{2n} + mg_x - \frac{\partial}{\partial t}(mu) - \rho_1 u_1(u_{1r})A_{1n} - \rho_2 u_2(u_{2r})A_{2n}, \quad (4.40)$$

where the signs of A_{1n} and A_{2n} are those of $(\mathbf{i} \cdot \mathbf{n}_1)$ and $(\mathbf{i} \cdot \mathbf{n}_2)$, respectively, and, of course, all u values are *algebraic*, i.e., must have their correct signs. Figure 4.5 depicts a one-dimensional single-input–single-output control volume together with the various terms of Eq. (4.40).

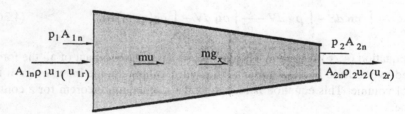

Figure 4.5 One-dimensional single-input–single-output control volume.

Example 4.5

Water enters a pipe bend with a uniform velocity, $q_1 = 5$ m/s. Referring to the bend shown in Fig. 4.6a, the other known data are

$$p_1 = 3.375 \times 10^5 \text{ Pa}, \qquad d_1 = 0.1 \text{ m},$$

$$p_2 = 1.5 \times 10^5 \,\text{Pa}, \qquad d_2 = 0.05 \,\text{m}.$$

The size of the bend is such that it contains 4 kg of water between its two flanges. Find the force, **R**, exerted on the bend by the water passing through it.

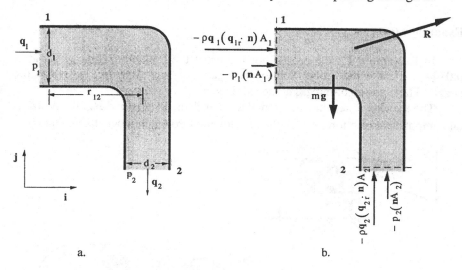

Figure 4.6 Pipe bend and forces.

Solution

The bend is a stationary SISO device; hence Eq. (4.38) may be used. The control volume is conveniently chosen to consist of the bend itself and the two planes of the flanges. The flow of the water is incompressible, and Eq. (4.21) yields

$$-q_1 A_1 = q_2 A_2,$$

or

$$q_2 = -q_1 \left(\frac{d_1}{d_2}\right)^2 = -\left(\frac{0.1}{0.05}\right)^2 = -20\,\text{m/s}.$$

For this uniform flow SISO device Eq. (4.38) yields, for an incompressible fluid of density ρ,

$$\mathbf{R} = -p_1(\mathbf{n}A_1) - p_2(\mathbf{n}A_2) + m\mathbf{g} - \rho q_1(\mathbf{q}_{1r} \cdot \mathbf{n}) - \rho q_2(\mathbf{q}_{2r} \cdot \mathbf{n})$$

$$= -p_1 A_1(-\mathbf{i}) - p_2 A_2(-\mathbf{j}) + m\mathbf{g}(-\mathbf{j}) + \mathbf{i}\rho q_1^2 A_1 + \mathbf{j}\rho q_2^2 A_2$$

$$= 2650.72\mathbf{i} + 294.52\mathbf{j} - 39.24\mathbf{j} + 196.35\mathbf{i} + 785.40\mathbf{j}.$$

The total force, **R**, exerted on the bend by the fluid is thus

$$R = (2847.07\mathbf{i} + 1040.68\mathbf{j})\ \text{N}.$$

The various forces involved are shown schematically in Fig. 4.6b.

Example 4.6

In Example 4.1 the cross-sectional area of the rocket nozzle at the exit is 0.0144 m², and the speed of the exhaust gas is $u_e = 1350$ m/s, relative to the nozzle. The pressure of the gas at the exit is $p_e = 1.1 \times 10^5$ Pa.

The rocket is fixed on a horizontal test bench and fired. Find $\mathbf{R}_H$, the horizontal component of the force which the gas flowing through the rocket applies to it.

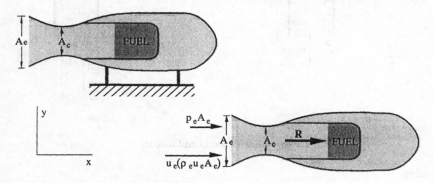

Figure 4.7 Rocket on test bench.

Solution

The coordinate system selected is fixed to the bench. The case is one-dimensional and Eq. (4.40) reduces to

$$R = p_e A_e + u_e(\rho_e u_e A_e).$$

Conservation of mass, Eq. (4.19), requires (see Example 4.1)

$$\rho_e u_e A_e = \rho_c u_c A_c = -\dot{m} = 1.4 \times 880 \times 0.01 = 12.32\ \text{kg/s}.$$

Thus,

$$R = p_e A_e + u_e(\rho_e u_e A_e). = 1.1 \times 10^5 \times 0.0144 + 1350 \times 12.32 = 18.216\ \text{N}.$$

Other Forces on the Control Volume

Equations (4.34) – (4.40) yield only **R**, i.e., the force exerted on the control volume by the fluid passing through it. For a control volume moving in space, with no atmosphere around it and where all body forces are negligible, **R** is indeed the only force which the control volume suffers. However, in most cases control volumes are located within the atmosphere and are subject to body forces. To obtain the total force which acts on a control volume two modifications are required: one which accounts for the environment pressure, and another which deals with body forces.

Body force modifications are considered first. Let the mass of the control volume itself, i.e., exclusive of the fluid inside it, be m_o. With the general body force still denoted by **g** this modification becomes

$$\mathbf{F}_g = m_o \mathbf{g}. \tag{4.41}$$

The environment atmosphere comes into contact with the outside surface of the control volume. The outside surface, exclusive of the openings, is denoted by S'', which may be quite different from S', the inner surface of the walls. The stress *at the openings* is already accounted for in **R**. In general the modification due to the stress on S'' is

$$\mathbf{F}_{S''} = \int_{S''} \mathbf{T} \, dS,$$

where **T** is the stress on the outside surface S''.

In many cases the approximation $\mathbf{T} = -p\mathbf{n}$ holds, and then

$$\mathbf{F}_{S''} = -\int_{S''} p\mathbf{n} \, dS \,.$$

We note that $S'' + S_o$, where S_o represents the openings, forms a closed surface, and we also note that with p_o constant,

$$\int p_o \mathbf{n} \, dS = p_o \int \mathbf{n} \, dS = 0,$$

on any surface. Therefore

$$-\int_{S''+S_o} p_o \mathbf{n} \, dS = 0, \qquad -\int_{S''} p_o \mathbf{n} \, dS = \int_{S_o} p_o \mathbf{n} \, dS,$$

and the modification due to the stress becomes

$$\mathbf{F}_{S''} = -\int_{S''} p\mathbf{n} \, dS = \int_{S_o} p_o \mathbf{n} \, dS = p_o \sum_i (\mathbf{n}A)_i, \tag{4.42}$$

where the summation is over all the openings.

We denote the total force which acts on a control volume by **F**, and then

$$\mathbf{F} = \mathbf{R} + \mathbf{F}_g + \mathbf{F}_{S''} = \mathbf{R} + m_o \mathbf{g} + \int_{S_o} p_o \mathbf{n} \, dS \tag{4.43}$$

Substitution of Eq. (4.36) into Eq. (4.43) results in an expression for the total force, **F**, which acts on the control volume,

$$F = -\int_{S_o} (p - p_o)\mathbf{n}\, dS + \int_V \rho \mathbf{g}\, dV + m_o \mathbf{g} - \frac{\partial}{\partial t}\int_V \rho \mathbf{q}\, dV - \int_{S_o} \rho \mathbf{q}(\mathbf{q_r} \cdot \mathbf{n})\, dS. \quad (4.44)$$

For the special case of a one-dimensional single-input–single-output control volume Eq. (4.40) may be modified to yield the total force as

$$F = -(p_1 - p_o)A_{1n} - (p_2 - p_o)A_{2n} + (m + m_o)g_x - \frac{\partial}{\partial t}(mu) - \rho_1 u_1 u_{1r} A_{1n} - \rho_2 u_2 u_{2r} A_{2n},$$
$$(4.45)$$

where the signs of the various terms are determined as in Eq. (4.40).

Example 4.7

Consider the pipe bend in Example 4.5. The mass of the steel walls of the bend is 6 kg, and the atmospheric pressure is 10^5 Pa. Find the total force which acts on the bend, i.e., that force that the bolts at the bend connections must support.

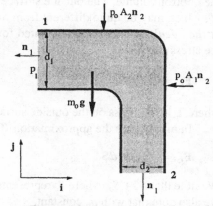

Figure 4.8 More forces on pipe bend.

Solution

From Eq. (4.41),

$$\mathbf{F}_g - m_o\mathbf{g} = -(6 \times 9.81)\mathbf{j} = -58.86\,\mathbf{j}.$$

By Eq. (4.42),

$$\mathbf{F}_{S''} = -\mathbf{i}p_o A_1 - \mathbf{j}p_o A_2 = -\mathbf{i}(10^5 \times 0.07854) - \mathbf{j}(10^5 \times 0.001963) = (-785.40\mathbf{i} - 196.35\mathbf{j}).$$

Thus, Eq. (4.43) yields

$$\mathbf{F} = \mathbf{R} + \mathbf{F}_g + \mathbf{F}_{S''} = \mathbf{i}(2{,}847.07 - 785.40) + \mathbf{j}(1{,}040.68 - 196.5 - 58.86)$$
$$= (2{,}061.67\mathbf{i} + 785.47\,\mathbf{j})\ \text{N}.$$

Example 4.8

We return to the rocket examples, Example 4.1 and Example 4.6. From Example 4.1 we know that the mass of the rocket and the fuel together is

$$m + m_o = 300 - 12.32t\ \text{kg}$$

and that

$$\frac{\partial m}{\partial t} = -12.32 \text{ kg/s}.$$

We also know that the cross-sectional area of the rocket nozzle at the exit is 0.0144 m^2, and the speed of the exhaust gas there is $u_e = 1350$ m/s, relative to the nozzle. The atmospheric pressure around the rocket is $p_o = 10^5$ Pa.

a. The rocket is fixed on a horizontal test bench and fired, as in Example 4.6. The pressure of the gas at the exit is $p_e = 1.1 \times 10^5$ Pa. The test bench is held in place by a dynamometer. Find the force read on the dynamometer.

b. Measurements in a wind tunnel show that when the wind speed is 200 m/s the pressure at the exit section A_e of the unfired rocket is $p_e = 0.9 \times 10^5$ Pa, and the total resistance force which keeps the rocket in place is $F = 16,776$ N. The rocket is then attached to an airplane which flies at 200 m/s and then releases the rocket. The rocket engine starts at the moment of release. Find whether the rocket accelerates.

c. The rocket is set with its nose upward and fired. Neglect the aerodynamic resistance and find its acceleration.

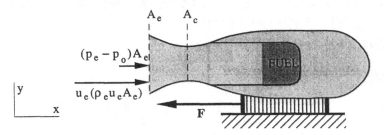

Figure 4.9 Total force on a rocket on a test bench.

Solution

a. We choose to solve this part anew, i.e., not using the results of Example 4.6. Defining the rocket as a control volume, we apply Eq. (4.45) with $A_{1n} = -A_e$, $p_1 = p_e$ and $u_1 = u_{1r} = -u_e$; and obtain the force read on the dynamometer

$$F = (p_e - p_o)A_e + u_e(\rho_e u_e A_e) = (1.1 - 1) \times 10^5 \times 0.0144 + 12.32 \times 1{,}350 = 16{,}776 \text{ N}.$$

Note that there is a vertical component of the body force, *downward*, of

$$(m + m_o)g = (300 - 12.32t) \times 9.81 = 2{,}943 - 120.86t.$$

b. We start by looking for the total resistance force acting on S'', i.e., the outer surface of the rocket not including its opening. Since the pressure at the exit section A_e pushes the rocket forward, the total resistance force becomes

$$F_{S''} = 16,776 + A_e \times 0.9 \times 10^5 = 16,776 + 0.0144 \times 0.9 \times 10^5 = 18,072 \, \text{N}.$$

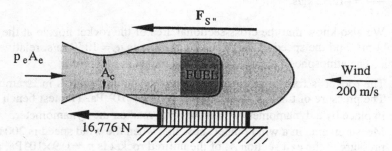

Figure 4.10 Unfired rocket in wind tunnel.

Next we assume the rocket to proceed with the constant speed of release, 200 m/s. If all forces acting on the rocket sum up to zero, it would mean that indeed this uniform motion is realized. A coordinate system convenient now is one which moves with the rocket. In this coordinate system R comes out to be exactly the same as in Example 4.6, i.e., $R = 18,216$ N. Equation (4.43) now gives the net force on the rocket as

$$F_{net} = R + F_{S''} = 18,216 - 18,072 = 144 \, \text{N}.$$

The rocket accelerates forward.

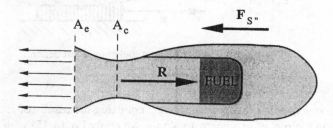

Figure 4.11 Fired rocket in flight.

c.   The sought acceleration is that of the whole rocket, i.e., that of the fuel and the structure together. Let this time-dependent acceleration be a. Let the coordinate system be chosen stationary with x pointing upward. This is also a one-dimensional case, and Eq. (4.45) states

$$F = (p_e - p_o)A_e - (m + m_o)g - \frac{\partial}{\partial t}(mu) + \rho_e(u_e - u)u_e A_e,$$

where u is the instantaneous velocity of the rocket, m is the mass of the fuel and m_o is the mass of the structure of the rocket:

$$\frac{\partial}{\partial t}(mu) = u\dot{m} + m\frac{\partial u}{\partial t} = u\dot{m} + ma;$$

hence, with $\dot{m} = -\rho_e u_e A_e$,

$$F = (p_e - p_o)A_e - (m + m_o)g - ma - \dot{m}u_e.$$

By Newton's second law

$$F = m_o a = (p_e - p_o)A_e - (m + m_o)g - ma - \dot{m}u_e.$$

Thus,

$$(m_o + m)a = (p_e - p_o)A_e - (m_o + m)g - \dot{m}u_e.$$

From Example 4.1,

$$m_o + m = 300 - 12.32t,$$

$$\dot{m} = -12.32t$$

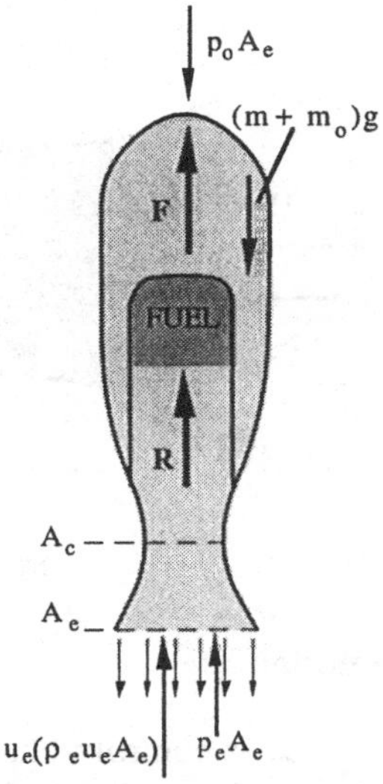

Figure 4.12 Vertical rocket.

and

$$a = \frac{0.1 \times 10^5 \times 0.0144}{300 - 12.32t} - g + \frac{12.32 \times 1{,}350}{300 - 12.32t} = \frac{16{,}776}{300 - 12.32t} - 9.81.$$

The velocity of the rocket may also be obtained as

$$u = \int_0^t a\,dt = \frac{16{,}776}{12.32}\ln\left(\frac{300}{300 - 12.32t}\right) - 9.81t.$$

This result, of course, is accurate only as long as aerodynamic resistance may still be neglected, i.e., for small velocities and hence for small t values.

Example 4.9

A single-engine jet airplane flies at the speed of $V = 280$ m/s. The atmospheric pressure is $p_a = 10^5$ Pa. The well-designed air intake of the engine has a cross-sectional area of $A_1 = 0.1$ m^2. The exit cross section of the jet nozzle has the area of $A_2 = 0.3$ m^2. The gas pressure at the exit is $p_2 = 1.1 \times 10^5$ Pa. The fuel-to-air ratio in the engine is $1 : 20$. The specific density of the air at the inlet is $\rho = 1.2$ kg/m^3. The gas speed at the exit relative to the airplane is $U = 650$ m/s. What is the horizontal force transferred by the bolts connecting the engine to the airplane body?

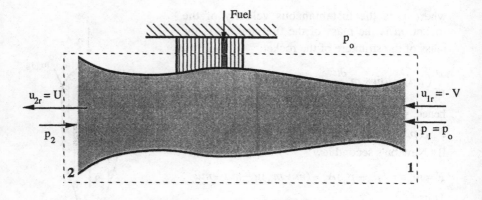

Figure 4.13 Jet engine and control volume.

Solution

Let a control volume be specified as consisting of the jet engine, Fig. 4.13. Now the jet engine is assumed to be of constant mass and to move in the x-direction with a constant speed. Moreover, in this case $p_1 = p_o$. The horizontal force transferred to the airplane body can be found from Eq. (4.45), which for this case simplifies to

$$F = -(p_2 - p_o)A_{2n} - \rho_1 u_1 u_{1r} A_{1n} - \rho_2 u_2 u_{2r} A_{2n} .$$

In our case the intake air moves with respect to the control volume with a speed $-V$; hence,

$$u_1 = u_{1r} = -V, \qquad A_{1n} = A_1, \qquad \rho_1 u_1 A_{1n} = -\rho_1 V A_1 = \dot{m}_1$$

and

$$u_{2r} = U, \qquad A_{2n} = -A_2 .$$

As the density of the exhaust gas is not given, we find the exit mass flowrate by using continuity, Eq. (4.17), and by accounting for the added mass of the fuel in the exit stream,

$$\rho_1 u_2 A_{2n} = \dot{m}_2 = -1.05\dot{m}_1 .$$

Substitution into Eq. (4.45) yields the force that the engine applies in the x-direction to the airplane body,

$$\begin{aligned}
F &= A_2(p_2 - p_o) - \dot{m}_1[1.05U - V] \\
&= 0.3 \times \left(1.1 \times 10^5 - 1 \times 10^5\right) - 1.2 \times 0.1 \times (-280) \times [1.05 \times 650 - 280] \\
&= 3{,}000\,\text{N} + 13{,}524\,\text{N} = 16{,}524\,\text{N}.
\end{aligned}$$

Example 4.10

The ancient Romans used a water-hammer pump, Fig. 4.14, to raise water above the level of their water sources, as they occurred in nature. They let water run in pipe A from the source. Once steady state was reached, they blocked the pipe with a wedge W. The water in pipe A had to decelerate, rising in pipe B up to the height h into an overflow-service tank C. Then they used pipe D to lead the water to their houses.

Find h, approximately, as a function of the length, L, of pipe A, and of the velocity, u_o, of the water in it at steady state, i.e., before the wedge is inserted. Neglect flow losses due to friction, and assume all the pipes to be of the same diameter.

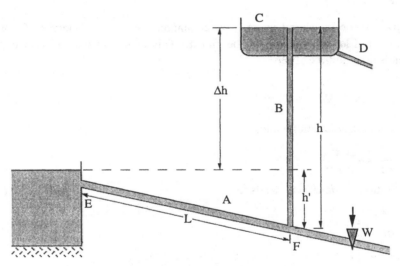

Figure 4.14 Roman water-hammer pump.

Solution

Let the pipe section between points E and F, Fig. 4.14, which is L meters long, be chosen as a control volume. Let the average velocity in this pipe at steady state and just before the wedge is inserted be u_o. The pressure above the water in pipe B is atmospheric, and so is, approximately, the pressure at point E. The difference of the water level, h', between the water source and point F is needed to accelerate the water to the velocity u.

Once the wedge is inserted, conservation of mass requires that as long as there is any flow, the water must rise in pipe B.

Denote the water level in pipe B by h. Then

$$\frac{dh}{dt} = u.$$

Since the control volume can have forces applied along its axis by shear stresses only, and since these are to be neglected, $\mathbf{R}_x = 0$ and   Eq. (4.36) yields

$$0 = -\int_{S_o} p\mathbf{n}\, dS + \left(\int_V \rho \mathbf{g}\, dV\right)_x - \frac{\partial}{\partial t}\int_V \rho \mathbf{q}\, dV - \int_{S_o} \rho \mathbf{q} u\, dS$$

$$= A_A(p_o - p_F) + 0 - LA_A\rho\frac{du}{dt} - 0$$

or

$$p_F - p_o = -\rho L\frac{du}{dt}.$$

Let $p_F - p_o$ be measured using a water manometer, i.e., the height of a water column, and let this manometer be pipe B. This is an approximation because h depends now on time. Then

$$p_F - p_o = \rho g h = -\rho L\frac{du}{dt}.$$

One more differentiation yields

$$\frac{dh}{dt} = -\frac{L}{g}\frac{d^2u}{dt^2}.$$

Substitution of $dh/dt = u$ leads to

$$\frac{d^2u}{dt^2} + \frac{g}{L}u = 0.$$

The solution of this linear differential equation yields

$$u = u_o \cos\left(\sqrt{\frac{g}{L}}\, t\right)$$

and

$$h = \int_0^t u\, dt = u_o\sqrt{\frac{L}{g}}\sin\left(\sqrt{\frac{g}{L}}\, t\right),$$

which attains its maximum at

$$\left(\sqrt{\frac{g}{L}}\, t\right) = \frac{\pi}{2}, \quad \text{i.e.,} \quad t_{max} = \frac{\pi}{2}\sqrt{\frac{L}{g}}.$$

Hence

$$h_{max} = u_o\sqrt{\frac{L}{g}}.$$

Examples are

	$u_o = 3$ m/s		$u_o = 5$ m/s	
L (m)	100	50	100	50
h_{max} (m)	9.5	6.7	16.0	11.2
h' (m)	0.5		1.3	

Here h' is the height needed to accelerate the flow to u_o. Its calculation is explained in Chapter 7.

Note: A simpler argument can be offered, although with the same level of approximation, by the use of the Bernoulli equation derived in Chapter 7. The Romans, however, did not know about this future glorious son of their land, Bernoulli, while their engineers appreciated very well the effect of inertia. Just regard their battering rams and note the similarity between a horizontally moving beam stopped abruptly by a wall and a horizontally moving water column stopped abruptly by a wedge.

Example 4.11

The nozzle shown in Fig. 4.15 ejects water at the rate of Q m³/s. The cross-sectional area of the water jet is A m². The jet meets a moving vane, B, flows along its surface and leaves it with the same relative velocity as at the point of their meeting. The vane moves away from the nozzle with the speed V. The pressure everywhere is atmospheric. You are an observer moving with the vane. Find the force exerted on the vane by the water jet; find the power transmitted and the velocity V at which this power becomes maximal.

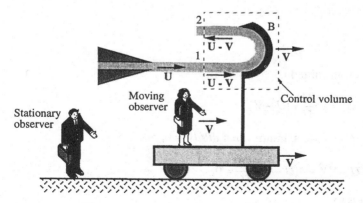

Figure 4.15 Water jet and moving vane.

Solution

To solve the problem, we use Eq. (4.40) for a control volume which moves with the vane at the constant speed of V.

For a fluid of constant density at steady state and a pressure uniform everywhere in the control volume, Eq. (4.40), which gives the force that the jet exerts on the vane, simplifies to

$$R = -\rho \left[u_1 u_{1r} A_{1n} + u_2 u_{2r} A_{2n} \right].$$

In our case

$$A_{1n} = A_{2n} = -A.$$

The velocity of the jet as it emerges out of the nozzle is $U = Q/A$, and this velocity is assumed to be conserved until the jet meets the vane. The jet velocity relative to the vane at point 1 is

$$u_{r1} = U - V.$$

The relative velocity at point 2, where the flow leaves the vane, is

$$u_{r2} = -(U - V).$$

As long as the jet velocity remains constant, continuity requires that its cross-sectional area be conserved. Thus

$$u_{r1} A_{1n} = -(U - V)A = -(U - V)\frac{Q}{U}, \qquad u_{r2} A_{2n} = (U - V)\frac{Q}{U}.$$

Now, in the inertial coordinate system moving with the vane

$$u_1 = u_{r1} = U - V, \qquad u_2 = u_{r2} - (U - V).$$

Substitution into Eq. (4.40) yields

$$R = 2\rho \frac{Q}{U}(U - V)^2.$$

The power transmitted is

$$P = RV = 2\rho \frac{Q}{U} V(U - V)^2.$$

This power has its maximum when $dP/dV = 0$, or

$$\frac{2Q}{U}(U - V)^2 - 4(U - V)\frac{QV}{U} = 0,$$

which leads to

$$V = U/3.$$

Example 4.12

Repeat Example 4.11 for an observer standing on the ground, i.e., stationary with respect to the nozzle.

Solution

The control volume is again chosen to move with the vane at the constant speed V. The velocity of the jet as it emerges out of the nozzle is $U = Q/A$, and this velocity is assumed to be conserved until the jet meets the vane. The velocities relative to the vane are still

$$u_{r1} = U - V, \qquad u_{r2} = -(U - V).$$

As long as the jet velocity remains constant, continuity requires that its cross-sectional area be conserved. Thus, once again,

$$u_{r1}A_{1n} = -(U - V)A = -(U - V)\frac{Q}{U}, \qquad u_{r2}A_{2n} = (U - V)\frac{Q}{U}.$$

Now, in the inertial coordinate system fixed to the ground,

$$u_1 = U, \qquad u_2 = -(U - V) + V = 2V - U.$$

Equation (4.40), again without the gravity and the pressure terms, yields

$$R = -\rho\left[u_1 u_{1r}A_{1n} + u_2 u_{2r}A_{2n}\right]$$

$$= -\rho\left[-U(U - V)\frac{Q}{U} + (2V - U)(U - V)\frac{Q}{U}\right] = 2\rho\frac{Q}{U}(U - V)^2.$$

The power transmitted is, again,

$$P = RV = 2\rho\frac{Q}{U}V(U - V)^2,$$

which again has its maximum at $V = U/3$.

Angular Momentum Theorem for a Control Volume

Many applications of fluid mechanics are through the use of rotating machines, such as turbines, turbopumps, turbocompressors, propellers, etc. A natural and convenient choice of the control volume for such a machine is the machine itself. Due to the large number of such applications, it is helpful to prepare a form analogous to Eq. (4.36), which is directly applicable to rotating control volumes. This form is known as the angular momentum theorem.

It is noted that the need for relations directly applicable to rotating coordinates exists in solid mechanics too, and analogous forms exist there. The use of these relations, which are valid for a thermodynamic system, simplifies the construction of the angular momentum theorem.

Reference is made to the control volume shown in Fig. 4.1. A center is chosen in the plane of the figure. This center may be any point, but once chosen, it must remain fixed. The radius vector from this center to any element of mass in the thermodynamic system is denoted $\mathbf{r}$.

The angular momentum of the thermodynamic system shown there at the time t is

$$\int_V \mathbf{r} \times \mathbf{q}(\rho \, dV).$$

It is noted that this expression for the angular momentum may be obtained by cross-multiplication of $\mathbf{r}$ by $\mathbf{q}$ in the expression for the linear momentum: Explicitly, since the angular momentum of the thermodynamic system is the sum of the angular momenta of the mass elements of the system, this cross-multiplication may be done before integration, i.e., under the integral sign. Proceeding to do this in Eq. (4.36), it becomes

$$\mathbf{M}_R = \int_{S_o} \mathbf{r} \times \mathbf{T} \, dS + \int_V \mathbf{r} \times \mathbf{g}(\rho \, dV) - \frac{\partial}{\partial t} \int_V [(\mathbf{r} \times \mathbf{q})\rho] \, dV - \int_{S_o} (\rho \mathbf{r} \times \mathbf{q})(\mathbf{q}_r \cdot \mathbf{n}) \, dS \quad (4.46)$$

where $\mathbf{M}_R$ is the reaction moment that the fluid exerts on the inner walls of the control volume.

Because the center from which $\mathbf{r}$ stems must be fixed, the direct application of Eq. (4.46) is limited to rotating systems with a fixed center of rotation. This fixed center is then the origin for the radius vector $\mathbf{r}$.

When the system is not rotating, any point may be chosen as a center, and Eq. (4.46) may then be used to calculate moments about that point.

Example 4.13

A small water turbine rotates at $n = 200$ rpm. The water enters at the center, at the rate of 2 kg/s, and leaves at the circumference with the relative velocity $\mathbf{q}_r = 17.2$ m/s, and with the angle $\beta = 25.8°$ between the relative velocity $\mathbf{q}_r$ and the circumferential velocity $\mathbf{q}_c$, Fig. 4.16. Note that β is determined by the geometry of the vanes of the turbine. The radius of the turbine rotor is $r = 0.2$ m. Find the absolute velocity at the exit, q_o, the moment of the turbine and the power.

Solution

$$|q_c| = \omega r = \frac{2\pi n}{60} \times 0.2 = 20.94 \times 0.2 = 4.2 \text{ m/s},$$

$$q_o = q_c + q_r.$$

Using the cosine theorem, Fig. 4.16,

$$q_o^2 = q_c^2 + q_r^2 - 2q_c q_r \cos\beta = 4.2^2 + 17.2^2 - 2 \times 4.2 \times 17.2 \cos 25.8°,$$

leading to

$$q_o = 13.54 \text{ m/s}.$$

By the sine theorem

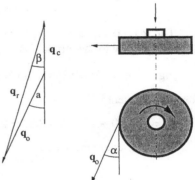

$$\sin(180° - \alpha) = \frac{q_r}{q_o}\sin\beta = \frac{17.2}{13.54}\sin 25.8°$$

and

$$\alpha = 33.6°.$$

Neglecting air friction, Eq. (4.46) yields

Figure 4.16 Water turbine.

$$M = \int_{S_o} \rho r \times q_o (q_r \cdot n) \, dS = 1000 \times 0.2 \times 13.54 \times \cos\alpha \times 17.2 A \sin\beta,$$

but

$$\rho(q_r \cdot n)A = 1000 \times 17.2 A \sin\beta = 2 \text{ kg/s}.$$

Hence,

$$M = 2 \times 0.2 \times 13.54 \times \cos 33.6° = 4.5 \text{ Nm},$$

and the power is

$$\text{Power} = |\omega \cdot M| = 20.94 \times 4.5 = 94.2 \text{ Nm/s} = 94.2 \text{ W}.$$

Example 4.14

For the pipe bend considered in Example 4.5, Fig. 4.6, what is the moment about point A just at the upper edge of flange number 1 for $r_{12} = 0.15$ m?

Solution

Using Eq. (4.42),

$$\int \mathbf{r} \times \mathbf{T}\, dS = (d_1/2)A_1(p_1 - p_2) + r_{12}A_2(p_2 - p_o)$$
$$= 0.05 \times \left(0.05^2\, \pi\right) \times (3.375 - 1.0) \times 10^5 + 0.15 \times \left(0.025^2\, \pi\right) \times (1.5 - 1.0) \times 10^5$$
$$= 108.00 \text{ Nm}$$

and

$$\int_{A_o} (\rho \mathbf{r} \times \mathbf{q})(\mathbf{q}_r \cdot \mathbf{n})\, dS = \frac{d_1}{2} \times 196.35 + r_{12} \times 785.4$$

$$= 0.05 \times 196.35 + 0.15 \times 785.4 = 127.63 \text{ Nm}.$$

The mass of the bend together with the water within it is 6 kg, and the contribution of gravity to the moment is thus

$$mgr_{12} = 6 \times 9.81 \times 0.15 = 8.83 \text{ Nm}.$$

The total moment is

$$\mathbf{M} = 108.00 + 127.63 - 8.83 = 226.8 \text{ Nm}.$$

Its direction is counterclockwise.

Example 4.15

Consider again the sprinkler of Example 4.3, Fig. 4.3. The angle of its short nozzle relative to the radial direction is 45°, and the projection of the arm is $r = 0.4$ m.

a. Find the moment acting on the sprinkler when the water is just turned on, neglecting friction and pressure effects.
b. Find the friction moment at steady state, given $\omega = 4$ rad/s.
c. Suppose friction disappeared altogether. Find now the angular velocity of the sprinkler.

Solution

The velocity triangles for the sprinkler are schematically shown in Fig. 4.17, with

$\mathbf{q}_o$ absolute velocity of water,
$\mathbf{q}_c$ circumferential velocity of sprinkler, and
$\mathbf{q}_r$ relative velocity of water with respect to sprinkler.

$$\mathbf{q}_o = \mathbf{q}_c + \mathbf{q}_r.$$

The mass flux through the two arms of the sprinkler is

$$\dot{m} = \frac{\pi}{4} 0.05^2 \times 0.2 \times 1000 = 0.393 \, \text{kg} / \text{s}.$$

a.   Equation (4.46) yields

$$\mathbf{M} = - \int_{A_o} (\mathbf{r} \times \mathbf{q}_o)(\rho \mathbf{q}_r \cdot \mathbf{n}) \, ds = -\mathbf{r} \times \mathbf{q}_o \dot{m},$$

$$q_c = 0, \qquad q_r = 10 \text{ m/s},$$

$$\mathbf{M} = -0.4 \times 10 \times \dot{m} \times \sin 45^\circ = -1.11 \text{ Nm}.$$

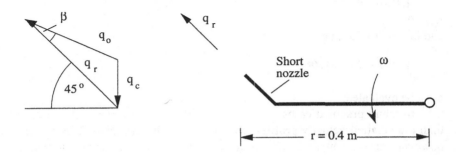

Figure 4.17 Velocity triangle for sprinkler.

b. $|\mathbf{q}_c| = \omega r = 4 \times 0.4 = 1.6 \, \text{m} / \text{s}, \qquad q_r = 10 \, \text{m} / \text{s},$
$q_o^2 = q_r^2 + q_c^2 - 2 q_r q_c \cos 45^\circ, \qquad q_o = 8.94 \, \text{m} / \text{s},$
$\sin \beta = (q_c / q_o) \sin 45^\circ, \qquad b = 7.27^\circ, \qquad 45^\circ - b = 37.73^\circ,$

$-\mathbf{r} \times \mathbf{q}_o = -0.4 \times 8.94 \sin 37.73^\circ = -2.19,$
$\mathbf{M}_f = -\mathbf{M} = 2.19 \dot{m} = 0.86 \, \text{Nm}.$

At steady state the moment of friction just balances the moment of the water.

c.   As friction disappears, so does the moment of the water. In this case $\mathbf{r} \times \mathbf{q}_o = 0$, or $\beta = 45^\circ$, which requires

$$q_c = q_r / \sqrt{2} = 7.07 \, \text{m} / \text{s} = \omega r,$$

$$\omega = 7.07 / 0.4 = 17.68 \text{ rad/s} = 169 \text{ rpm}.$$

Correction Factor for Average Velocity in Momentum Theorem

To use control volume considerations for the conservation of mass, the integral in Eq. (4.14) of the mass flux through the opening must be computed, i.e.,

$$\int_{A_o} \rho(\mathbf{q} \cdot \mathbf{n}) \, dS.$$

To use control volume considerations for the momentum theorem, both the linear and the angular one, the integrals in Eq. (4.36) , i.e.,

$$\int_{A_o} \rho \mathbf{q}(\mathbf{q}_r \cdot \mathbf{n}) \, dS,$$

and in   Eq. (4.46) , i.e.,

$$\int_{A_o} \rho(\mathbf{r} \times \mathbf{q})(\mathbf{q}_r \cdot \mathbf{n}) \, dS,$$

must be evaluated.

In many practical cases it seems desirable to use average velocities, rather than the detailed velocity profiles, which may not be available. This possibility, however, raises a new problem. Let the term average, as applied here to velocities, mean that in considerations of conservation of mass, both integration of the detailed velocity profile and calculations based on the average velocities yield exactly the same mass flux. Obviously, conservation of mass then needs no correction factors at all. But suppose that the same average velocities are used to calculate the flux of momentum into the control volume. Are the results thus obtained the same as those obtained by integrating the velocity profiles? In general the results are different, and those results obtained using average velocities must be multiplied by *correction factors* to yield the correct results.

Let the correction factor for the momentum be denoted β_M, and let its evaluation be restricted to cases where $\mathbf{q}$, $\mathbf{q}_r$ and $\mathbf{n}$ are parallel.

The momentum flux in terms of average velocities is

$$A_i \rho_i \bar{q}_1^2, \tag{4.47}$$

with $\bar{q}_i$ being the average velocity, while the flux obtained by integration is

$$\left| \int_{A_o} \rho \mathbf{q}(\mathbf{q}_r \cdot \mathbf{n}) dS \right| = \int_{A_o} \rho q^2 dS. \tag{4.48}$$

Hence the *correction factor for the momentum flux* becomes

$$(4.49)$$

$$\beta_M = \frac{1}{\rho_i A_i \bar{q}_i^2} \int_A \rho q^2 dS.$$

The correction factors obtained in Examples 4.16 and 4.17 can be used to estimate errors when the exact velocity profiles are not known.

Example 4.16

The velocity profile in laminar flow through a circular pipe is

$$w = w_{max}\left[1 - (r/R)^2\right].$$

Find the momentum correction factor for this flow.

Solution

We first find for this flow the average velocity, i.e., the volumetric flowrate Q per unit cross-sectional area A,

$$\overline{w} = \frac{Q}{A} = \frac{1}{\pi R^2} = \int_A w \, dA = \frac{1}{\pi R^2} = \int_0^R w_{max}\left[1 - \left(\frac{r}{R}\right)^2\right] 2\pi r \, dr$$

or

$$\overline{w} = \frac{w_{max}}{2}.$$

The correction factor is obtained from Eq. (4.49),

$$\beta_M = \frac{1}{\pi R^2 \left(\dfrac{w_{max}}{2}\right)^2} \int_0^R w_{max}^2 \left[1 - \left(\frac{r}{R}\right)^2\right]^2 2\pi r \, dr = \frac{4}{3}.$$

Example 4.17

The velocity profile in turbulent flow through a circular pipe is

$$w = w_{max}\left[1 - r/R\right]^{1/7}.$$

Find the correction factor for the momentum for this flow.

Solution

The average velocity is here

$$\overline{w} = \frac{Q}{A} = \frac{1}{\pi R^2} \int_A w \, dA = \frac{1}{\pi R^2} \int_0^R w_{max}\left(1 - \frac{r}{R}\right)^{\frac{1}{7}} 2\pi r \, dr = 0.817 w_{max}.$$

The correction factor is found from Eq. (4.49),

$$\beta_M = \frac{1}{\pi R^2 (0.817 w_{max})^2} \int_0^R w_{max}^2 \left[1 - \frac{r}{R}\right]^{\frac{2}{7}} 2\pi r \, dr = 1.020.$$

Hence, for turbulent flow the correction factor can usually be neglected.

Problems

4.1   A system of pipes is shown in Fig. P4.1. All pipes have the same diameter $d = 0.1$ m. The average flow velocity in pipe number 1 is $q_1 = 10$ m/s to the right. In pipe 2, $q_2 = 6$ m/s to the right. Find q_3. Can there be a flow with q_2 to the left? What is q_3 then?

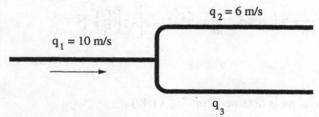

Figure P4.1 Three pipes.

4.2   The average air velocity in the intake duct of an air conditioner is 3 m/s. The intake air temperature is 35°C. The air comes out of the air conditioner at 20°C and flows through a duct, also at 3 m/s. The pressure is atmospheric at 10^5 Pa everywhere. What is the ratio of the cross-sectional areas of the ducts?

Assuming that some water has condensed from the air in the air conditioner, how does it affect the velocity in the outlet duct?

4.3   A pressure vessel is equipped with a small piston of area A_S and a large piston of area A_L, Fig. P4.3. The vessel is filled with oil, and a force **F** is applied to the small piston. Neglect friction and differences of heights of oil and show that conservation of mass and conservation of energy lead to

$\mathbf{G} = \mathbf{F}(A_L/A_S)$. This is the principle of the hydraulic lift. For $\mathbf{F} = 10\,\text{N}$ and $A_L/A_S = 1{,}000$, find $\mathbf{G}$.

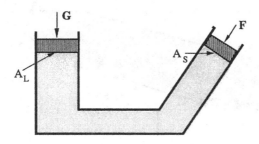

Figure P4.3 Hydraulic lift.

4.4 Compressed air is introduced into the ballast tank of a submarine and drives the water out of there with the rate of 2 m³/s when the submarine is at the depth of 10 m. The atmospheric pressure is 10^5 Pa. What is the flowrate of the ejected water when the same mass flux of compressed air is introduced at the depth of 100 m? Can there be a situation in which, because of depth and downward motion of the submarine, the rate of water ejection becomes zero or even negative?

4.5 A steam boiler is fed with water at the rate of 1 kg/s. It supplies steam at atmospheric pressure and 105°C at the same rate. The steam speed in the pipe is 10 m/s. What is the pipe diameter?

4.6 A water container has a hole in its side, Fig. P4.6. The water flows through this hole with the velocity of $q_h = 5$ m/s. The effective size of the hole is $A = 0.1\,\text{m}^2$. The pressure around the container is atmospheric. What horizontal force acts on the container because of the hole?

Figure P4.6 Container with hole.

4.7 A fireman hose ends with a nozzle, as shown in Fig. P4.7. Measurements show $A_F = 0.01$ m², $A_E = 0.0025$ m², $q_F = 8$ m/s, $p_F = 578{,}000$ N/m², $p_E = 98{,}000$ N/m² and so is the atmospheric pressure. The hose is flexible and the nozzle is connected to it by a flange at G. What are the magnitude and direction of the forces acting on the flange? If the hose is not held, which way will it move?

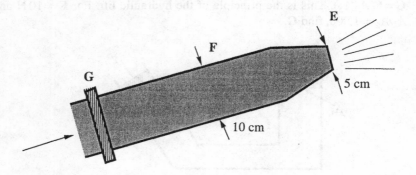

Figure P4.7 Fireman hose.

4.8   A jet of fluid of absolute velocity q_i emerges from a nozzle and hits a
moving vane and then turns, Fig. P4.8. The speed of the fluid leaving the
vane relative to the vane, q_{er}, equals the speed of the fluid hitting the vane
relative to the vane, q_{ir}. The vane moves at the absolute speed V_v. Derive
expressions for the force that the fluid applies to the vane and for the power
developed. Find V_v for which the force is the largest and one for which the
power is maximized.

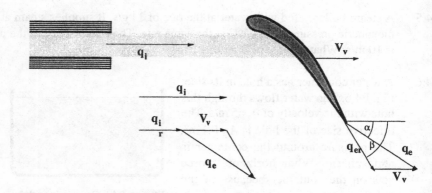

Figure P4.8 Jet, vane and velocities.

4.9   A nozzle moves with the velocity $\mathbf{V}_n$, Fig. P4.9. It ejects a jet which has the
velocity $\mathbf{q}_{er}$, relative to the moving nozzle. The nozzle can swivel to form
different angles β between $\mathbf{V}_n$ and $\mathbf{q}_{er}$. Find an expression for the force driving the nozzle in the direction of $\mathbf{V}_n$, and the power obtained by this force.
Are there $\mathbf{V}_n$ values which maximize this force, or this power, as in Problem
4.8? Are there β values which do this?

4.10   The nozzle in Problem 4.9 is fixed
       to the circumference of a rotating
       disk, such that it still moves with
       V_n. Find the turning moment and
       the power.

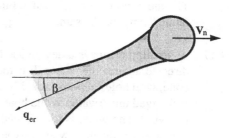

4.11   The sprinkler of Examples 4.3 and
       4.15 (Figs. 4.3 and 4.17) is set
       inside a high-pressure water tank
       such that now water enters
       through the nozzles at its arms and
       exits through its central pipe. The
       water flow through this central pipe is at 0.6 m/s.

Figure P4.9 Moving nozzle and jet.

       a. Which way does the sprinkler turn?
       b. What is the maximum power obtained at 60 rpm?

4.12   A rocket is released from a flying vehicle at a horizontal velocity of
       300 m/s. The rocket engine ignites and sends backward a horizontal jet of
       gas at a relative velocity of 280 m/s. The flux of the gas is 3 kg/s, and the
       mass of the rocket is 30 kg. The flight takes place at such a height that
       atmospheric pressure and aerodynamic resistance are negligible.
       a. What is the acceleration of the rocket 1 s after release?
       b. What is the velocity of the rocket's jet relative to the ground?

4.13   A ram-jet airplane flies horizontally with a velocity of $V = 600$ m/s, which
       is also the velocity with which it sees air coming into its engine. Fuel is
       burnt at the rate of 0.12 kg/kg air. The combustion gas leaves the nozzle at
       $q_r = 700$ m/s relative to the airplane. The air intake area is 0.25 m², and the
       nozzle exit area is 0.75 m². The density of the air is $\rho_i = 1.2$ kg/m³. Find the
       driving force and the power used.

4.14   A steel pipe of $d = 0.150$ m, wall thickness of 6 mm and 1,000 m long is
       used to supply water. The water flows at 10 m/s, and the nominal stress
       allowed in the pipe wall is 5,000 N/cm² (these pipes get rusty). To prevent
       high stresses due to water hammer, the valve at the end of the pipe must not
       be closed faster than at a certain rate. What is the minimal time required to
       close the valve?

4.15   What are the estimated correction factors for Problems 4.6, 4.7, 4.8, 4.9,
       4.10, 4.13 and 4.14 ?

4.16   The flow rate in the water turbine of Example 4.13 is doubled, to 4 kg/s. The moment remains the same. Find q_o, n and the power of the turbine. Hint: β remains the same, and $|q_r|$ is doubled.

4.17   A floating anchor is a device used in lifeboats to keep the nose of the boat directed against the waves. It is made of heavy cloth and has the shape of a cone, with holes at both ends, Fig. P4.17. It is tied to the rear of the boat and is dragged underwater by the boat, which is itself dragged by the waves and the wind. The water is thus forced into the anchor at its larger opening and comes out at the narrow end. Experiments show that the pressure at the exit, i.e., the narrow end, is slightly below the hydrostatic pressure for this depth, that the pressure at the wide end is above hydrostatic by about $0.4\rho V^2$, where V is the speed of the anchor relative to the water, and a good assumption is that the water does not leave the anchor at a relative speed higher than V.

For $d_i = 1$ m, $d = 0.4$ m, estimate the resistance force of this anchor to relative speeds of 1 m/s, 3 m/s, 10 m/s, 15 m/s, 20 m/s.

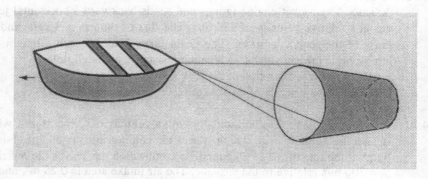

Figure P4.17 Floating anchor.

4.18   A shallow water boat is propelled by a pump as shown in Fig. P4.18. The inlet pipe, A, has a diameter of 0.4 m, and the average velocity in it is 10 m/s. The outlet pipe, B, has the diameter of 0.2 m. The centers of both pipes are 1.5 m below the water level and the pressures at these centers are approximately hydrostatic. Find the force that the pump and its pipes apply to the boat.

An inventor who did not study fluid mechanics but had such a boat modified it by taking the outlet pipe 1.5 m into the air, as shown in configuration C in Fig. P4.18. Is this a good idea?

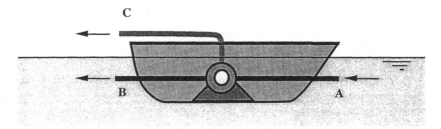

Figure P4.18 A boat with a pump.

4.19 A boat moves in the water at a speed of 10 m/s propelled by a pump which takes in water at the front (point 1) through a suction pipe with the diameter of $D_1 = 0.25$ m and discharges the water at the rear (point 2) through a discharge pipe with the diameter $D_2 = 0.2$ m; $H_1 = 2$ m, $H_2 = 2.5$ m.

The pressures p_1 and p_2 are the same as the hydrostatic pressures outside the boat, and the relative velocity V_1 is that of the boat, i.e., $V_1 = 10$ m/s.

a. Find the total driving force applied to the boat (i.e., the force which is transferred through the bolts which connect the pump to the boat).

b. Someone suggests to connect a short converging cone at point 2 such that the exit of the water becomes through an opening of $D_3 = 0.15$ m. The pump operates as in a. Will this increase or decrease the speed of the boat?

c. Someone suggests to connect a short diverging cone at point 2 such that the exit of the water becomes through an opening of $D_4 = 0.25$ m. The pump operates as in a. Will this increase or decrease the speed of the boat?

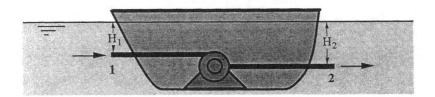

Figure P4.19 A boat propelled by a pump.

4.20 The pressure at the entrance to pipe 1 in Problem 4.1 is 200,000 Pa, and that at the exit of pipes 2 and 3 is 100,000 Pa, which is also the atmospheric pressure. Find what force is needed to keep the pipe structure from moving.

4.21 A jet of water has the diameter of 0.04 m and the average velocity of 8 m/s. The water hits a stationary flat vane, as shown in Fig. P4.21. The water is assumed to spread at the point of impact with cylindrical symmetry, and its velocity is conserved. The pressure everywhere outside the water jet is atmospheric, and so it is inside the jet well before it hits the vane. Find the force acting on the vane and the power extracted by the vane.

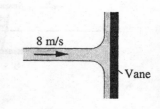

Figure P4.21

4.22 The vane in Problem 4.21 is tilted by the angle $\pi/4$, Fig. P4.22. The water still conserves its velocity after hitting the vane. Now, however, conservation of momentum in the direction tangent to the vane decides how the flow is divided at the point of impact. Assume the flow two-dimensional and find the force acting on the vane and the power extracted by the vane.

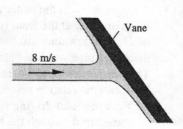

Figure P4.22

4.23 The vane in Problem 4.21 now recedes from the water jet with the speed of 2 m/s, as shown in Fig. P4.23. The water now conserves its velocity relative to the vane after hitting it. Find the force acting on the vane and the power extracted by the vane.

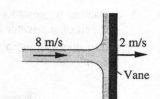

Figure P4.23

4.24 The vane in Problem 4.22 now recedes from the jet with the speed of 2 m/s, Fig. P4.24. The water still conserves its velocity relative to the vane after hitting it. But now conservation of relative momentum in the direction tangent to the moving vane determines how the flow is divided at the point of impact. Assume the flow two-dimensional and find the force acting on the vane and the power extracted by the vane.

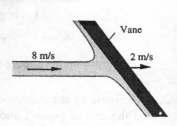

Figure P4.24

4.25   A water container is open at its top. It is fitted with an
       inlet pipe controlled by valve A and with an outlet
       pipe and valve B, as shown in Fig. P4.25. When a
       valve is opened, the flow in the corresponding pipe is
       1 m³/s at the mean velocity of 10 m/s. When both
       valves are closed, the container transfers to the floor a
       force of 100,000 N. Find what forces are transferred
       to the floor a short time after:
       a. Valve A is opened.
       b. Valve B is opened.
       c. Both valves are opened.

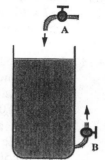

Figure P4.25

4.26   The angle between the course of a sailboat and the wind is $\alpha + \beta$. The
       velocity of the wind is 15 m/s. The sail has an area of 20 m² and is set with
       the angle β to the wind direction, as shown in Fig. P4.26. Assume that the
       wind is completely turned by the sail so as to become tangent to its back-
       ward direction and that the speed of the wind relative to the sail is con-
       served. The density of the blowing air is 1.15 kg/m³, and the speed of the
       boat is 0.5 m/s.
       a. Find the force vector the mast transfers to the boat. Note that only the
           component in the sailing direction of this force vector serves to advance
           the boat.
       b. Find β which makes this forward component the largest.
       c. For $\alpha + \beta = 60°$ find numerical values.

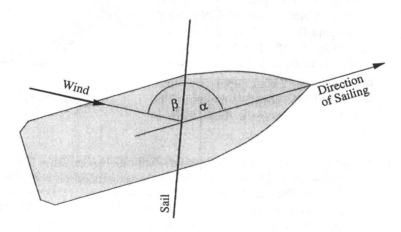

Figure P4.26 A sailboat.

4.27 Referring to Problem 4.26, show that a sailboat can sail upwind, i.e., when $\alpha+\beta<\pi/2$. For a given wind, show that the greatest speed that the boat may achieve is sometimes not directly downwind, i.e., not for $\alpha+\beta=\pi$.

4.28 A pipe bend is shown in Fig. P4.28. The inner diameter at the entrance flange, point A, is 0.5 m, and that at the outlet, point B, is 0.25 m. The mean velocity of the water at point A is 5.0 m/s. The manometric pressure in the water at point A is 187,500 Pa, and the atmospheric pressure, which is also the pressure at point B, is 100,000 Pa. The mass of the

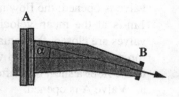

Figure P4.28

pipe section between points A and B is estimated as 10% of that of the body of water inside the pipe. For bend angles of $\alpha=\pi/6$ and $\alpha=\pi/4$, find the mass of water in the bend section which makes the force in the bolts of the flange at A just horizontal, i.e., there is no vertical component. Does this mean that there are no moments at point A?

4.29 For the dimensions, pressures and velocities given in Problem 4.28 find the angle α for which the horizontal force in the bolts of flange A becomes the largest. Is there an angle α for which this horizontal force vanishes? Is there an angle for which the bend section pushes to the left against the water supply pipe?

4.30 The pipe which supplies the water to the bend in Problems 4.28 and 4.29 is made from some flexible material and looks as in Fig. P4.30. The bend angle is $\alpha=0$, i.e., no bend, just a cone. Now the bend and the flexible supply pipe must be held to prevent their motion. Find the force necessary to hold them in place.

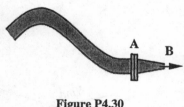

Figure P4.30

4.31 Figure P4.31 shows a water sprinkler with dissimilar arms. The inlet pipe has the diameter of 0.02 m, and the mean velocity of the water there, point A, is 15 m/s. The outlets are nozzles with diameters of 0.003 m. The length of the longer arm, AB, is 0.3 m, and that of the shorter one, AC, is 0.2 m. The sprinkler is designed to run at 120 rpm. The water is assumed to divide equally between the two arms, and the power obtained from the rotor is used to drive the sprinkler over the field. Find the angles α_1 and α_2 by which the nozzles must be set to maximize the power obtained. Calculate this power.

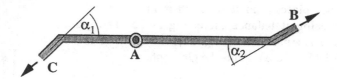

Figure P4.31 Asymmetric sprinkler.

4.32 A propeller has a diameter of 1.0 m, and when it rotates, it spans a disk, sometimes referred to as an actuating disk. It is designed for a nominal operation where a stream of air at atmospheric pressure enters the actuating disk at 400 km/h and leaves it at 750 km/h, still at atmospheric pressure.
  a. Find the thrust of the propeller operating on an airplane flying at 400 km/h.
  b. Find the power delivered by it to the airplane.

4.33 A ventilation scoop, Fig. P4.33, is used to ventilate compartments in ships. One such scoop has an opening with the diameter of 1 m. The winds the ship is expected to encounter are not faster than 100 m/s. Find the upward force for which the scoop must be designed.

Figure P4.33

4.34 A cheap blower has the vanes in its rotor made of flat strips of metal, Fig. P4.34. The air enters centrally and comes out at the circumference, with its velocity relative to the rotor tangent to the vanes. The angle between the tangent to the vane at the circumference and the radius is $\alpha = \pi/6$. The inlet conduit has a diameter of $r = 0.04$ m, and the velocity of the air there is 20 m/s. The outer diameter of the rotor is 0.10 m, and the width of the rotor at the outlet is 0.01 m. The rotor rotates at 3000 rpm. The density of the air may be taken as constant, at 1.1 kg/m^3. Find the power needed to run the blower.

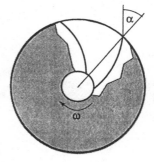

Figure P4.34

4.35 A round jet of water comes straight up from a nozzle in a water fountain. The jet diameter as it comes out of the nozzle is 0.002 m, and its velocity there is 20 m/s. A little boy places a small glass sphere in the jet and enjoys seeing it balance there, Fig. P4.35. The glass sphere has a mass of 0.01 kg. Find the diameter of the water jet just before it hits the glass sphere.

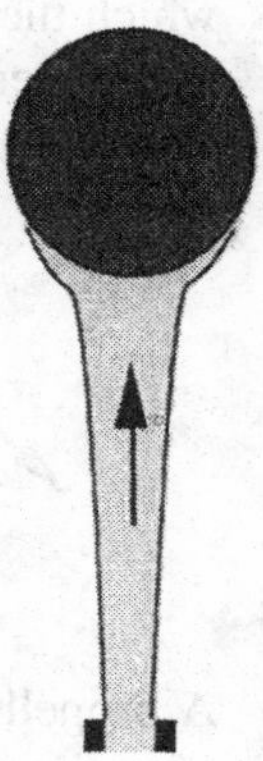

4.36 A light airplane is used to spray cotton fields. The spray nozzles are directed toward the rear of the airplane and the spray comes out at a rate of 100 kg/s and at a speed of 20 m/s relative to the airplane.

   a. Draw the velocity vectors of the spray relative to the airplane and as seen by an observer on the ground.

Figure P4.35

   b. Find the thrust added to the airplane by the spray.
   c. Find the velocity and the thrust when the nozzles are directed downward.

5. FLUIDS IN MOTION – DIFFERENTIAL ANALYSIS

Differential Representation

Two conditions must be satisfied to make the integral analysis considered in the previous chapter useful: The sought information must be such that no details are required; and enough information must be known a priori to supply numerical values for all the symbols in the equations of the integral analysis. In many situations details are meaningful and necessary, and in very many cases, particularly engineering ones, no a priori information is given.

The difficulty mentioned in Chapter 4, that of identification of fluid particles, cannot be alleviated any more by an integral approach and the concept of the *field* must be introduced.

The field point of view, which is traditionally called the Eulerian approach, uses the same properties associated with fluid particles, e.g., density, velocity, temperature, etc., but assigns them to the field. One then considers the field density, i.e., the density at a particular point (x,y,z) and at a particular time, or one considers the field velocity or the field temperature. There is then a *flow field*, with all its properties at each point and for any time, and the problem of identification seems to have been successfully circumvented.

There is still, however, an intermediate step necessary before field analysis can be applied: the basic laws dealing with mass and momentum, which are formulated for well-defined, and therefore identifiable, thermodynamic systems, must be translated into expressions which contain field concepts only. An analogous step has been taken in going from thermodynamic systems to control volumes in Chapter 4. The next step is, therefore, to go from macroscopic thermodynamic system considerations to differential relations of the field properties, i.e., to obtain the differential equations of the field.

The most important field properties are the density, the velocity, the pressure and the temperature, all of which become now dependent variables, with the independent variables being the coordinates x, y, z and the time t. The differential equations of the field now contain these dependent and independent variables and express the basic laws of physics.

The two differential equations sought are that which expresses conservation of mass and the other which represents Newton's second law of motion.

We note that both physical laws involved contain rates of change: *the rate of change of the mass of a thermodynamic system is nil*, and *the rate of change of the momentum of a thermodynamic system equals the sum total of the forces acting on the system*. We also note that the rates of change in both cases must be evaluated "while following the system."

Now, such problems have already been addressed in the integral analysis, using Reynolds transport theorem. We would therefore like to use some results already obtained for control volumes. We observe that in the integral analysis all the examples involved nonfluid geometries, such as pipes, airplanes, rockets, etc., which conveniently served as control volumes. To apply our integral results to regions of pure fluid, we must find there natural control volumes. We therefore start our analysis by showing that indeed the flow field itself provides all the necessary control volumes. We then take up the concept of the material derivative, briefly mentioned in Chapter 4, and define it using field terms. Once this has been achieved, we are ready to derive the differential equations that express the laws of conservation of mass and change of momentum.

Streamlines, Stream Sheets and Stream Tubes

The velocity in a continuum constitutes a vectorial field, which is defined everywhere in the fluid, at any moment.

A sufficiently smooth vectorial field admits field lines, defined as continuous lines tangent to the field vector everywhere. The velocity field lines, i.e., those lines tangent everywhere to the velocity vector, are called *streamlines*. For steady flows the streamline pattern is also steady. In time-dependent flows the streamlines change in time. Streamlines can intersect only at points of zero velocity, as otherwise a finite velocity vector would have more than one direction at a point. Also, because the velocity vector is tangent to the streamline, there is no velocity component normal to it and, therefore, no flow through it.

To enhance the geometric interpretation of the concept of streamlines, two auxiliary concepts are introduced: *pathlines* and *streaklines*.

A *pathline* is a line traced by a particle in the fluid. A small particle, such as a dust speck moving in a clear fluid, can be photographed using a movie camera. The time history of a single fluid particle obtained this way is the particle pathline.

By seeding a fluid with many small particles, a large number of pathlines can be recorded simultaneously.

A *streakline* is the locus of all the particles which have passed through a particular point. A thin hollow needle, held in place in the fluid domain, slowly releasing a continuous stream of dye, would cause all particles passing at the point of the needle to be dyed. These particles then form a dyed streakline in the fluid.

In steady flows both pathlines and streaklines coincide with streamlines, and this is their main usefulness: they both can be used to make streamlines visible. We now return to streamlines and remember that because the field has the velocity vector defined everywhere, each and every point has a streamline passing through it.

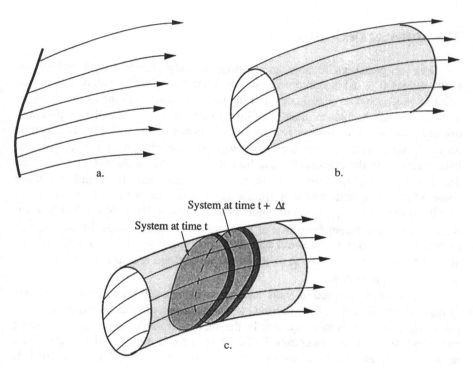

Figure 5.1 a. Stream sheet formed by streamlines intersecting a curve.
b. Stream tube formed by streamlines intersecting a closed curve.
c. Thermodynamic system moving through a stream tube.

Consider a general curve drawn in the fluid domain, Fig. 5.1a. This curve is chosen not to be a streamline itself; hence it is continuously intersected by streamlines. All of these intersecting streamlines constitute a *stream sheet*, which is a three-dimensional surface with no flow through it. When the originally drawn curve is closed, i.e., forms a ring, the stream sheet forms a tube with no flow through its walls, called a *stream tube*, Fig. 5.1b. The concepts of streamlines, stream sheets and stream tubes are thus very useful in the description of flows. It is noted that because there is no flow through the "walls" of a stream tube, it can conveniently serve as a wall of a control volume. The flow field is thus full of such partitions available for use whenever necessary. In particular consider the thermodynamic system shown in Fig. 5.1c. It is shown at the time t and at the time $t + dt$, and because its motion has been affected by the velocity field, it has moved in a stream tube, i.e., through a control volume.

The Material Derivative

Let two problems be considered simultaneously: that of a steel ball located at the point (x, y, z) and that of a similar "sphere of fluid" located at the same point in another set of coordinates. Let the rates of change in time of the densities of both systems be investigated. The investigator of the steel ball will probably use the notation $d\rho/dt$ to describe the rate of change of the ball density. There is no doubt whatsoever as to which density is considered, and even when several balls take part in the process, the addition of a subscript, i.e. $(d\rho/dt)_i$, will suffice. The investigator of the moving fluid, however, cannot use this notation without some additional qualification. A simple $d\rho/dt$ could mean the rate of change of density at the point (x_1, y_1, z_1) brought about because at the time $t + \Delta t$ it is occupied by a fluid different from that of the time t; or it could mean the time rate of change of the density of the small sphere of fluid which was resident at (x_1, y_1, z_1) at time t but went elsewhere at $t + \Delta t$. Both possibilities are meaningful, but with quite different meanings.

The notation adopted in fluid mechanics is such that when a rate of change *at a point* is considered, $\partial/\partial t$ is used, while for a rate of change observed *while following the fluid system* (system in the thermodynamic sense, i.e., the same particles), the notation becomes D/Dt. Thus $\partial\rho/\partial t$ is the rate of change of the density at a point, and it vanishes for time-independent flows. The term $D\rho/Dt$ is the rate of change of the density of a thermodynamic system; this rate need not vanish for time-independent flows.

The differences between the various derivatives can be explained in a more formal manner as follows: Consider a fluid particle moving with a local velocity:

$$q = iu + jv + kw , \tag{5.1}$$

and let the change of the property $b = b\,(x,y,z,t)$ of the particle be investigated.

The change in b with time and position may be expressed as

$$db = \left(\frac{\partial b}{\partial t}\right)dt + \left(\frac{\partial b}{\partial x}\right)dx + \left(\frac{\partial b}{\partial y}\right)dy + \left(\frac{\partial b}{\partial z}\right)dz, \qquad (5.2)$$

and the rate of change of b in time, db/dt, is

$$\frac{db}{dt} = \left(\frac{\partial b}{\partial t}\right) + \left(\frac{\partial b}{\partial x}\right)\frac{dx}{dt} + \left(\frac{\partial b}{\partial y}\right)\frac{dy}{dt} + \left(\frac{\partial b}{\partial z}\right)\frac{dz}{dt}. \qquad (5.3)$$

Now dx_i/dt is the rate of change in the coordinate x_i of the particle and is therefore equal to q_i. This leads to the definition of the *material derivative*, i.e., the derivative taken while following the fluid motion,

$$\frac{Db}{Dt} = \frac{\partial b}{\partial t} + u\frac{\partial b}{\partial x} + v\frac{\partial b}{\partial y} + w\frac{\partial b}{\partial z}, \qquad (5.4)$$

which may also be written, using vector notation, as

$$\frac{Db}{Dt} = \frac{\partial b}{\partial t} + \mathbf{q}\cdot\nabla b. \qquad (5.5)$$

For the case of a static fluid, Eq. (5.5) becomes

$$\frac{Db}{Dt} = \frac{\partial b}{\partial t} \qquad (5.6)$$

while for time-independent flows it takes the form

$$\frac{Db}{Dt} = \mathbf{q}\cdot\nabla b = u\frac{\partial b}{\partial x} + v\frac{\partial b}{\partial y} + w\frac{\partial b}{\partial z}. \qquad (5.7)$$

Equation (5.5) is quite general and holds in any system of coordinates. Its detailed form, using cartesian coordinates, is Eq. (5.4). Analogous forms in other systems of coordinates may be obtained by either coordinate transformations, from the cartesian to the desired ones, starting with Eq. (5.4), or by the direct operation of $\mathbf{q}\cdot\nabla b$ in the new coordinates. Thus Eq. (5.5) in cylindrical and spherical coordinates, respectively, becomes

$$\frac{Db}{Dt} = \frac{\partial b}{\partial t} + q_r\frac{\partial b}{\partial r} + q_\theta\frac{\partial b}{r\partial\theta} + q_z\frac{\partial b}{\partial z} \qquad (5.8)$$

and

$$\frac{Db}{Dt} = \frac{\partial b}{\partial t} + q_r\frac{\partial b}{\partial r} + q_\theta\frac{\partial b}{r\partial\theta} + q_\phi\frac{\partial b}{r\sin\theta\,\partial\phi}. \qquad (5.9)$$

The basic laws used in thermodynamics and mechanics are formulated for thermodynamic systems. When these laws are applied in fluid mechanics, the point of view must still be that of the system, i.e., while following the particle, and the operator D/Dt is used. This operator is defined by Eq. (5.5), in which b may be any scalar property.

The material derivative may also be defined for a vector quantity. Formally one may write for a vector $\mathbf{A}$

$$\frac{D\mathbf{A}}{Dt} = \frac{\partial \mathbf{A}}{\partial t} + (\mathbf{q}\cdot\nabla)\mathbf{A} \qquad (5.10)$$

in which the operator $(\mathbf{q}\cdot\nabla)$ is taken literally as meaning: perform the scalar product of the vector $\mathbf{q}$ and the operator ∇ in the particular coordinate system and use the resulting terms as an operator on the vector written to the right of the parentheses. The form of Eq. (5.10) is quite similar to that of Eq. (5.5), expressing the material derivative of a scalar. However, $\mathbf{q}\cdot\nabla b$ in Eq. (5.5) is a common vectorial operation; whereas $(\mathbf{q}\cdot\nabla)\mathbf{A}$ had to be specifically defined here.

For the particular case of $\mathbf{A} = \mathbf{q}$, Eq. (5.10) can also be written in terms of common vectorial operations, yielding an expression for the material derivative of the velocity vector of a fluid particle,

$$\frac{D\mathbf{q}}{Dt} = \frac{\partial \mathbf{q}}{\partial t} + \frac{1}{2}\nabla(\mathbf{q}\cdot\mathbf{q}) - \mathbf{q}\times(\nabla\times\mathbf{q}). \qquad (5.11)$$

In cartesian coordinates this vector may be written as

$$\frac{D\mathbf{q}}{Dt} = \frac{\partial \mathbf{q}}{\partial t} + u\frac{\partial \mathbf{q}}{\partial x} + v\frac{\partial \mathbf{q}}{\partial y} + w\frac{\partial \mathbf{q}}{\partial z}. \qquad (5.12)$$

In dynamics of solids the time rate of change of the velocity of a particle, taken while following the particle, i.e., the material derivative, has the simple interpretation of the acceleration vector of the particle. Considerations of Newton's second law of motion, applied to a system, will show this interpretation to be indeed valid in fluid dynamics too, i.e., to give the correct acceleration vector in the field.

The operator D/Dt has already been used in the formulation of the Reynolds transport theorem, in Chapter 4, with the same interpretation, i.e., that of *while following the system*. We note, however, that there it has been operating on extensive quantities, i.e., on integrals over the thermodynamic systems. Here it applies to local properties, and its detailed forms are therefore different.

Example 5.1

The price of fruit, P, is given by

$$P = C + Ax - Bt,$$

where x is the distance from its place of growth and t is the time that has elapsed since it was picked. Find the rate at which the price of fruit changes while it is being transported at a velocity v.

Solution

The rate of change in P is

$$\frac{DP}{Dt} = \frac{\partial P}{\partial t} + \frac{dx}{dt}\frac{\partial P}{\partial x} = -B + \frac{dx}{dt}A.$$

In our case $dx/dt = v$ and thus, $DP/Dt = -B + vA$. For the special case when no transportation takes place, $v = 0$ and

$$\frac{DP}{Dt} = \frac{\partial P}{\partial t} = -B$$

while for fruit which does not spoil easily, $B \approx 0$ and $DP/Dt = vA$.

Conservation of Mass – The Equation of Continuity

The law of conservation of mass has already been presented in a form applicable to a control volume, as Eq. (4.12), which may be rewritten as

$$0 = \int_V \frac{\partial \rho}{\partial t}\, dV + \int_S \rho \mathbf{q} \cdot \mathbf{n}\, dS. \tag{5.13}$$

Application of the divergence theorem to the surface integral

$$\int_S \rho \mathbf{q} \cdot \mathbf{n}\, dS = \int_V \nabla \cdot (\rho \mathbf{q})\, dV \tag{5.14}$$

transforms Eq. (5.13) into

$$\int_V \left[\frac{\partial \rho}{\partial t} + \nabla \cdot (\rho \mathbf{q}) \right] dV = 0. \tag{5.15}$$

Equation (5.15) must be satisfied by *any* thermodynamic system chosen in the domain of the fluid. In other words the integration region V is arbitrary. The integral of Eq. (5.15) can vanish over any arbitrary region inside a given domain only if its integrand vanishes at all points in that domain; hence

$$\frac{\partial \rho}{\partial t} + \nabla \cdot (\rho \mathbf{q}) = 0. \tag{5.16}$$

Equation (5.16) is known as the equation of continuity. It is the differential form of the law of conservation of mass, written in terms of the flow field.

Equation (5.16) is now rewritten in detail in the three most commonly used coordinate systems.

In cartesian coordinates:

$$\frac{\partial \rho}{\partial t} + \frac{\partial(\rho u)}{\partial x} + \frac{\partial(\rho v)}{\partial y} + \frac{\partial(\rho w)}{\partial z} = 0. \tag{5.17}$$

In cylindrical coordinates:

$$\frac{\partial \rho}{\partial t} + \frac{1}{r}\frac{\partial(r\rho q_r)}{\partial r} + \frac{1}{r}\frac{\partial(\rho q_\theta)}{\partial \theta} + \frac{\partial(\rho q_z)}{\partial z} = 0. \tag{5.18}$$

In spherical coordinates:

$$\frac{\partial \rho}{\partial t} + \frac{1}{r^2}\frac{\partial(r^2 \rho q_r)}{\partial r} + \frac{1}{r\sin\theta}\frac{\partial(\rho q_\theta \sin\theta)}{\partial \theta} + \frac{1}{r\sin\theta}\frac{\partial(\rho q_\phi)}{\partial \phi} = 0. \tag{5.19}$$

In some particular cases the equation of continuity assumes simpler forms, given here in cartesian coordinates.

a. Time-independent flows – steady flows, $\partial \rho / \partial t = 0$:

$$\nabla \cdot (\rho \mathbf{q}) = 0,$$

$$\frac{\partial(\rho u)}{\partial x} + \frac{\partial(\rho v)}{\partial y} + \frac{\partial(\rho w)}{\partial z} = 0. \tag{5.20}$$

b. Incompressible flows, $\rho = const$:

$$\nabla \cdot \mathbf{q} = 0,$$

$$\frac{\partial u}{\partial x} + \frac{\partial v}{\partial y} + \frac{\partial w}{\partial z} = 0. \tag{5.21}$$

Example 5.2

Using the equation of continuity, show that for any specific property b the Reynolds transport theorem, Eq. (4.10), may be put in the form

$$\frac{D}{Dt}\int_V \rho b\, dV = \int_V \rho \frac{Db}{Dt}\, dV.$$

Solution

Using the form of Eq. (4.10) of the Reynolds transport theorem,

$$\frac{D}{Dt}\int_V (\rho b)dV = \int_V \frac{\partial(\rho b)}{\partial t}dV + \int_S \rho b\mathbf{q}\cdot\mathbf{n}dS,$$

we apply the divergence theorem to the surface integral on the right-hand side of the equation and obtain

$$\int_S \rho b\mathbf{q}\cdot\mathbf{n}\,dS = \int_V \nabla\cdot(\rho b\mathbf{q})dV.$$

Substitution into the Reynolds transport theorem results in

$$\frac{D}{Dt}\int_V (\rho b)dV = \int_V \left[\frac{\partial(\rho b)}{\partial t} + \nabla\cdot(\rho b\mathbf{q})\right]dV$$

or

$$\frac{D}{Dt}\int_V \rho b\,dV = \int_V \left[\rho\left(\frac{\partial b}{\partial t} + \mathbf{q}\cdot\nabla b\right) + b\left(\frac{\partial\rho}{\partial t} + \nabla\cdot(\rho\mathbf{q})\right)\right]dV.$$

Now the first term inside the integral on the right-hand side of the equation yields

$$\rho\left(\frac{\partial b}{\partial t} + \mathbf{q}\cdot\nabla b\right) = \rho\frac{Db}{Dt}$$

while the equation of continuity, Eq. (5.16), makes the second term vanish,

$$b\left(\frac{\partial\rho}{\partial t} + \nabla\cdot(\rho\mathbf{q})\right) = 0.$$

Hence

$$\frac{D}{Dt}\int_V \rho b\,dV = \int_V \rho\frac{Db}{Dt}dV.$$

Example 5.3

Air flows at steady state in a square duct of constant cross section, Fig. 5.2. Measurements at two points, 1 and 2, which are 30 m apart indicate uniform velocities of $u_1 = 30$ m/s and $u_2 = 130$ m/s, respectively. The flow is assumed one dimensional. Is the flow compressible?

Solution

The flow is one dimensional and time independent. Equation (5.20) thus becomes

$$\frac{d(\rho u)}{dx} = 0 \qquad \text{or} \qquad \rho \frac{du}{dx} + u \frac{d\rho}{dx} = 0.$$

Since du/dx is different from zero so must be dp/dx and the flow is compressible. Furthermore $d(\rho u)/dx = 0$ implies $\rho u = \text{const}$. Hence,

$$\frac{\rho_2}{\rho_1} = \frac{u_1}{u_2} = \frac{30}{130} = 0.23 .$$

Example 5.4

Water flows in a square duct of constant cross section, Fig. 5.2. Measurements show that $v = w = 0$. A man who did not study fluid mechanics claims that "because of friction," he expects the water to slow down along the duct. Is he right?

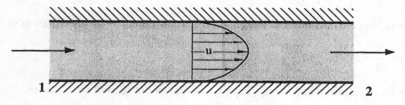

Figure 5.2 Duct flow.

Solution

Water is incompressible. Thus Eq. (5.21) applies for this case. Because the flow is in the x-direction only, $v = w = 0$. Hence Eq. (5.21) simplifies to

$$\frac{\partial u}{\partial x} = 0, \qquad \rho u = \text{const},$$

which implies that $u \neq u(x)$. Therefore, the velocity of the fluid does not change in the x-direction. On the other hand, the velocity may depend on y or z. Indeed, integration of the continuity equation, given above for this case, leads in the most general case to

$$u = u(y, z) .$$

Newton's Second Law of Motion

Newton's second law of motion states that the rate of change of the momentum of a thermodynamic system equals the sum total of the forces acting on this system. Thus, the change in the momentum of any thermodynamic system subject to both body forces and surface forces, as defined in Chapter 1, is

$$\frac{D}{Dt} \int_v \rho \mathbf{q} \, dV = \int_v \mathbf{g}\rho \, dV + \int_s \mathbf{T} \, dS, \tag{5.22}$$

where $\mathbf{g}$ is a general body force per unit mass and $\mathbf{T}$ is the stress at the system boundary.

Newton's second law has already been applied to a small cube in Chapter 2, resulting in the three components of the momentum equations (2.13), (2.14), (2.15), i.e.,

$$\rho a_x = \rho g_x + \frac{\partial T_{xx}}{\partial x} + \frac{\partial T_{yx}}{\partial y} + \frac{\partial T_{zx}}{\partial z}, \tag{2.13}$$

$$\rho a_y = \rho g_y + \frac{\partial T_{xy}}{\partial x} + \frac{\partial T_{yy}}{\partial y} + \frac{\partial T_{zy}}{\partial z}, \tag{2.14}$$

$$\rho a_z = \rho g_z + \frac{\partial T_{xz}}{\partial x} + \frac{\partial T_{yz}}{\partial y} + \frac{\partial T_{zz}}{\partial z}, \tag{2.15}$$

where a_i are the three components of the acceleration vector of the small cube, which we now suspect to be given by Eqs. (5.11) or (5.12).

Equation (5.22) holds in inertial coordinate systems only. In such systems the cartesian unit vectors may be considered constant. The unit vectors in cartesian coordinates do not undergo differentiation and behave like constant multipliers.

Let the x-component of Eq. (5.22) be first considered:

$$\frac{D}{Dt} \int_V \rho u \, dV = \int_V g_x \rho \, dV + \int_S T_{nx} \, dS. \tag{5.23}$$

The Reynolds transport theorem may now be applied to the left-hand side of this equation (see Example 5.2), to yield

$$\frac{D}{Dt} \int_V \rho u \, dV = \int_V \rho \frac{Du}{Dt} dV = \int_V \rho \left[\frac{\partial u}{\partial t} + u \frac{\partial u}{\partial x} + v \frac{\partial u}{\partial y} + w \frac{\partial u}{\partial z} \right] dV. \tag{5.24}$$

The stress term T_x inside the surface integral of Eq. (5.28) is now rewritten in terms of its components as given by Eq. (2.29) to yield

$$\int_S T_{nx} \, dS = \int_S \left[T_{xx}\mathbf{i} + T_{yx}\mathbf{j} + T_{zx}\mathbf{k} \right] \cdot \mathbf{n} \, dS = \int_V \left[\frac{\partial T_{xx}}{\partial x} + \frac{\partial T_{yx}}{\partial y} + \frac{\partial T_{zx}}{\partial z} \right] dV, \tag{5.25}$$

where the divergence theorem has been used again.

Substitution of Eqs. (5.24) and (5.25) into Eq. (5.23) yields

$$\int_V \left\{ \rho\left[\frac{\partial u}{\partial t} + u\frac{\partial u}{\partial x} + v\frac{\partial u}{\partial y} + w\frac{\partial u}{\partial z} - g_x \right] - \left[\frac{\partial T_{xx}}{\partial x} + \frac{\partial T_{yx}}{\partial y} + \frac{\partial T_{zx}}{\partial z} \right] \right\} dV = 0.$$

(5.26)

But because Newton's second law applies to *any* thermodynamic system, V is arbitrary and for the integral to vanish the integrand must vanish at any point of the domain of integration, i.e.,

$$\rho\left[\frac{\partial u}{\partial t} + u\frac{\partial u}{\partial x} + v\frac{\partial u}{\partial y} + w\frac{\partial u}{\partial z} \right] = \rho g_x + \frac{\partial T_{xx}}{\partial x} + \frac{\partial T_{yx}}{\partial y} + \frac{\partial T_{zx}}{\partial z}.$$

(5.27)

Similarly for the y- and z-components

$$\rho\left[\frac{\partial v}{\partial t} + u\frac{\partial v}{\partial x} + v\frac{\partial v}{\partial y} + w\frac{\partial v}{\partial z} \right] = \rho g_y + \frac{\partial T_{xy}}{\partial x} + \frac{\partial T_{yy}}{\partial y} + \frac{\partial T_{zy}}{\partial z},$$

(5.28)

$$\rho\left[\frac{\partial w}{\partial t} + u\frac{\partial w}{\partial x} + v\frac{\partial w}{\partial y} + w\frac{\partial w}{\partial z} \right] = \rho g_z + \frac{\partial T_{xz}}{\partial x} + \frac{\partial T_{yz}}{\partial y} + \frac{\partial T_{zz}}{\partial z}.$$

(5.29)

Equations (5.27) - (5.29) relate rates of change of momentum to body forces and to stresses existing in the fluid. It is enlightening to compare these equations with Eqs. (2.13) - (2.15). The progress we made in attaining the Reynolds transport theorem and the concept of the control volume made it possible to express the acceleration, which in Eqs. (2.13) - (2.15) was just a letter **a**, in field terms that are meaningful and correct. We have, however, paid a heavy price: Newton's second law is linear, and Eqs. (2.13) - (2.15) look linear, but Eqs. (5.27) - (5.29), which represent this law in field terms, are nonlinear.

As they stand, Eqs. (5.27) - (5.29) cannot be solved because they contain too many dependent variables. Therefore, some additional relations connecting stresses and velocities must be added to make this system of equations complete.

Newton's Law of Viscosity

Consider the fluid between the lower stationary plate and the upper moving plate, shown in Fig. 1.4. Experiments show that the force per unit area required to move the upper plate is proportional to the velocity of this plate, V, and inversely proportional to the gap between the parallel plates, h:

$$\frac{F}{A} \propto \frac{V}{h}.$$

(5.30)

The force per unit area exerted on the upper plate is equal to shear stress T_{yx} applied to the fluid by the upper plate. The proportionality constant that converts Eq. (1.17) into an equality is called the *dynamic* or *absolute viscosity* μ. Thus

$$T_{yx} = \mu \frac{V}{h}.$$
(5.31)

Equation (5.31) may be generalized for the case of two adjacent layers of fluid separated by a distance dy, both moving parallel to the x-direction with the velocities u and $u + du$, respectively. Here the shear stress exerted by one layer of fluid on the other is proportional to its velocity relative to that of the other layer and inversely proportional to the distance between them. Hence

$$T_{yx} = \mu \frac{du}{dy}.$$
(5.32)

Equation (5.32) is known as Newton's law of viscosity, and a fluid obeying it is called a *Newtonian fluid*. Equation (5.32) states that in unidirectional flow the shear stress in a Newtonian fluid is directly proportional to the transverse velocity gradient, *du/dy*, also known as the rate of shear strain or the rate of shear deformation.

There is no obvious reason why real fluids should obey Newton's law of viscosity, Eq. (5.32). As a matter of fact, there are more fluids that do not obey Eq. (5.32) than those that do. Such fluids are called *non-Newtonian*. Fortunately, the three most abundant fluids, air, water and petroleum, obey Newton's law of viscosity quite closely. Typical non-Newtonian fluids are paints, polymer solutions and melts, blood and many liquid food products, such as soups, jellies, etc.

*Analysis of Deformation

Equation (5.32), which defines a Newtonian fluid, can be applied to unidirectional flows only. However, the definition of a Newtonian fluid as one in which the stress depends linearly on the rate of deformation may be generalized to three-dimensional flows. To obtain this general relationship, the rate of deformation has to be considered in some more detail. We start by noting two points:

a. A fluid has been defined as a continuum which cannot support shear stress while at rest with respect to any coordinate system. Because coordinate systems can move and rotate, a fluid whose only motion is like that of a rigid body does not suffer shear stresses. As already shown, the stress tensor for such motions reduces to a diagonal matrix, with the three equal normal stresses, i.e., the pressure, occupying the diagonal. The magnitude of this pressure, however, does depend on the fluid motion. The diagonal terms in the stress tensor thus seem to contain a part which does not vanish in a rigid-body-like motion, i.e., with a zero rate of deformation. To exhibit shear stresses, the fluid must undergo a motion which cannot look like rest to any observer. We call such a motion deformation, and its rate, the rate of deformation. The shear terms in the stress tensor are taken as proportional to these rates of deformation; the normal stresses are also modified, with the modifications proportional to the rates of deformations. These proportionality relations are the extension of the Newtonian fluid definition to three-dimensional flows.

b.   Suppose our analysis is finished, and we have those desired relations between stress and deformation. When our system of coordinates rotates the stress components transform as tensor components should. However, they must remain proportional to the deformation. We should therefore expect the deformation to transform accordingly. One way to satisfy this requirement is for the deformation to be a tensor, a symmetrical one with its principal directions coinciding with those of the stress tensor. Indeed, in such a case the transformation of a stress component would be just the transformation of a deformation component multiplied by the proportionality constant.

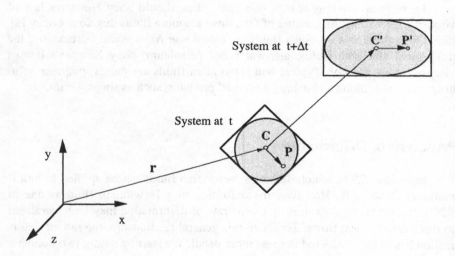

Figure 5.3 System undergoing deformation.

With these two points in mind we consider a fluid system that undergoes deformation as shown in Fig. 5.3. In order to describe this deformation, we first seek the motion of point P relative to C. The difference between the x-components of the velocity u at points P and C which are very close together is

$$du = \frac{\partial u}{\partial x}dx + \frac{\partial u}{\partial y}dy + \frac{\partial u}{\partial z}dz. \tag{5.33}$$

In general, for the i-th velocity component,

$$dq_i = \frac{\partial q_i}{\partial x_j} dx_j.$$

(5.34)

The motion of P relative to C depends, therefore, on the nine components dq_i/dx_j and may be written as

$$dq = \begin{vmatrix} \dfrac{\partial u}{\partial x} & \dfrac{\partial u}{\partial y} & \dfrac{\partial u}{\partial z} \\ \dfrac{\partial v}{\partial x} & \dfrac{\partial v}{\partial y} & \dfrac{\partial v}{\partial z} \\ \dfrac{\partial w}{\partial x} & \dfrac{\partial w}{\partial y} & \dfrac{\partial w}{\partial z} \end{vmatrix} \cdot dr.$$

(5.35)

The relative motion described by Eqs. (5.33) - (5.35) results from the combined effects of rotation and deformation. A rotating rigid body which does not deform still exhibits relative motion between its points. To eliminate the effect of rotation, we subtract from the relative motion of Eq. (5.35) the part that corresponds to a rigid body rotation.

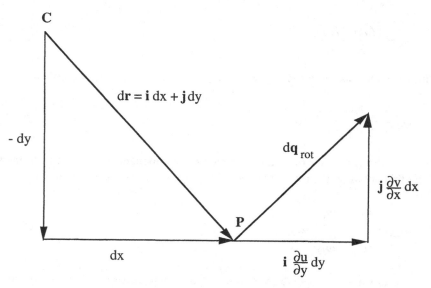

Figure 5.4 Rigid body rotation in x-y plane.

Similar relations can be obtained for ω_x and ω_y:

To obtain this rotation, we consider Fig. 5.4 which describes a rigid body rotating in the x - y plane. The rotational velocity $d\mathbf{q}_{rot}$ is given by

$$d\mathbf{q}_{rot} = \mathbf{i}\frac{\partial u}{\partial y}dy + \mathbf{j}\frac{\partial v}{\partial x}dx. \tag{5.36}$$

Rotation in the x - y plane may also be described as $\boldsymbol{\omega} = \mathbf{k}\omega_z$, and thus

$$d\mathbf{q}_{rot} = \boldsymbol{\omega} \times d\mathbf{r} = -\mathbf{i}\omega_z\, dy + \mathbf{j}\omega_z\, dx. \tag{5.37}$$

Comparing Eqs. (5.36) and (5.37), one obtains

$$\omega_z = -\frac{\partial u}{\partial y}, \qquad \omega_z = \frac{\partial v}{\partial x}, \tag{5.38}$$

which, upon addition and division by 2, attains the more convenient form

$$\omega_z = \frac{1}{2}\left(\frac{\partial v}{\partial x} - \frac{\partial u}{\partial y}\right). \tag{5.39}$$

$$\omega_x = \frac{1}{2}\left(\frac{\partial w}{\partial y} - \frac{\partial v}{\partial z}\right), \tag{5.40}$$

$$\omega_y = \frac{1}{2}\left(\frac{\partial u}{\partial z} - \frac{\partial w}{\partial x}\right). \tag{5.41}$$

With the additional notation $\omega_{xy} = \omega_z$, $\omega_{yz} = \omega_x$ and $\omega_{zx} = \omega_y$, we obtain

$$\omega_{ij} = \frac{1}{2}\left(\frac{\partial q_j}{\partial x_i} - \frac{\partial q_i}{\partial x_j}\right). \tag{5.42}$$

The subscript ij denotes the plane in which the component of rotation is defined. Using this notation, the rotational part of Eqs. (5.34) and (5.35) may be put as

$$d\mathbf{q}_{rot} = \begin{vmatrix} 0 & \omega_{12} & \omega_{13} \\ \omega_{21} & 0 & \omega_{23} \\ \omega_{31} & \omega_{32} & 0 \end{vmatrix} \cdot d\mathbf{r}. \tag{5.43}$$

We are now in a position to subtract the rotational relative velocity $d\mathbf{q}_{rot}$ from the total relative motion, $d\mathbf{q}$, Eq. (5.35). We thus obtain the relative motion caused by deformation, $d\mathbf{q}_{def}$, and the tensor which gives the rate of deformation $\boldsymbol{\varepsilon}$. Hence,

$d\mathbf{q}_{def} = d\mathbf{q} - d\mathbf{q}_{rot}$

$$= \begin{vmatrix} \dfrac{\partial u}{\partial x} & \dfrac{\partial u}{\partial y} & \dfrac{\partial u}{\partial z} \\[2mm] \dfrac{\partial v}{\partial x} & \dfrac{\partial v}{\partial y} & \dfrac{\partial v}{\partial z} \\[2mm] \dfrac{\partial w}{\partial x} & \dfrac{\partial w}{\partial y} & \dfrac{\partial w}{\partial z} \end{vmatrix} \cdot d\mathbf{r} - \dfrac{1}{2} \begin{vmatrix} 0 & \left(\dfrac{\partial u}{\partial y}-\dfrac{\partial v}{\partial x}\right) & \left(\dfrac{\partial u}{\partial z}-\dfrac{\partial w}{\partial x}\right) \\[2mm] \left(\dfrac{\partial v}{\partial x}-\dfrac{\partial u}{\partial y}\right) & 0 & \left(\dfrac{\partial v}{\partial z}-\dfrac{\partial w}{\partial y}\right) \\[2mm] \left(\dfrac{\partial w}{\partial x}-\dfrac{\partial u}{\partial z}\right) & \left(\dfrac{\partial w}{\partial y}-\dfrac{\partial v}{\partial z}\right) & 0 \end{vmatrix} \cdot d\mathbf{r}$$

$$= \begin{vmatrix} \dfrac{\partial u}{\partial x} & \dfrac{1}{2}\left(\dfrac{\partial u}{\partial y}+\dfrac{\partial v}{\partial x}\right) & \dfrac{1}{2}\left(\dfrac{\partial u}{\partial z}+\dfrac{\partial w}{\partial x}\right) \\[2mm] \dfrac{1}{2}\left(\dfrac{\partial v}{\partial x}+\dfrac{\partial u}{\partial y}\right) & \dfrac{\partial v}{\partial y} & \dfrac{1}{2}\left(\dfrac{\partial v}{\partial z}+\dfrac{\partial w}{\partial y}\right) \\[2mm] \dfrac{1}{2}\left(\dfrac{\partial w}{\partial x}+\dfrac{\partial u}{\partial z}\right) & \dfrac{1}{2}\left(\dfrac{\partial w}{\partial y}+\dfrac{\partial v}{\partial z}\right) & \dfrac{\partial w}{\partial z} \end{vmatrix} \cdot d\mathbf{r} = \boldsymbol{\varepsilon} \cdot d\,\mathbf{r}. \qquad (5.44)$$

The rate of deformation tensor, $\boldsymbol{\varepsilon}$, is noted to be symmetrical, as we have anticipated. The components of this rate of deformation tensor may be put in the form

$$\varepsilon_{ij} = \frac{1}{2}\left(\frac{\partial q_i}{\partial x_j}+\frac{\partial q_j}{\partial x_i}\right). \qquad (5.45)$$

This rate of deformation must now be related to the stress in the fluid. The relation has to be consistent with two premises: It must reduce to the elementary definition of a Newtonian fluid for unidirectional flow, Eq. (5.32), and it must simplify to the hydrostatic stress equation for the case of no deformation, Eq. (3.2). The relations that indeed satisfy these requirements are presented in the following section.

Newtonian Fluids

The definition of a Newtonian fluid, which so far related to one-dimensional relations of the form of Eq. (5.32), may now be extended to three-dimensional relations using the so-called rate of deformation tensor, Eq. (5.45), i.e.,

$$\varepsilon_{ij} = \frac{1}{2}\left(\frac{\partial q_i}{\partial x_j}+\frac{\partial q_j}{\partial x_i}\right). \qquad (5.45)$$

We, therefore, redefine the Newtonian fluid as one that satisfies

$$T_{ij} = -p\delta_{ij} + 2\mu\varepsilon_{ij} \qquad (5.46)$$

in which the Kronecker delta δ_{ij} equals unity for $i=j$ and vanishes for $i \neq j$. The stress T_{ij} thus reduces to its pressure terms, Eq.(3.2), for vanishing strain ε_{ij}. Equation (5.46) is written in detail in cartesian coordinates as

$$T_{xx} = -p + 2\mu \frac{\partial u}{\partial x}, \qquad T_{yy} = -p + 2\mu \frac{\partial v}{\partial y}, \qquad T_{zz} = -p + 2\mu \frac{\partial w}{\partial z},$$

$$T_{xy} = \mu \left(\frac{\partial u}{\partial y} + \frac{\partial v}{\partial x} \right), \qquad T_{xz} = \mu \left(\frac{\partial u}{\partial z} + \frac{\partial w}{\partial x} \right), \qquad T_{yz} = \mu \left(\frac{\partial v}{\partial z} + \frac{\partial w}{\partial y} \right). \tag{5.47}$$

Equation (5.46) suggests a definition for the pressure in a moving fluid. Addition of the three normal components of the stress matrix results in

$$T_{xx} + T_{yy} + T_{zz} = -p(1+1+1) + 2\mu \nabla \cdot \mathbf{q} = -3p. \tag{5.48}$$

The $\nabla \cdot \mathbf{q}$ term vanishes for incompressible flows. Hence, the thermodynamic pressure may be defined for an incompressible fluid as the average normal stress:

$$p = -\frac{T_{xx} + T_{yy} + T_{zz}}{3}. \tag{5.49}$$

It is customary to separate out the pressure terms from the total stress, Eq. (5.46). The remainder of T_{ij}, i.e., the part expressing the deviation of the stress from pure pressure, is called the *deviatoric stress* and is usually denoted by τ_{ij}. Thus

$$T_{ij} = -p\delta_{ij} + \tau_{ij} \tag{5.50}$$

and

$$\tau_{ij} = 2\mu\varepsilon_{ij}. \tag{5.51}$$

Equation (5.50) is rewritten in tensor form as

$$\mathbb{T} = -\mathbb{P} + \mathfrak{T} \tag{5.52}$$

where $\mathbb{P}$ is the diagonal tensor

$$\mathbb{P} = \begin{vmatrix} p & 0 & 0 \\ 0 & p & 0 \\ 0 & 0 & p \end{vmatrix}. \tag{5.53}$$

Equation (5.50) is used to modify the momentum equations, Eqs. (5.27) – (5.29), to

$$\rho \left[\frac{\partial u}{\partial t} + u \frac{\partial u}{\partial x} + v \frac{\partial u}{\partial y} + w \frac{\partial u}{\partial z} \right] = -\frac{\partial p}{\partial x} + \rho g_x + \frac{\partial \tau_{xx}}{\partial x} + \frac{\partial \tau_{yx}}{\partial y} + \frac{\partial \tau_{zx}}{\partial z}, \tag{5.54}$$

$$\rho \left[\frac{\partial v}{\partial t} + u \frac{\partial v}{\partial x} + v \frac{\partial v}{\partial y} + w \frac{\partial v}{\partial z} \right] = -\frac{\partial p}{\partial y} + \rho g_y + \frac{\partial \tau_{xy}}{\partial x} + \frac{\partial \tau_{yy}}{\partial y} + \frac{\partial \tau_{zy}}{\partial z}, \tag{5.55}$$

$$\rho \left[\frac{\partial w}{\partial t} + u \frac{\partial w}{\partial x} + v \frac{\partial w}{\partial y} + w \frac{\partial w}{\partial z} \right] = -\frac{\partial p}{\partial z} + \rho g_z + \frac{\partial \tau_{xz}}{\partial x} + \frac{\partial \tau_{yz}}{\partial y} + \frac{\partial \tau_{zz}}{\partial z}, \tag{5.56}$$

which may be put in the symbolic more compact form

$$\rho \frac{Dq}{Dt} = -\nabla p + \rho g + \nabla \cdot \tau. \qquad (5.57)$$

Equations (5.54)-(5.57) are quite general. For Newtonian fluids the stress components are given by Eq. (5.51). Where the relations between τ_{ij} and ε_{ij} are nonlinear, the fluid is non-Newtonian, but Eqs. (5.54) - (5.57) still hold. Equation (5.51) expresses a fundamental relation between stress and rate of strain and is independent of the coordinate system used. The expressions for the stress and the rate of strain components in several coordinate systems are now written down:

In cartesian coordinates $q = iu + jv + kw$

$$\varepsilon_{xy} = \frac{1}{2}\left(\frac{\partial u}{\partial y} + \frac{\partial v}{\partial x} \right), \quad \varepsilon_{yz} = \frac{1}{2}\left(\frac{\partial v}{\partial z} + \frac{\partial w}{\partial y} \right), \quad \varepsilon_{xz} = \frac{1}{2}\left(\frac{\partial u}{\partial z} + \frac{\partial w}{\partial x} \right),$$

$$\tau_{xy} = \mu\left(\frac{\partial u}{\partial y} + \frac{\partial v}{\partial x} \right), \quad \tau_{yz} = \mu\left(\frac{\partial v}{\partial z} + \frac{\partial w}{\partial y} \right), \quad \tau_{xz} = \mu\left(\frac{\partial u}{\partial z} + \frac{\partial w}{\partial x} \right),$$

$$\varepsilon_{xx} = \frac{\partial u}{\partial x}, \qquad \varepsilon_{yy} = \frac{\partial v}{\partial y}, \qquad \varepsilon_{zz} = \frac{\partial w}{\partial z}, \qquad (5.58)$$

$$\tau_{xx} = 2\mu \frac{\partial u}{\partial x}, \qquad \tau_{yy} = 2\mu \frac{\partial v}{\partial y}, \qquad \tau_{zz} = 2\mu \frac{\partial w}{\partial z}.$$

In cylindrical coordinates $q = e_r q_r + e_\theta q_\theta + e_z q_z$

$$\varepsilon_{r\theta} = \frac{1}{2}\left[\frac{1}{r}\frac{\partial q_r}{\partial \theta} + r\frac{\partial}{\partial r}\left(\frac{q_\theta}{r} \right) \right], \quad \tau_{r\theta} = \mu\left[\frac{1}{r}\frac{\partial q_r}{\partial \theta} + r\frac{\partial}{\partial r}\left(\frac{q_\theta}{r} \right) \right],$$

$$\varepsilon_{rz} = \frac{1}{2}\left[\frac{\partial q_r}{\partial z} + \frac{\partial q_z}{\partial r} \right], \qquad \tau_{rz} = \mu\left[\frac{\partial q_r}{\partial z} + \frac{\partial q_z}{\partial r} \right],$$

$$\varepsilon_{\theta z} = \frac{1}{2}\left[\frac{\partial q_\theta}{\partial z} + \frac{1}{r}\frac{\partial q_z}{\partial \theta} \right], \qquad \tau_{\theta z} = \mu\left[\frac{\partial q_\theta}{\partial z} + \frac{1}{r}\frac{\partial q_z}{\partial \theta} \right], \qquad (5.59)$$

$$\varepsilon_{rr} = \frac{\partial q_r}{\partial r}, \quad \varepsilon_{\theta\theta} = \left(\frac{1}{r}\frac{\partial q_\theta}{\partial \theta} + \frac{q_r}{r} \right), \quad \varepsilon_{zz} = \frac{\partial q_z}{\partial z},$$

$$\tau_{rr} = 2\mu \frac{\partial q_r}{\partial r}, \quad \tau_{\theta\theta} = 2\mu\left(\frac{1}{r}\frac{\partial q_\theta}{\partial \theta} + \frac{q_r}{r} \right), \quad \tau_{zz} = 2\mu \frac{\partial q_z}{\partial z}.$$

In spherical coordinates $\mathbf{q} = \mathbf{e}_r q_r + \mathbf{e}_\theta q_\theta + \mathbf{e}_\phi q_\phi$

$$\varepsilon_{R\theta} = \frac{1}{2}\left[\frac{1}{R}\frac{\partial q_R}{\partial \theta} + R\frac{\partial}{\partial R}\left(\frac{q_\theta}{R}\right)\right], \qquad \tau_{R\theta} = \mu\left[\frac{1}{R}\frac{\partial q_R}{\partial \theta} + R\frac{\partial}{\partial R}\left(\frac{q_\theta}{R}\right)\right],$$

$$\varepsilon_{R\phi} = \frac{1}{2}\left[\frac{1}{R\sin\theta}\frac{\partial q_R}{\partial \theta} + R\frac{\partial}{\partial R}\left(\frac{q_\theta}{R}\right)\right], \qquad \tau_{R\phi} = \mu\left[\frac{1}{R\sin\theta}\frac{\partial q_R}{\partial \theta} + R\frac{\partial}{\partial R}\left(\frac{q_\theta}{R}\right)\right],$$

$$\varepsilon_{\theta\phi} = \frac{1}{2}\left[\frac{1}{R\sin\theta}\frac{\partial q_\theta}{\partial \phi} + \frac{\sin\theta}{R}\frac{\partial}{\partial \theta}\left(\frac{q_\phi}{\sin\theta}\right)\right], \quad \tau_{\theta\phi} = \mu\left[\frac{1}{R\sin\theta}\frac{\partial q_\theta}{\partial \phi} + \frac{\sin\theta}{R}\frac{\partial}{\partial \theta}\left(\frac{q_\phi}{\sin\theta}\right)\right],$$

$$\varepsilon_{RR} = \frac{\partial q_R}{\partial R}, \qquad\qquad\qquad \tau_{RR} = 2\mu\frac{\partial q_R}{\partial R},$$

$$\varepsilon_{\theta\theta} = \left(\frac{1}{R}\frac{\partial q_\theta}{\partial \theta} + \frac{q_r}{R}\right), \qquad\qquad \tau_{\theta\theta} = 2\mu\left(\frac{1}{R}\frac{\partial q_\theta}{\partial \theta} + \frac{q_r}{R}\right),$$

$$\varepsilon_{\phi\phi} = \left(\frac{1}{R\sin\phi}\frac{\partial q_\phi}{\partial \phi} + \frac{q_R}{R} + \frac{q_\theta\cot\theta}{R}\right), \qquad \tau_{\phi\phi} = 2\mu\left(\frac{1}{R\sin\phi}\frac{\partial q_\phi}{\partial \phi} + \frac{q_R}{R} + \frac{q_\theta\cot\theta}{R}\right).$$

$$(5.60)$$

The Navier–Stokes Equations

Equations (5.58) - (5.60) may be used to eliminate the stress components from the differential momentum equations (5.54) - (5.56). The result is

$$\rho\left(\frac{\partial u}{\partial t} + u\frac{\partial u}{\partial x} + v\frac{\partial u}{\partial y} + w\frac{\partial u}{\partial z}\right) = \rho g_x - \frac{\partial p}{\partial x} + \mu\left(\frac{\partial^2 u}{\partial x^2} + \frac{\partial^2 u}{\partial y^2} + \frac{\partial^2 u}{\partial z^2}\right),$$

$$\rho\left(\frac{\partial v}{\partial t} + u\frac{\partial v}{\partial x} + v\frac{\partial v}{\partial y} + w\frac{\partial v}{\partial z}\right) = \rho g_y - \frac{\partial p}{\partial y} + \mu\left(\frac{\partial^2 v}{\partial x^2} + \frac{\partial^2 v}{\partial y^2} + \frac{\partial^2 v}{\partial z^2}\right), \qquad (5.61)$$

$$\rho\left(\frac{\partial w}{\partial t} + \frac{\partial w}{\partial x} + v\frac{\partial w}{\partial y} + w\frac{\partial w}{\partial z}\right) = \rho g_z - \frac{\partial p}{\partial z} + \mu\left(\frac{\partial^2 w}{\partial x^2} + \frac{\partial^2 w}{\partial y^2} + \frac{\partial^2 w}{\partial z^2}\right).$$

These momentum equations are called the Navier–Stokes equations. They constitute a system of three nonlinear second order partial differential equations. Together with the continuity equation they form a set of four equations which is complete for incompressible Newtonian flows, i.e., in principle they are sufficient to solve for the four dependent variables $p, u, v,$ and w.

The Navier–Stokes equations require for their solution initial conditions as well as boundary conditions. The proper boundary conditions for the velocity on a rigid boundary are

$$q_n = q_t = 0, \qquad (5.62)$$

where q_n is the normal component of the velocity relative to the solid boundary, and q_t is its tangential component. These conditions are also termed the *no-penetration* $(q_n = 0)$ and *no-slip* $(q_t = 0)$ *viscous boundary conditions*. When the region occupied by the fluid is not closed, i.e., the fluid is not completely confined, additional conditions are still required on some surfaces which completely enclose the domain of the solution. These may represent some real physical surfaces or they may be chosen quite arbitrarily, provided the velocity on them is known.

The pressure, which is also a dependent variable, requires boundary conditions too. More is said on this point later, e.g., in Examples 5.11 - 5.14.

We now proceed to express the Navier–Stokes equations in other coordinate systems. To do this, we first write these equations in their vectorial form as

$$\rho \frac{Dq}{Dt} = -\nabla p + \rho g - \mu \nabla \times (\nabla \times q) = -\nabla p + \rho g + \mu \nabla^2 q. \qquad (5.63)$$

The first form, i.e., the one containing $\nabla \times (\nabla \times q)$, is a standard vectorial form for the incompressible Navier–Stokes equations and can be shown to be correct by expansion in cartesian coordinates. The other form is symbolic and must be taken literally, i.e., as if ∇^2 is the Laplacian operator applied to the velocity vector in cartesian coordinates. Expanding $\nabla \times (\nabla \times q)$ in cylindrical polar coordinates and using the equation of continuity, $\nabla \cdot q = 0$ for incompressible fluids, we obtain

$$\rho \left(\frac{\partial q_r}{\partial t} + q_r \frac{\partial q_r}{\partial r} + q_\theta \frac{\partial q_r}{r \partial \theta} + q_z \frac{\partial q_r}{\partial z} - \frac{q_\theta^2}{r} \right)$$
$$= \rho g_r - \frac{\partial p}{\partial r} + \mu \left[\frac{\partial}{\partial r} \left(\frac{1}{r} \frac{\partial}{\partial r} (r q_r) \right) + \frac{1}{r^2} \frac{\partial^2 q_r}{\partial \theta^2} + \frac{\partial^2 q_r}{\partial z^2} - \frac{2}{r^2} \frac{\partial q_\theta}{\partial \theta} \right], \qquad (5.64)$$

$$\rho \left(\frac{\partial q_\theta}{\partial t} + q_r \frac{\partial q_\theta}{\partial r} + q_\theta \frac{\partial q_\theta}{r \partial \theta} + q_z \frac{\partial q_\theta}{\partial z} + \frac{q_r q_\theta}{r} \right)$$
$$= \rho g_\theta + \frac{\partial p}{r \partial \theta} + \mu \left[\frac{\partial}{\partial r} \left(\frac{1}{r} \frac{\partial}{\partial r} (r q_\theta) \right) + \frac{1}{r^2} \frac{\partial^2 q_\theta}{\partial \theta^2} + \frac{\partial^2 q_\theta}{\partial z^2} + \frac{2}{r^2} \frac{\partial q_r}{\partial \theta} \right], \qquad (5.65)$$

$$\rho \left(\frac{\partial q_z}{\partial t} + q_r \frac{\partial q_z}{\partial r} + q_\theta \frac{\partial q_z}{r \partial \theta} + q_z \frac{\partial q_z}{\partial z} \right)$$
$$= \rho g_z - \frac{\partial p}{\partial z} + \mu \left[\frac{1}{r} \frac{\partial}{\partial r} \left(r \frac{\partial q_z}{\partial r} \right) + \frac{1}{r^2} \frac{\partial^2 q_z}{\partial \theta^2} + \frac{\partial^2 q_z}{\partial z^2} \right]. \qquad (5.66)$$

Repeating the process for spherical coordinates, we obtain

$$
\rho\left(\frac{\partial q_R}{\partial t}+q_R\frac{\partial q_R}{\partial R}+\frac{q_\theta}{R}\frac{\partial q_R}{\partial \theta}+\frac{q_\phi}{R\sin\theta}\frac{\partial q_R}{\partial \phi}-\frac{q_\theta^2+q_\phi^2}{R}\right)
$$

$$
=\rho g_R-\frac{\partial p}{\partial R}+\mu\left[\frac{1}{R^2}\frac{\partial}{\partial R}\left(R^2\frac{\partial q_R}{\partial R}\right)+\frac{1}{R^2\sin\theta}\frac{\partial}{\partial \theta}\left(\sin\theta\frac{\partial q_R}{\partial \theta}\right)\right]
$$

$$
+\mu\left[\frac{1}{R^2\sin^2\theta}\frac{\partial^2 q_R}{\partial \phi^2}-\frac{2q_R}{R^2}-\frac{2}{R^2}\frac{\partial q_\theta}{\partial \theta}-\frac{2q_\theta\cot\theta}{R^2}-\frac{2}{R^2\sin\theta}\frac{\partial q_\phi}{\partial \phi}\right], \tag{5.67}
$$

$$
\rho\left(\frac{\partial q_\theta}{\partial t}+q_R\frac{\partial q_\theta}{\partial R}+\frac{q_\theta}{R}\frac{\partial q_\theta}{\partial \theta}+\frac{q_\phi}{R\sin\theta}\frac{\partial q_\theta}{\partial \phi}-\frac{q_R q_\theta}{R}-\frac{q_\phi^2\cot\theta}{R}\right)
$$

$$
=\rho g_\theta-\frac{1}{R}\frac{\partial p}{\partial \theta}+\mu\left[\frac{1}{R^2}\frac{\partial}{\partial R}\left(R^2\frac{\partial q_\theta}{\partial R}\right)+\frac{1}{R^2\sin\theta}\frac{\partial}{\partial \theta}\left(\sin\theta\frac{\partial q_\theta}{\partial \theta}\right)\right]
$$

$$
+\mu\left[\frac{1}{R^2\sin^2\theta}\frac{\partial^2 q_\theta}{\partial \phi^2}+\frac{2}{R^2}\frac{\partial q_R}{\partial \theta}-\frac{q_\theta}{R^2\sin^2\theta}-\frac{2\cos\theta}{R^2\sin^2\theta}\frac{\partial q_\phi}{\partial \phi}\right], \tag{5.68}
$$

$$
\rho\left[\frac{\partial q_\phi}{\partial t}+q_R\frac{\partial q_\phi}{\partial R}+\frac{q_\theta}{R}\frac{\partial q_\phi}{\partial \theta}+\frac{q_\phi}{R\sin\theta}\frac{\partial q_\phi}{\partial \phi}+\frac{q_\phi q_R}{R}+\frac{q_\theta q_\phi\cot\theta}{R}\right]
$$

$$
=\rho g_\phi-\frac{1}{R\sin\theta}\frac{\partial p}{\partial \phi}+\mu\left[\frac{1}{R^2}\frac{\partial}{\partial R}\left(R^2\frac{\partial q_\phi}{\partial R}\right)+\frac{1}{R^2\sin\theta}\frac{\partial}{\partial \theta}\left(\sin\theta\frac{\partial q_\phi}{\partial \theta}\right)\right]
$$

$$
+\mu\left[\frac{1}{R^2\sin^2\theta}\frac{\partial^2 q_\phi}{\partial \phi^2}-\frac{q_\phi}{R^2\sin^2\theta}+\frac{2}{R^2\sin^2\theta}\frac{\partial q_r}{\partial \phi}+\frac{2\cos\theta}{R^2\sin^2\theta}\frac{\partial q_\phi}{\partial \phi}\right]. \tag{5.69}
$$

The Euler Equations

Substitution of $\mu=0$ in the Navier–Stokes equations (5.61) - (5.69) reduces them to a form called the *Euler equations:*

$$
\rho\frac{D\mathbf{q}}{Dt}=\rho\mathbf{g}-\nabla p. \tag{5.70}
$$

Historically the Euler equations were formulated earlier than the Navier–Stokes equations and were considered an approximation. It is noted that the Euler equations are of the first order and cannot in general satisfy both boundary conditions Eq. (5.62). It is therefore formally concluded that the Euler equations do not form a good approximation near a rigid boundary. Far from a boundary, and where $\mu\approx 0$ is a fair estimate, they have an important role as approximations and are generally easier to solve than the full Navier–Stokes equations. The Euler equations are further considered in later chapters.

Solutions for Two-dimensional Flows – The Stream Function

Solutions of the Navier–Stokes equations result in velocity vectors, **q**, and pressures, p, which satisfy both the momentum equations and the continuity equation. Given such a combination, [**q**, p], one can check whether it constitutes a solution by substitution into the equations. How to find such a solution is another matter, and any general step leading toward this goal is useful. For two-dimensional flows it is possible to eliminate the continuity equation from the system of equations by using only functions which satisfy the continuity equation. This elimination is a formal step toward a solution, and the functions which affect this elimination are the *stream functions*.

A flow is defined as two dimensional when its description in cartesian coordinates shows no z-component of the velocity and no dependence on the z-coordinate. Such a flow can be described in the $z = 0$ plane, with the velocity vector and the streamlines lying in this plane. Furthermore, $z = C$ planes, which are parallel to the $z = 0$ plane, show a flow pattern identical to that in the $z = 0$ plane. The $z = 0$ plane is therefore called the *representative plane*.

Figure 5.5 shows a representative plane for a two-dimensional flow, with four streamlines denoted by the letters A, B, C, D. The whole pattern may be shifted in the z-direction parallel to itself. Thus the streamlines also represent stream sheets, i.e., barriers which are not crossed by the flow. The mass flux entering at the left, between, say, streamlines A and B must therefore come out at the right side without change. Because the distance between the two streamlines accommodating this mass flux seems in the drawing to increase, the mass flux per unit cross section, $\rho \mathbf{q}$, must decrease from left to right. There is therefore some relation between the convergence and divergence of streamlines and the vector $\rho \mathbf{q}$. Furthermore, because stream sheets are not crossed by the flow, each sheet represents a certain mass flux per unit depth of stream sheet taking place "below it," i.e., flowing between it and some particular stream sheet representing zero flux. This mass flux is called the stream function and is denoted by ψ.

Let the stream function corresponding to streamline C in Fig. 5.5 be ψ kg/m. Let streamline D be close to it and have the stream function $\psi + d\psi$. Obviously,

$$d\psi = (dy)(u\rho)$$
$$= (-dx)(v\rho).$$

From which follows

$$u\rho = \frac{\partial \psi}{\partial y}, \qquad v\rho = -\frac{\partial \psi}{\partial x}. \tag{5.71}$$

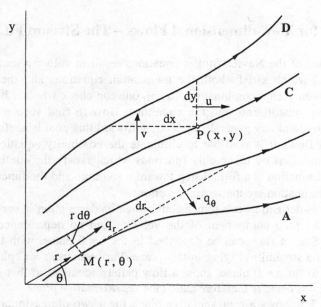

Figure 5.5 Two-dimensional representative plane and streamlines.

Using plane polar coordinates in the representative plane and letting $\psi_B = \psi_A + d\psi$,

$$d\psi = (rd\theta)(q_r\rho)$$
$$= (dr)(-q_\theta\rho),$$

from which follows

$$q_r\rho = \frac{1}{r}\frac{\partial\psi}{\partial\theta}, \qquad q_\theta\rho = -\frac{\partial\psi}{\partial r}. \tag{5.72}$$

It is noted that substitution of Eq. (5.71) or of Eq. (5.72) in Eq. (5.17) satisfies the continuity equation identically. Indeed, conservation of mass which generated the equation of continuity is also the basis for the derivation of Eqs. (5.71) and (5.72).

For incompressible flows ρ appears in the definition of the stream function as a constant coefficient, and so it does in Eqs. (5.71) and (5.72). It is customary to drop this constant for incompressible flows and to define ψ as a measure of the volumetric flowrate, i.e., to have the dimensions of $[m^3/(m\cdot s)]$. The analog of Eqs. (5.71) and (5.72) for incompressible steady flows is

$$u = \frac{\partial\psi}{\partial y}, \qquad v = -\frac{\partial\psi}{\partial x} \tag{5.73}$$

and

$$q_r = \frac{1}{r}\frac{\partial \psi}{\partial \theta}, \qquad q_\theta = -\frac{\partial \psi}{\partial r}, \tag{5.74}$$

and the relations, of course, satisfy Eq. (5.21) identically.

It is important to realize that the use of stream functions is equivalent to the inclusion of the continuity equation in the considerations. Crudely speaking, in steady two-dimensional flows the use of stream functions is equivalent to "one differentiation and one substitution"; this is so because further reference to one first order differential equation, the continuity equation, becomes unnecessary, but the remaining equations increase their order by 1. It is also helpful to review the manner in which this equivalence has been established. Conservation of mass has led to both the continuity equation and the stream function, and therefore their mathematical consequences must be consistent.

Finally it is noted that the tangent to the streamline,

$$ds = \mathbf{i}\,dx + \mathbf{j}\,dy + \mathbf{k}\,dz,$$

is parallel to the velocity vector,

$$\mathbf{q} = \mathbf{i}u + \mathbf{j}v + \mathbf{k}w,$$

which implies that their components are proportional to one another, i.e.,

$$\frac{dx}{u} = \frac{dy}{v} = \frac{dz}{w} \tag{5.75}$$

or, for two-dimensional flows, $v\,dx - u\,dy = 0$. Thus there is some function which is conserved on the streamlines. One may check whether the form written above is a total differential, i.e., whether

$$\frac{\partial v}{\partial y} = -\frac{\partial u}{\partial x} ,$$

which is indeed satisfied by continuity. Denoting this total differential as $-d\psi$, we obtain

$$d\psi = -v\,dx + u\,dy = 0 \quad \text{on streamlines.}$$

This is an alternative way to introduce the concept of the stream function.

Example 5.5

A two-dimensional source of intensity Q, shown in Fig. 5.6, is defined as a singular point out of which a fluid flows at the constant rate of Q kg/s·m, distributed symmetrically in all angular directions.

Write the equation of continuity in cartesian and in polar coordinates and choose the convenient form for this case. Find the velocity field and the stream function. Assume incompressible flow.

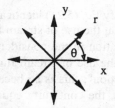

Figure 5.6 Two-dimensional source flow.

Solution

The flow is independent of time, and the continuity equation becomes

$$\nabla \cdot \mathbf{q} = 0.$$

In cartesian coordinates

$$\nabla \cdot \mathbf{q} = \frac{\partial u}{\partial x} + \frac{\partial v}{\partial y} + \frac{\partial w}{\partial z} = 0.$$

In polar coordinates

$$\nabla \cdot \mathbf{q} = \frac{1}{r}\frac{\partial}{\partial r}(rq_r) + \frac{1}{r}\frac{\partial}{\partial \theta}q_\theta + \frac{\partial}{\partial z}q_z = 0.$$

For two-dimensional flow $w = q_z = 0$. Because of the angular symmetry, $q_\theta = 0$. The polar form, therefore, is the more convenient, retaining one term only:

$$\frac{d}{dr}(rq_r) = 0, \qquad rq_r = \text{const.}$$

The intensity of the source Q is related to q_r by $Q = 2\pi rq_r$; hence, $q_r = Q/2\pi r$. In polar coordinates

$$\frac{1}{r}\frac{\partial \psi}{\partial \theta} = q_r \, .$$

Integration yields

$$\psi = \frac{Q\theta}{2\pi} + f(r).$$

Also

$$-\frac{\partial \psi}{\partial r} = -f' = q_\theta = 0.$$

Hence f is a constant which just determines the location of $\theta = 0$. The stream function is constant along radii.

Example 5.6

Parallel flow at a constant velocity U is shown in Fig. 5.7. The flow is formally defined by

$$u = U, \qquad v = 0.$$

Find the stream function for this case in cartesian and in polar coordinates.

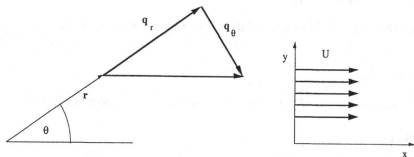

Figure 5.7 Parallel flow at constant velocity U.

Solution

Substitution of $u = U$ into the definition of the two-dimensional stream function, Eq. (5.73) yields

$$\frac{\partial \psi}{\partial y} = U,$$

which upon integration leads to

$$\psi = Uy + f(x),$$

where f is found by the substitution of $v = 0$ into Eq. (5.73),

$$v = -\frac{\partial \psi}{\partial x} = -f' = 0, \qquad f = \text{const.}$$

Hence, the cartesian stream function is given by

$$\psi = Uy + C.$$

Since $y = r\sin\theta$, the polar form becomes

$$\psi = Ur\sin\theta + C.$$

One may try to first express the velocity in polar form and then obtain ψ. The polar decomposition of U, shown in Fig. 5.7, yields the velocity components

$$q_r = U\cos\theta, \qquad q_\theta = U\sin\theta.$$

Substitution of the radial velocity component into Eq. (5.74) yields

$$\frac{\partial \psi}{\partial \theta} = Ur\cos\theta,$$

$$\psi = Ur\sin\theta - f(r).$$

The function f is obtained from the tangential velocity component

$$\frac{\partial \psi}{\partial r} = U\sin\theta - f' = -q_\theta,$$

leading to $f' = 0$, and $f =$ const; and the same result has been obtained.

Example 5.7

What flow results of the superposition of source flow and parallel flow?

Solution

Addition of the stream functions of source flow and parallel flow, obtained in Examples 5.5 and 5.6, respectively, yields

$$\psi = \frac{Q}{2\pi}\theta + Ur\sin\theta.$$

Making use of Eq. (5.74) leads to

$$q_r = \frac{1}{r}\frac{\partial \psi}{\partial \theta} = \frac{Q}{2\pi r} + U\cos\theta,$$

$$q_\theta = -\frac{\partial \psi}{\partial r} = -U\sin\theta.$$

For very large r the flow approaches that of the parallel flow, while for very small r it becomes the source flow. The flow is sketched in Fig. 5.8. As seen in the figure, at point A the streamline splits and the velocity vector seems to have four different directions. This can happen only if the velocity vanishes there, because only a vector of zero length may have several directions without indicating a contradiction. Such a point of zero velocity is denoted a *stagnation point*.

Indeed, for $\theta = \pi$, i.e., on $\psi = Q/2$,

$$q_\theta = 0, \text{ and}$$

$$q_r = \frac{Q}{2\pi r} - U.$$

Thus for $r = Q/(2\pi U)$, $q_r = 0$, and the stagnation point has been obtained.

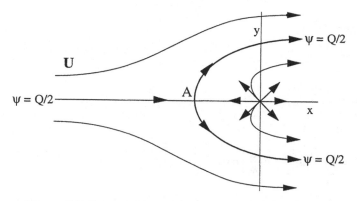

Figure 5.8 Superposition of source flow and parallel flow.

Example 5.8

The flow between two parallel flat plates, with the lower one fixed and the upper one moving, as shown in Fig. 2.6, is given by

$$u = V\frac{y}{h}, \qquad v = w = 0.$$

Does the flow satisfy the continuity equation? Find the stream function for this flow.

Solution

The continuity Eq. (5.21) states

$$\frac{\partial u}{\partial x} + \frac{\partial v}{\partial y} + \frac{\partial w}{\partial z} = 0$$

and is therefore satisfied by this flow. Now $u = \partial\psi/\partial y$, the flow is two-dimensional and hence

$$\frac{\partial \psi}{\partial y} = y\frac{V}{h}, \qquad \psi = y^2\frac{V}{2h} + C.$$

Example 5.9

The steady flow between two parallel plates shown in Fig. 2.6, and considered in Example 5.8, has the velocity vector

$$q = iu = V\frac{y}{h}.$$

Find Dq/Dt.

Solution

From Eq. (5.10)

$$\frac{Dq}{Dt} = \frac{\partial q}{\partial t} + \left(u\frac{\partial q}{\partial x} + v\frac{\partial q}{\partial y} \right) = 0.$$

Example 5.10

The source flow shown in Fig. 5.6, and considered in Example 5.5, has the velocity vector

$$q = e_r q_r = \frac{e_r}{r}\left(\frac{Q}{2\pi} \right).$$

Find Dq/Dt.

Solution

From Eq. (5.10)

$$\frac{Dq}{Dt} = \frac{\partial q}{\partial t} + q\left[\frac{\partial q_r}{\partial r} + 0 \right]$$

$$= 0 + \left(\frac{Q}{2\pi} \right)^2 \frac{e_r}{r} \frac{\partial}{\partial r}\left(\frac{1}{r} \right) = -\frac{e_r}{r^3}\left(\frac{Q}{2\pi} \right)^2.$$

Or, another way, using Eq. (5.11),

$$\frac{Dq}{Dt} = \frac{\partial q}{\partial t} + \frac{1}{2}\nabla(q \cdot q) - q \times (\nabla \times q),$$

where

$$\nabla \times q = \begin{vmatrix} \dfrac{e_r}{r} & \dfrac{e_\theta}{r} & \dfrac{e_z}{r} \\[2mm] \dfrac{\partial}{\partial r} & \dfrac{\partial}{\partial \theta} & \dfrac{\partial}{\partial z} \\[2mm] \dfrac{Q}{2\pi r} & 0 & 0 \end{vmatrix} = 0$$

and

$$\mathbf{q}\cdot\mathbf{q} = \left(\frac{Q}{2\pi}\right)^2 \frac{1}{r^2}.$$

Hence,

$$\frac{D\mathbf{q}}{Dt} = \frac{1}{2}\nabla(\mathbf{q}\cdot\mathbf{q}) = \frac{1}{2}\mathbf{e}_r\frac{\partial}{\partial r}(\mathbf{q}\cdot\mathbf{q}) = -\left(\frac{Q}{2\pi}\right)^2\frac{\mathbf{e}_r}{r^3}.$$

The Pressure in the Momentum Equations

We already know that the momentum equations are nonlinear. On the other hand the relations between the stream function and the velocity components, Eqs. (5.71) - (5.74), are linear, and so is the continuity equation. Having two reasonable flow fields, such as the source flow, Example 5.5, and the parallel flow, Example 5.6, we have attempted superposition in Example 5.7, which has resulted in a flow field which is indeed the superposition of those two fields. The continuity equation is, obviously, satisfied. However, because of the nonlinearity of the momentum equation, the pressure gradient, obtained from the substitution of the combined velocity field, is not the superposition of the pressure gradients corresponding to the two individual fields. This point is considered in the following examples.

Example 5.11

Consider the source flow discussed in Example 5.5. The velocity vector found there has been obtained without the use of the Navier–Stokes equations. Substitute this velocity vector into the equations and evaluate the pressure field.

Solution

The velocity vector for the source flow is

$$\mathbf{q} = \mathbf{e}_r q_r = \mathbf{e}_r\frac{Q}{2\pi r}.$$

The form convenient in this case is the expression of the Navier–Stokes equations in cylindrical coordinates, Eqs. (5.64) - (5.66), which yield

$$\rho\left(q_r\frac{\partial q_r}{\partial r}\right) = -\frac{\partial p}{\partial r} + \rho g_r + \mu\frac{\partial}{\partial r}\left(\frac{1}{r}\frac{\partial}{\partial r}(rq_r)\right),$$

$$0 = -\frac{\partial p}{\partial \theta} + \rho g_\theta,$$

or taking $g = 0$,

$$\rho \frac{Q^2}{(2\pi)^2} \cdot \frac{1}{r^3} = \frac{dp}{dr}.$$

Integration yields

$$p = p_o - \rho \frac{Q^2}{8\pi^2} \left[\frac{1}{r^2} - \frac{1}{r_0^2} \right].$$

The Navier–Stokes equations are satisfied, and we now know the resulting pressure field. This flow can exist only if the obtained pressure is possible. An acceptable boundary condition may be

$$p = p_\infty = \text{const} \qquad \text{at} \qquad r \to \infty,$$

which then implies

$$p = p_\infty - \frac{\rho Q^2}{8\pi^2} \cdot \frac{1}{r^2}.$$

We also note that in the solution for the pressure there is no trace of the viscosity. This pressure, therefore, also satisfies the Euler equation (5.70).

Example 5.12

Consider the parallel flow, Example 5.6. Check whether the velocity vector obtained there satisfies the Navier–Stokes equations.

Solution

The velocity vector obtained is $\mathbf{q} = \mathbf{i}U$, and we choose, therefore, the cartesian form of the Navier–Stokes equations, Eq. (5.61). These yield

$$0 = \rho g_x - \frac{\partial p}{\partial x},$$

$$0 = \rho g_y - \frac{\partial p}{\partial y}, \quad \text{or, neglecting } g, \quad p = \text{const}.$$

An acceptable boundary condition for the pressure is

$$p = p_\infty = \text{const} \qquad \text{at} \qquad r \to \infty.$$

Again we find that the same velocity and pressure also satisfy the Euler equation.

Example 5.13

Consider the superposition of parallel flow and source flow, Example 5.7. Find what pressure comes out when the combined velocity vector is substituted in the Navier–Stokes equations.

Solution

The combined velocity vector is $\mathbf{q} = \mathbf{e}_r q_s + \mathbf{e}_\theta q_\theta$,

$$q_r = \frac{Q}{2\pi r} + U\cos\theta,$$

$$q_\theta = -U\sin\theta.$$

Substitution in Eqs. (5.64) and (5.65), with g neglected, yields

$$\rho\left[-\left(\frac{Q}{2\pi r}+U\cos\theta\right)\frac{Q}{2\pi r^2}+U\sin\theta\cdot\frac{U}{r}\sin\theta-\frac{U^2}{r}\sin^2\theta\right]$$

$$=-\frac{\partial p}{\partial r}+\mu\left[-\frac{U}{r^2}\cos\theta-\frac{U}{r^2}\cos\theta+2\frac{U}{r^2}\cos\theta\right],$$

$$\rho\left[U\sin\theta\frac{1}{r}U\cos\theta-\frac{1}{r}\left(\frac{Q}{2\pi r}+U\cos\theta\right)U\sin\theta\right]$$

$$=-\frac{\partial p}{r\,\partial\theta}+\mu\left[\frac{U}{r^2}\sin\theta+\frac{U}{r^2}\sin\theta-\frac{2}{r^2}U\sin\theta\right],$$

or

$$-\rho\frac{Q^2}{(2\pi)^2 r^3}-\rho\frac{QU}{2\pi r^2}\cos\theta=-\frac{\partial p}{\partial r},$$

$$\rho\frac{QU}{2\pi r^2}\sin\theta=\frac{1}{r}\frac{\partial p}{\partial\theta}.$$

The first equation may be integrated with respect to r to yield

$$-\frac{p}{\rho}=\frac{Q^2}{(2\pi)^2 2r^2}+\frac{QU}{2\pi r}\cos\theta+f(\theta).$$

Differentiation with respect to θ, division by r and substitution into the second equation yield

$$-\frac{QU}{2\pi r^2}\sin\theta+\frac{1}{r}\frac{df}{d\theta}=-\frac{QU}{2\pi r^2}\sin\theta,$$

or

$$\frac{df}{d\theta} = 0$$

and

$$f = \text{const.}$$

Thus finally,

$$p = p_o - \rho \left[\frac{Q^2}{8\pi^2 r^2} + \frac{QU}{2\pi r}\cos\theta \right],$$

which is rather different from the superposition of the pressures obtained in Examples 5.11 and 5.12. We remember, however, that the Navier–Stokes equations are not linear, and there is no reason for the superimposed pressures from Examples 5.11 and 5.12 to equal this pressure.

An acceptable boundary condition for the pressure is

$$p = p_\infty = \text{const} \qquad \text{at} \qquad r \to \infty.$$

Again we note that the viscous terms do not appear here and that the solution satisfies the Euler equation too.

Example 5.14

Consider the shear flow of Example 5.8. Find what pressure is obtained when this flow is substituted in the Navier–Stokes equations.

Solution

The velocity vector for the shear flow is

$$q = iu = iV\frac{y}{h}, \qquad v = w = 0.$$

Equation (5.61) states, with g neglected,

$$\frac{\partial p}{\partial x} = \frac{\partial p}{\partial y} = 0;$$

hence, p is constant.

The Navier–Stokes equations are satisfied, and so are the Euler equations.

Problems

5.1   Find the two-dimensional flow field described by

  a. $\psi = U(y-x)$ (Parallel flow)
  b. $\psi = Ur^2$ (Rigid body rotation)
  c. $\psi = -Q\theta/(2\pi)$ (Sink flow)
  d. $\psi = Q\theta/(2\pi)$ (Source flow)
  e. $\psi = Uy^2$ (Shear flow)

  Find the x- and y-components of the velocity field and draw, free-hand, lines of flow. Are these streamlines? Pathlines? Streaklines?

5.2   Check explicitly if all the flows in Problem 5.1 satisfy the continuity equation. Suppose one flow did not satisfy the continuity equation. Is this possible?

5.3   Do all the flows in Problem 5.1 satisfy the Navier–Stokes equations? How do you check this? Are there sufficient boundary conditions? If not, add the missing ones.

5.4   A function $F = F(x,y)$ is continuous and has at least three partial derivatives. Can such a function always be considered a stream function? Does it necessarily satisfy the equation of continuity? The Navier–Stokes equations? Try some polynomials of various orders.

5.5 A two-dimensional source of intensity Q is located at $(0,0)$. A sink of the same intensity is located at $(5,0)$. Find the velocity field and sketch the streamlines. What are the shapes of the streamlines?

5.6   Consider superposition of parallel flow and the source–sink combination of Problem 5.5. How does the flow look? Can the flow field so obtained be considered a flow around a rigid oval body? Explain.

5.7   The flow in a round pipe is given by
$$w = w_{max}\left[1-\left(\frac{r}{R}\right)^2\right].$$

  Check if this flow satisfies the continuity equation and find the pressure distribution.

5.8   For the flow in the round pipe given in Problem 5.7:

a. Find the shear stress at the wall and the total shear force at the circumference. Compute the pressure gradient necessary to balance this force.

b. Substitute the velocity vector in the Navier–Stokes equations and obtain the pressure gradient. Compare with a above.

5.9   An elliptical pipe has the inner contour

$$\frac{x^2}{a^2} + \frac{y^2}{b^2} = 1.$$

The velocity in the pipe is suggested as $\mathbf{q} = \mathbf{k}w$, with

$$w = w_{max}\left(1 - \frac{x^2}{a^2} - \frac{y^2}{b^2}\right).$$

Does it satisfy the continuity equation? The Navier–Stokes equation? The viscous boundary conditions? Find the pressure distribution in the flow.

5.10   A circular pipe, Problems 5.7 and 5.8, and an elliptical pipe, Problem 5.9, have the same w_{max}. Find relations between a, b and R such that the longitudinal pressure drops are the same.

5.11   A can of milk completely full with milk such that there is no air bubble in it is set on a turntable and rotated with ω. After some time the milk rotates like a solid body. Write the velocity vector in the milk and check whether it satisfies the continuity equation, the Navier–Stokes equation and the viscous boundary conditions. Find the pressure distribution.

5.12   A square can half filled with water is set on a turntable and rotated with ω. Once it reaches solid body rotation, does the velocity field satisfy continuity, momentum and boundary conditions?

5.13   Does any velocity field which looks stationary to any observer (not necessarily in an inertial coordinate system) satisfy continuity, momentum and boundary conditions?

5.14   A two-dimensional source of strength $Q = 4$ m³/s·m is located at the origin, point $(0;0)$, and a sink of the same strength is located at point $(5;0)$. Also given are points A $(-3;1)$, B $(-3;-1)$ and C $(0;4)$.

a. Calculate the volumetric flow [m³/m·s] between points A and B, between points A and C and between points B and C.

b. A parallel flow,

$$\mathbf{q} = \mathbf{i}u = 9\mathbf{i} \ [\text{m}/\text{s}],$$

has been added to the flow field. Find the stagnation points. Find the velocity at the point (2.5; 2.5).

5.15   The stream function in a certain region of a flow may be approximated by

$$\psi = 4\left(x^2 - y^2\right).$$

a.   Find the velocity vector and the pressure distribution in this region.
b.   The flow is viewed by an observer who moves with the velocity **V** = 3**i**. Find the velocity vector of the field, as seen by this observer.
Note that the moving system of the observer is also an inertial one.

5.16   Find the shear stress distribution in the region described in Problem 5.15.

5.17.   The velocity field in a certain region of a flow is approximated by

$$\mathbf{q} = \mathbf{i}y - \mathbf{j}x.$$

Find the pressure distribution and the stream function in this region.

5.18   Find the shear stress distribution in the region described in Problem 5.17.

5.19   A rectangular conduit has the sides 0.05 m by 0.025 m and is oriented such that its sides are parallel to the z-axis, Fig. P5.19. Water flows through the conduit with a velocity which may be approximated by

$$\mathbf{q} = \mathbf{k}w = 3\mathbf{k}\sin\frac{\pi x}{0.05}\sin\frac{\pi y}{0.025}.$$

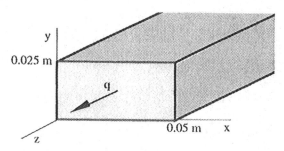

Figure P5.19 Flow in a rectangular channel.

a.   Is the flow one, two or three dimensional?
b.   Are the viscous boundary conditions satisfied by this approximation?
c.   Is the continuity equation satisfied by this approximation?
d.   Find the shear stresses on the sides of the conduit.

e. Find the approximate pressure distribution in the flow. Note that the shear forces on the sides must be balanced by the pressure forces.

5.20   The side of the conduit in Fig. P5.19 located at $x = 0.05\,\mathrm{m}$ now slides in the direction of the z-axis with the velocity $2\,\mathrm{m/s}$. The velocity field is now approximated by

$$q = kw = k\left(2\sin\frac{\pi x}{0.05}\times\sin\frac{\pi y}{0.025}+2\frac{x}{0.05}\times\sin\frac{\pi y}{0.025}\right).$$

a. Find the shear stress distribution on the sides of the conduit and the approximate pressure distribution.
b. Is the continuity equation satisfied by this approximation?
c. Are the viscous boundary conditions satisfied?

5.21   The velocity in a certain region of a flow field is approximated by

$$q = iu = i(4\sin y).$$

a. Is the flow there two dimensional?
b. Find the stream function there, the pressure distribution and the shear stress distribution.
c. Draw the streamlines and the velocity distribution.

6. EXACT SOLUTIONS OF THE NAVIER–STOKES EQUATIONS

The Navier–Stokes equations, obtained in the previous chapter, are nonlinear and submit to no general method of solution. Each new problem must be carefully formulated as to geometry and proper boundary conditions, and then some scheme of attack may be chosen, with the hope of reaching a solution. In most cases all attempts to obtain an exact solution fail, and approximate solutions must suffice. In a few cases exact solutions can be found. Some of these solvable cases are presented now as examples.

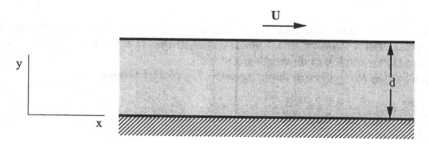

Figure 6.1 Parallel plates and coordinates.

Flows between Parallel Plates

Two parallel plates with the gap d between them in the y-direction are shown in Fig. 6.1. The plates and the incompressible fluid between them are assumed to extend very far in the $\pm x$-direction, and the flow field is considered two-dimensional, i.e., nothing depends on the z-coordinate. The upper plate has the velocity $\mathbf{i}U$, where U may be a function of time. The lower plate is stationary. The velocity of the fluid, $\mathbf{q} = \mathbf{i}u + \mathbf{j}v$, where $u = u\,(x, y, t)$, $v = v\,(x, y, t)$, must satisfy the boundary conditions

$$u(x,0,t) = 0, \tag{6.1}$$
$$u(x,d,t) = U(t), \tag{6.2}$$
$$v(x,0,t) = v(x,d,t) = 0. \tag{6.3}$$

The continuity equation, Eq. (5.21), and the Navier–Stokes equations, Eqs. (5.61), for this case are

$$\frac{\partial u}{\partial x} + \frac{\partial v}{\partial y} = 0, \tag{6.4}$$

$$\rho\left(\frac{\partial u}{\partial t} + u\frac{\partial u}{\partial x} + v\frac{\partial u}{\partial y}\right) = -\frac{\partial p}{\partial x} + \rho g_x + \mu\left(\frac{\partial^2 u}{\partial x^2} + \frac{\partial^2 u}{\partial y^2}\right), \tag{6.5}$$

$$\rho\left(\frac{\partial v}{\partial t} + u\frac{\partial v}{\partial x} + v\frac{\partial v}{\partial y}\right) = -\frac{\partial p}{\partial y} + \rho g_y + \mu\left(\frac{\partial^2 v}{\partial x^2} + \frac{\partial^2 v}{\partial y^2}\right). \tag{6.6}$$

The boundary conditions, Eq. (6.3), together with the assumption that the plates and the fluid extend very far in the $\pm x$-direction, bring to mind the possibility that perhaps the flow is unidirectional, i.e.,

$$v(x,y,t) = 0. \tag{6.7}$$

Equation (6.7) is not an assumption. Rather it is an intuitive guess we pursue until we either find a solution or become convinced that it leads to no solution, in which case we mark it as an unsuccessful trial.

With Eq. (6.7), the continuity equation, Eq. (6.4), becomes

$$\frac{\partial u}{\partial x} = 0, \tag{6.8}$$

which upon integration yields

$$u = u(y,t).$$

Hence, the velocity u depends on y and t only. Whatever the velocity profile is at some *x-coordinate*, it repeats itself for other x values. Such a flow is denoted *fully developed*. In going along a constant y line we find the same u values, i.e., no further development. With Eqs. (6.7) and (6.8), Eqs. (6.5) and (6.6) become

$$\frac{\partial u}{\partial t} = -\frac{1}{\rho}\frac{\partial p}{\partial x} + g_x + v\frac{\partial^2 u}{\partial y^2}, \tag{6.9}$$

$$0 = -\frac{1}{\rho}\frac{\partial p}{\partial y} + g_y .$$

(6.10)

At this point the Navier–Stokes equations, (6.9) and (6.10), may be simplified by combining the gravitational force **g** with the effect of the pressure p. We define the distance from some reference plane *upward*, i.e., in the direction opposed to that of gravity, as

$$h = h(x,y,z)$$

and let

$$\mathbf{g} = -g\,\nabla h$$

(6.11)

or

$$g_x = -g\frac{\partial h}{\partial x}, \qquad g_y = -g\frac{\partial h}{\partial y}, \qquad g_z = -g\frac{\partial h}{\partial z}.$$

For example, if h coincides with the coordinate y in Fig. 6.1 (implying that g is directed *downward*), one would obtain from Eq. (6.11)

$$g_y = -g, \qquad g_x = g_z = 0.$$

Substitution of Eq. (6.11) into Eq. (5.63) yields for the case of constant ρ and g

$$\rho\frac{D\mathbf{q}}{Dt} = -\nabla(p + \rho g h) - \mu\nabla\times(\nabla\times\mathbf{q})$$

(6.12)

One may now define a *modified pressure*,

$$P = p + \rho g h,$$

(6.13)

thus simplifying Eq. (6.12) to

$$\rho\frac{D\mathbf{q}}{Dt} = -\nabla P - \mu\nabla\times(\nabla\times\mathbf{q})$$

(6.14)

and Eqs. (6.9) and (6.10) to

$$\frac{\partial u}{\partial t} = -\frac{1}{\rho}\frac{\partial P}{\partial x} + v\frac{\partial^2 u}{\partial y^2},$$

(6.15)

$$0 = \frac{\partial P}{\partial y} .$$

(6.16)

The modified pressure, P, defined by Eq. (6.13), is most useful in hydrodynamics. In a fluid at rest the modified pressure is constant throughout, and a variation in P indicates that the fluid is moving. However, the use of this modified pressure is limited to cases in which ρ is constant and the absolute pressure p does not affect the boundary conditions. Thus, it may be used in any internal flow of an incompressible fluid. In flows with free surfaces the use of the static pressure may be more convenient.

Elimination of **g** from the equations makes the solutions independent of the orientation of gravity. Thus the solution for a vertical tube is the same as for a horizontal tube provided both flows are subject to the same modified pressure gradient, ∇P.

Returning to our problem, we note from Eq. (6.16) that P does not depend on y. Now let Eq. (6.15) be rearranged,

$$\mu \frac{\partial^2 u}{\partial y^2} - \rho \frac{\partial u}{\partial t} = \frac{\partial P}{\partial x}. \tag{6.17}$$

In this form the right-hand side of the equation does not depend on y, while its left-hand side does not depend on x. Thus both sides can depend only on t, i.e.,

$$\frac{\partial P}{\partial x} = F(t). \tag{6.18}$$

Equation (6.18) indicates that a necessary condition for a fully developed flow is a uniform pressure gradient all through the fluid, i.e.,

$$\frac{\partial P}{\partial x} = \frac{\Delta P}{\Delta x} = F(t). \tag{6.19}$$

We now proceed to solve Eq. (6.15) under various sets of boundary conditions.

Example 6.1

A fluid of density $\rho = 1,000$ kg/m³ flows between horizontal parallel plates, Fig 6.2a. The x- and y-coordinates are as shown, and the respective dimensions are
$$x_1 = 6 \text{ m}, \qquad x_2 = 10 \text{ m}, \qquad x_3 = 14 \text{ m}, \qquad d = 1.5 \text{ m}.$$
The pressure is measured at points 1, 2 and 3, as shown, and the measurements are
$$p_1 = 170,000 \text{ Pa}, \qquad p_2 = 145,285 \text{ Pa}, \qquad p_3 = 150,000 \text{ Pa}.$$
Can the flow be fully developed? Find the modified pressure at points 1, 2 and 3.

Solution

Let points 1 and 3 have the height zero. The modified pressures are
$$P_1 = p_1 = 170,000 \text{ Pa},$$
$$P_2 = p_2 + \rho g h = 145,285 + 1,000 \times 9.81 \times 1.5 = 160,000 \text{ Pa},$$
$$P_3 = p_3 = 150,000 \text{ Pa}.$$

The pressure gradient is

$$\left(\frac{\Delta P}{\Delta x}\right)_{12} = \frac{10,000}{4} = 2,500 \ \frac{\text{Pa}}{\text{m}} = \left(\frac{\Delta P}{\Delta x}\right)_{23}.$$

As shown, the pressure gradient is uniform. Hence, the flow may be fully developed.

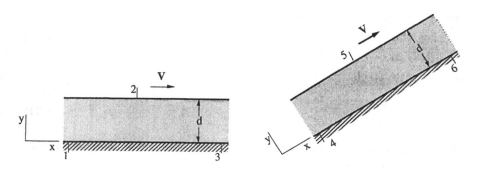

Figure 6.2   Flow between parallel plates.
a. Horizontal. b. Inclined by 30°.

Example 6.2

The channel of Example 6.1 is now tilted by 30°, as shown in Fig. 6.2b. Once again x_4= 6 m, x_5 = 10 m , x_6 = 14 m, with x measured along the channel axis. The pressure at point 4 is p_4 = 130,000 Pa. The flow is fully developed and the same flowrate as in Example 6.1 is measured. Find p_5 and p_6.

Solution

Let point 4 have the height zero. Then
$$P_4 = p_4 = 130,000 \text{ Pa}.$$
To have the same flowrate, the modified pressure must have the same gradient as in Example 6.1. Therefore

$$P_5 = P_4 + 2,500 \times 4 = 140,000 \text{ Pa},$$
$$P_6 = P_5 + 10,000 = 150,000 \text{ Pa}.$$

Hence

$$p_5 = P_5 - \rho g h_5 = 140,000 - 1,000 \times 9.81 \times \left(4\sin 30° + 1.5\cos 30°\right) = 107,636 \text{ Pa},$$

$$p_6 = P_6 - \rho g h_6 = 150,000 - 1,000 \times 9.81 \times 8 \sin 30^\circ = 110,760 \text{ Pa.}$$

Time-independent Flows

Let us first consider time-independent flows. For this case Eq. (6.19) becomes

$$\frac{\partial P}{\partial x} = \frac{\Delta P}{\Delta x} = F = \text{const.} \qquad (6.20)$$

and Eq. (6.17) yields

$$\frac{\Delta P}{\Delta x} = \mu \frac{\partial^2 u}{\partial y^2}. \qquad (6.21)$$

The velocity u is not a function of x, and this equation can be integrated directly to yield the general solution

$$u = \frac{\Delta P}{\Delta x} \frac{d^2}{2\mu} \left(\frac{y}{d}\right)^2 + C_1 \left(\frac{y}{d}\right) + C_2 \qquad (6.22)$$

using the boundary conditions

$$y = 0, \qquad u = 0,$$
$$y = d, \qquad u = U, \qquad (6.23)$$

One obtains

$$u = \left(-\frac{\Delta P}{\Delta x}\right) \frac{d^2}{2\mu} \left[\frac{y}{d} - \left(\frac{y}{d}\right)^2\right] + U\left(\frac{y}{d}\right). \qquad (6.24)$$

The special case of $U = 0$ results in

$$u = \left(-\frac{\Delta P}{\Delta x}\right) \frac{d^2}{2\mu} \left[\frac{y}{d} - \left(\frac{y}{d}\right)^2\right], \qquad (6.25)$$

which is known as plane Poiseuille flow, while the case of $\Delta P/\Delta x = 0$ results in

$$u = U \frac{y}{d}, \qquad (6.26)$$

which is known as *shear flow* or *plane Couette flow*.

Several velocity profiles given by Eq. (6.24), including the plane Poiseuille and the plane Couette flows, are shown in Fig. 6.3.

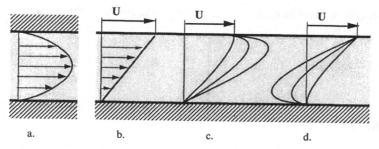

Figure 6.3 Parallel flow, Eq. (6.24):
a. Plane Poiseuille flow.   b. Shear flow
c. Flows with $\Delta P/\Delta x < 0$.   d. Flows with $\Delta P/\Delta x > 0$.

Example 6.3

Find the maximum velocity, the net flowrate and the average velocity for plane Poiseuille flow.

Solution

The velocity profile is given by Eq. (6.25) as

$$u = \left(-\frac{\Delta P}{\Delta x}\right)\frac{d^2}{2\mu}\left[\frac{y}{d} - \left(\frac{y}{d}\right)^2\right].$$

This expression attains its maximum at $y = d/2$. The maximum velocity, u_o, is therefore

$$u_o = \left(-\frac{\Delta P}{\Delta x}\right)\frac{d^2}{8\mu}.$$

The velocity distribution may be expressed in terms of u_o as

$$u = 4u_o\left[\frac{y}{d} - \left(\frac{y}{d}\right)^2\right].$$

The flowrate per unit channel width, Q, is now obtained,

$$Q = \int_0^d u\,dy = 4u_o\int_0^d\left[\frac{y}{d} - \left(\frac{y}{d}\right)^2\right]dy = \frac{2}{3}u_od,$$

and by substituting the value of u_o, this becomes

$$Q = \left(-\frac{\Delta P}{\Delta x}\right)\frac{d^3}{12\mu}.$$

The average velocity is defined as the flowrate divided by the distance between the plates, d, i.e.,

$$\bar{u} = \frac{Q}{d} = \frac{2}{3}u_o.$$

Example 6.4

Find the net flowrate and the average velocity for plane Couette flow.

Solution

From Eq. (6.26),

$$y = U\frac{y}{d},$$

$$Q = \int_0^d u\,dy = U\int_0^d \frac{y}{d}\,dy = \frac{Ud}{2},$$

$$\bar{u} = \frac{Q}{d} = \frac{U}{2}.$$

Example 6.5

The viscosity of glycerin at 10°C is $\mu = 2\,\text{Ns/m}^2$. A gap of 0.05 m between two plates is filled with glycerin. One plate is at rest and the other moves in the x-direction with the velocity $U = 2$ m/s. Because of an adverse pressure gradient, there is no net flow in the gap (see Fig. 6.3d). Find the pressure gradient.

Solution

The equation which governs this flow, Eq. (6.21), is linear, and therefore superposition is permissible. Thus the flowrate for the flow under the combined effect of pressure gradient and plate movement is the sum of flowrates for plane Poiseuille flow (Example 6.3) and plane Couette flow (Example 6.4):

$$Q = \left(-\frac{\Delta P}{\Delta x}\right)\frac{d^3}{12\mu} + \frac{Ud}{2}.$$

For no net flow, $Q = 0$, and the required pressure gradient is

$$\frac{\Delta P}{\Delta x} = \frac{6\mu U}{d^2} = \frac{6\times 2\times 2}{0.05^2} = 9,600\ \text{Pa / m}.$$

Note: The expression for the flowrate may be alternatively obtained by the integration of the velocity, Eq. (6.24), over the flow gap.

Time-dependent Flow – The Rayleigh Problem

Let us now return to Eq. (6.15) but consider flows for which $\partial P/\partial x$ vanishes. Equation (6.15) now assumes the form

$$\frac{\partial u}{\partial t} = v\frac{\partial^2 u}{\partial y^2}.$$ (6.27)

Consider an infinite flat plate with an infinite domain of fluid on its upper side, Fig. 6.4. The fluid and the plate are at rest. At the time $t = 0$ the plate is impulsively set into motion with the velocity U and continues to move at that speed. Thus the initial condition and the boundary conditions are, respectively,

$$u(y,0) = 0,$$
$$u(\infty,t) = 0,$$ (6.28)
$$u(0,t) = U, \quad \text{for} \quad t > 0.$$

This is known as the *Rayleigh Problem*, and again a solution is sought, in which $v = 0$ everywhere and u satisfies Eqs. (6.27) and (6.28).

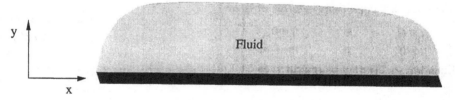

Figure 6.4 Rayleigh flow.

To obtain this solution, we try to find a new independent variable η in the form

$$\eta = Byt^n$$ (6.29)

and then to express u as a function of this single variable. Such a method of solution is called a *similarity transformation*. Thus if $u = u(\eta)$, then

$$\frac{\partial u}{\partial t} = \frac{du}{d\eta}\frac{\partial\eta}{\partial t} = \frac{n}{t}\eta\frac{du}{d\eta},$$ (6.30)

$$\frac{\partial u}{\partial y} = \frac{1}{y}\eta\frac{du}{d\eta},$$ (6.31)

$$\frac{\partial^2 u}{\partial y^2} = \frac{1}{y^2}\eta^2\frac{d^2 u}{d\eta^2}.$$ (6.32)

Substitution into Eq. (6.27) yields

$$\frac{n}{t}\frac{du}{d\eta} = \frac{v}{y^2}\eta\frac{d^2 u}{d\eta^2}$$ (6.33)

or

$$\frac{d^2 u}{d\eta^2} - \frac{n}{\eta}\left(\frac{y^2}{vt}\right)\frac{du}{d\eta} = 0.$$ (6.34)

If u is to be expressed as a function of η only, no y or t terms should remain in the differential equation, Eq. (6.34). We therefore choose η such that the combination y^2/t is proportional to η^2 and select a convenient B such that Eq. (6.34) becomes

$$\frac{d^2 u}{d\eta^2} + 2\eta\frac{du}{d\eta} = 0.$$ (6.35)

Comparison with Eq. (6.34) requires

$$\eta = \frac{y}{2\sqrt{vt}}, \qquad n = -\tfrac{1}{2}, \qquad B = \frac{1}{2\sqrt{v}}.$$

The boundary conditions in terms of η are

$$\begin{aligned} u &= U && \text{at} && \eta = 0, \\ u &= 0 && \text{at} && \eta \to \infty. \end{aligned}$$ (6.36)

Equation (6.35) may be rewritten as

$$\frac{u''}{u'} = -2\eta,$$

which is integrated to

$$\ln u' = -\eta^2 + \ln C_1$$

or

$$\frac{du}{d\eta} = C_1 e^{-\eta^2}.$$

Another integration yields

$$u = C_1\int e^{-\eta^2}d\eta + C_2.$$ (6.37)

The boundary conditions, Eqs. (6.36), then give

$$u = U\left[1 - \frac{2}{\sqrt{\pi}} \int_0^\eta e^{-\eta^2} d\eta\right],$$ (6.38)

which can also be written in terms of the error function as

$$u = U\left[1 - \operatorname{erf}\eta\right] = U\left[1 - \operatorname{erf}\left(\frac{y}{2\sqrt{vt}}\right)\right].$$ (6.39)

Obviously, for a given η, u is uniquely determined. However, there are infinite pairs of y and t which give the same η - all those satisfying $\eta = y/(2\sqrt{vt})$. The same u is thus obtained for all points on the $y/(2\sqrt{vt}) = $ const curve in the y-t plane. In this sense all these points are similar, and hence the term similarity solution.

Example 6.6

The kinematic viscosity of air at 25ºC is $v_a = 1.5\times10^{-5}$ m²/s, and that of water is $v_w = 10^{-6}$ m²/s. A large plate immersed in water is suddenly set into motion at $U = 5$ m/s, and the resulting flow is similar to that of the Rayleigh problem.

How long will it take a water layer 1 m away from the plate to reach the velocity of 2 m/s? Repeat for 3 m away and for air.

η	0.1	0.2	0.3	0.4	0.5	0.6	0.7	0.8	0.9	1.0
erf(η)	0.117	0.223	0.329	0.428	0.520	0.604	0.678	0.742	0.797	0.843
η	1.1	1.2	1.3	1.4	1.5	1.6	1.7	1.8	1.9	2.0
erf(η)	0.880	0.910	0.934	0.952	0.966	0.976	0.984	0.989	0.993	0.995

Table 6.1 Short table of the error function.

Solution

Equation (6.39) states

$$u = U\left[1 - \operatorname{erf}\eta\right]$$

where $\eta = y/(2\sqrt{vt})$. In our case $u = 2$ and $U = 5$. Hence,

$$\text{erf}(\eta) = 1 - \frac{u}{U} = 0.6$$

and from Table 6.1, $\eta = 0.6 = y / (2\sqrt{vt})$. Therefore

$$t_w = \frac{(y/\eta)^2}{4v} = \frac{(1/0.6)^2}{4\times10^6} = 6.94\times10^5 \text{s} = 193 \text{ h}.$$

For air, the same yields

$$t_a = t_w \frac{v_w}{v_a} = 0.067 t_w = 12.9 \text{ h}.$$

For 3 m,

$$t_{w_3} = t_{w_1} \times 3^2 = 1,737 \text{ h} \quad \text{and} \quad t_{a_3} = t_{a_1} \times 3^2 = 116.1 \text{ h}.$$

Steady Flow in a Round Tube

We consider a round tube with the z-coordinate as its axis of symmetry. The tube is filled with an incompressible fluid and extends very far in the $\pm z$-directions. The boundary conditions are

$$\mathbf{q} = 0 \qquad \text{at} \qquad r = R, \tag{6.40}$$

and there is complete circular symmetry about the z-axis.

As in the case of flows between parallel plates, we start by trying the possibility

$$q_r = q_\theta = 0 \qquad \text{everywhere.} \tag{6.41}$$

The continuity equation in polar cylindrical coordinates reads

$$\nabla \cdot \mathbf{q} = \frac{1}{r}\frac{\partial}{\partial r}(rq_r) + \frac{\partial}{r\partial\theta}q_\theta + \frac{\partial}{\partial z}q_z = 0, \tag{6.42}$$

and by Eq. (6.41)

$$\frac{\partial q_z}{\partial z} = 0.$$

Hence

$$q_z = w = w(r,\theta). \tag{6.43}$$

However, because of the assumed complete axisymmetry of the flow,

$$w = w(r)\text{only}, \tag{6.44}$$

w does not change with z and the velocity profile is the same everywhere. This flow is, therefore, also fully developed. The Navier–Stokes equations (5.64) - (5.66) become for this case of steady flow

$$0 = \frac{\partial P}{\partial r},$$ (6.45)

$$0 = \frac{1}{r}\frac{\partial P}{\partial \theta},$$ (6.46)

$$0 = -\frac{\partial P}{\partial z} + \frac{\mu}{r}\frac{\partial}{\partial r}\left(r\frac{\partial w}{\partial r}\right),$$ (6.47)

where $P = p + \rho g h$ is the modified pressure. The first two equations indicate that P is a function of z only, and in a manner similar to the flow between parallel plates, one may show that

$$\frac{\partial P}{\partial z} = \text{const} = \frac{\Delta P}{\Delta z}.$$ (6.48)

Thus the flow equation becomes

$$\frac{1}{\mu}\frac{\Delta P}{\Delta z} = \frac{1}{r}\frac{\partial}{\partial r}\left(r\frac{\partial w}{\partial r}\right),$$ (6.49)

which upon integration results in a general solution,

$$w = \frac{1}{4\mu}\left(\frac{\Delta P}{\Delta z}\right)r^2 + C_1 + C_2 \ln r.$$ (6.50)

For a flow in a tube of radius R, with the boundary condition of Eq. (6.40) and finite velocity at the tube centerline, one obtains

$$w = -\left(\frac{\Delta P}{\Delta z}\right)\frac{R^2}{4\mu}\left[1 - \left(\frac{r}{R}\right)^2\right].$$ (6.51)

The minus sign indicates that the flow is in the direction of decreasing modified pressure.

The centerline velocity w_o is obtained by letting $r = 0$,

$$w_o = -\frac{\Delta P}{\Delta z}\frac{R^2}{4\mu},$$ (6.52)

and combining Eqs. (6.51) and (6.52) results in

$$w = w_o\left[1 - \left(\frac{r}{R}\right)^2\right].$$ (6.53)

The volumetric flowrate may now be obtained from

$$Q = \int_0^R w(r)2\pi r\,dr = \int_0^R w_o\left[1 - \left(\frac{r}{R}\right)^2\right]2\pi r\,dr,$$ (6.54)

which upon integration yields

$$Q = \frac{1}{2}\pi R^2 w_o \tag{6.55}$$

or in combination with Eq. (6.52)

$$Q = \frac{\pi R^4}{8\mu}\left(-\frac{\Delta P}{\Delta z}\right). \tag{6.56}$$

Equation (6.56) is known as the *Poiseuille equation*.

The average velocity $\overline{w}$ is found from Eq. (6.54) or Eq. (6.55) as

$$\overline{w} = \frac{Q}{A} = \frac{\frac{1}{2}\pi R^2 w_o}{\pi R^2} = \frac{w_o}{2} \tag{6.57}$$

or

$$\overline{w} = -\frac{\Delta P}{\Delta z}\frac{R^2}{8\mu}. \tag{6.58}$$

Example 6.7

A heavy oil of viscosity $\mu = 1.5$ Ns/m^2 flows in a pipe whose diameter is $d = 0.01$ m, with a mean velocity of $\overline{w} = 0.1$ m/s. Find the pressure gradient. Find the shear stress at the pipe wall, and show that the pressure drop is balanced by the shear forces.

Solution

The pressure gradient is found from Eq. (6.58) ,

$$-\frac{\Delta P}{\Delta z} = \frac{8\overline{w}\mu}{(d/2)^2} = \frac{8\times 0.1\times 1.5}{0.005^2} = 48,000 \ \frac{N/m^2}{m}.$$

The wall shear stress is found from Eq. (5.59) and Eq. (6.53),

$$\varepsilon_{rz} = \frac{1}{2}\left(\frac{\partial q_r}{\partial z} + \frac{\partial q_z}{\partial r}\right) = \frac{1}{2}\frac{\partial w}{\partial r} = -2w_o \left.\frac{r}{2R^2}\right|_{r=R} = -\frac{w_o}{R}$$

From Eq. (5.51)

$$\tau_{rz} = 2\mu\varepsilon_{rz} = -\frac{2\mu w_o}{R}.$$

The total shear force per unit length of pipe is

$$F_s = 2\pi R\tau_{rz} = -4\pi\mu w_o = -8\pi\mu\overline{w} = -8\pi\times 1.5\times 0.1 = -3.77\text{N}.$$

The pressure drop over this length, i.e., for $\Delta z = 1$, is just 48,000 N/m^2, and the force exerted by this pressure drop is

$$F_p = \frac{\pi}{4} d^2 \Delta P = \frac{\pi}{4} 0.01^2 \times 48,000 = 3.77 \text{N} = F_s.$$

Flow in an Axisymmetric Annulus

Starting with the same assumptions as in the considerations of the Poiseuille flow, Eq. (6.50) is obtained for the flow in a cylindrical annulus too. The boundary conditions for the annulus are, however, different. As in the flow between plates let us consider first a flow for which there is no axial pressure drop. The inner cylinder does slide axially relative to the outer one, which is assumed to be at rest. This kind of motion may be encountered in dies for coating of wire with plastic material.

The boundary conditions for this case are

$$\begin{aligned} w &= 0 \quad \text{at} \quad r = R_o, \\ w &= V_i \quad \text{at} \quad r = R_i, \end{aligned} \tag{6.59}$$

and with $\Delta P / \Delta z = 0$, Eq. (6.50) yields the velocity

$$w = V_i \frac{\ln(r / R_o)}{\ln(R_i / R_o)}, \tag{6.60}$$

which is the analog of the velocity distribution in plane shear flow.

The volumetric flowrate is obtained as

$$Q = \int_{R_1}^{R_o} w(r) 2\pi r \, dr = \tfrac{1}{2} \pi R_o^2 V_i \left[\frac{1 - R_i / R_o}{\ln(R_o / R_i)} - 2\left(\frac{R_i}{R_o}\right)^2 \right]. \tag{6.61}$$

Next let us consider a case where both cylinders are stationary but there is an axial pressure drop. The boundary conditions are now

$$\begin{aligned} w &= 0 \quad \text{at} \quad r = R_o, \\ w &= 0 \quad \text{at} \quad r = R_i. \end{aligned} \tag{6.62}$$

For this case Eq. (6.50) yields the velocity distribution,

$$w = \left(-\frac{\Delta P}{\Delta z}\right) \frac{1}{4\mu} \left[(R_o^2 - r^2) - (R_o^2 - R_i^2) \frac{\ln(r / R_o)}{\ln(R_i / R_o)} \right], \tag{6.63}$$

which is the analog of the plane Poiseuille flow.

The maximum velocity is obtained at $dw/dr = 0$, i.e.,

$$-\frac{2r}{R_o^2}-\left[1-\left(\frac{R_i}{R_o}\right)^2\right]\left(\frac{1/r}{\ln(R_o/R_i)}\right)=0$$

or

$$\left(\frac{r}{R_o}\right)_{max}^2=\frac{1-(R_i/R_o)^2}{2\ln(R_o/R_i)},\qquad(6.64)$$

which can now be substituted in Eq. (6.63) to give the maximum velocity, w_o,

$$w_o=\left(-\frac{\Delta P}{\Delta z}\right)\frac{R_o^2}{4\mu}\left(1-\frac{1-(R_i/R_o)^2}{2\ln(R_o/R_i)}\left[1-\ln\frac{1-(R_i/R_o)^2}{2\ln(R_o/R_i)}\right]\right).\qquad(6.65)$$

Again the volumetric flowrate is obtained as

$$Q=\int_{R_i}^{R_o}w(r)2\pi r\,dr=\left(-\frac{\Delta P}{\Delta z}\right)\frac{\pi R_o^4}{8\mu}\left(1-\left(\frac{R_i}{R_o}\right)^4-\frac{\left[1-(R_i/R_o)^2\right]^2}{\ln(R_o/R_i)}\right)\qquad(6.66)$$

and the average velocity comes out as

$$\overline{w}=\frac{Q}{A}=\frac{Q}{\pi\left(R_o^2-R_i^2\right)}=\left(-\frac{\Delta P}{\Delta z}\right)\frac{R_o^2}{8\mu}\left[1+\left(\frac{R_i}{R_o}\right)^2-\frac{1-(R_i/R_o)^2}{\ln(R_o/R_i)}\right].\qquad(6.67)$$

Superposition of the solutions Eqs. (6.60) and (6.63) yields flows corresponding to combinations where one of the cylinders moves and an axial pressure gradient exists. Such flows are schematically shown in Fig. 6.5.

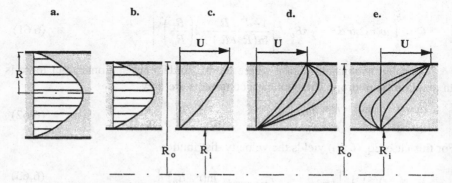

Figure 6.5 Velocity profiles in pipe and annulus flows.
a. Poiseuille flow.   **b.** Poiseuille annulus flow.
c. Annulus shear flow.   **d.** Shear plus axial pressure gradient.
e. Shear minus axial pressure gradient.

Example 6.8

An experimental system consists of a long tube of radius R_o filled with glycerin. The pressure gradient along the tube $\Delta P/\Delta z = dP/dz$ is given, and the flow is fully developed. The temperature in the glycerin at the center of the tube is sought, and a suggestion is made to stretch a thin wire along the tube axis, as shown in Fig. 6.6. It is claimed that since the wire is very thin, the flow field is only slightly modified by the presence of the wire. Assess the modification.

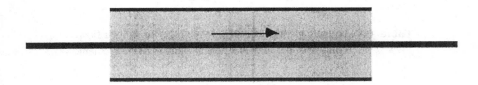

Figure 6.6 Tube with wire at its axis.

Solution

With the wire in the middle, the flow is an annular flow. To assess the modification caused by the wire, we consider the ratio of maximum velocity of the annular flow to that of the tube flow and the ratio of the average velocity of annular flow to that of the tube flow.

We use the subscript p for the pipe without the wire and w for the tube with the wire, i.e., for the annulus. The ratio of maximum velocities is found by the division of Eq. (6.65) by Eq. (6.52) resulting in

$$\left(\frac{w_w}{w_p}\right)_{max} = 1 - \frac{1-\left(R_i/R_o\right)^2}{2\ln\left(R_o/R_i\right)}\left[1-\ln\frac{1-\left(R_i/R_o\right)^2}{2\ln\left(R_o/R_i\right)}\right].$$

The ratio of the average velocities is found from Eqs. (6.67) and (6.58),

$$\frac{\overline{w}_w}{\overline{w}_p} = 1 + \left(\frac{R_i}{R_o}\right)^2 - \frac{1-\left(R_i/R_o\right)^2}{\ln\left(R_o/R_i\right)}.$$

Numerically computed values of the location of the maximum velocity, Eq.(6.64), values of $(w_w/w_p)_{max}$ and values of w_w/w_p are given in Table 6.2 for several ratios of R_i/R_o. As seen from the table, the modification of the flow is not small.

$\dfrac{R_i}{R_o}$	$\left(\dfrac{r}{R}\right)_{\max}$	$\left(\dfrac{w_w}{w_p}\right)_{\max}$	$\left(\dfrac{w_w}{w_p}\right)_{\text{mean}}$
0.01	0.32949	0.65038	0.78297
0.02	0.35744	0.60936	0.74488
0.05	0.40803	0.53503	0.66953
0.10	0.46365	0.45456	0.58005
0.20	0.54611	0.34093	0.44352
0.30	0.61475	0.25434	0.33417
0.50	0.73553	0.12664	0.16798
0.70	0.84554	0.04516	0.06013
0.90	0.94956	0.00500	0.00667

Table 6.2   Numerical values for $w_{\max}$ and w_{mean} for annular flow in comparison to flow in a pipe at equal pressure gradient.

Example 6.9

A metal wire of radius $R_i = 2$ mm is pulled vertically upward with the speed V_i through a long pipe of radius $R_o = 5$ mm, Fig. 6.7. The gap between the wire and the pipe is filled with molten plastic material of density ρ and viscosity μ. As the wire comes out of the pipe, it carries on its surface a layer of plastic which cools and solidifies. The thickness of the solid layer is 0.1 mm. Find the speed with which the wire is pulled upward.

Solution

The net flowrate of the molten plastic between the wire and the pipe must supply the coating. The volumetric flowrate of the 0.1-mm-thick solidified plastic coating moving upward with the wire at a speed V_i is

$$Q = AV_i,$$

where A is the cross-sectional area of the annulus formed by the plastic coating,

$$A = \pi\left[\left(R_i + 10^{-4}\right)^2 - R_i^2\right] = 1.288 \times 10^{-6}\,\text{m}^2.$$

Figure 6.7
Wire coating.

This flow must equal the sum of the flow induced by the motion of the wire alone, Eq. (6.61), and that of the flow downward due to gravity. The latter can be obtained from Eq. (6.66) written for a flow under the influence of a modified pressure drop $\Delta P / \Delta z$. In our case of constant pressure p_o, we have

$$P = p_o + \rho g h$$

and

$$-\frac{\Delta P}{\Delta z} = -\rho g \frac{\Delta h}{\Delta z} = -\rho g.$$

This expression for pressure drop is substituted into Eq. (6.66), which is combined with Eq. (6.61) to yield the expression for the flowrate induced by the motion of the wire coupled with the effects of gravity,

$$Q = A V_i = \frac{\pi R_o^2 V_i}{8} \left(4 \left[\frac{1 - \dfrac{R_i}{R_o}}{\ln \dfrac{R_o}{R_i}} - 2 \left(\frac{R_i}{R_o} \right)^2 \right] - \frac{\rho g R_o^2}{\mu V_i} \left[1 - \left(\frac{R_i}{R_o} \right)^4 - \frac{\left(1 - (R_i / R_o)^2 \right)^2}{\ln \dfrac{R_o}{R_i}} \right] \right).$$

The term containing V_i is extracted, leading to

$$\frac{\mu V_i}{\rho g R_o^2} = \frac{1 - \left(\dfrac{R_i}{R_o} \right)^4 - \dfrac{\left[1 - (R_i / R_o)^2 \right]^2}{\ln (R_o / R_i)}}{8 \left[\dfrac{1 - R_i / R_o}{2 \ln (R_o / R_i)} - \left(\dfrac{R_i}{R_o} \right)^2 - \dfrac{A}{\pi R_o^2} \right]}$$

and substitution of $R_i = 0.002$ m, $R_o = 0.005$ m, $A = 1.288 \times 10^{-6}$ m^2 gives

$$\frac{\mu V_i}{\rho g R_o^2} = 0.16915$$

or

$$V_i = \frac{0.16915 \rho g R_o^2}{\mu}.$$

For $\rho = 1{,}000$ kg/m³, $g = 9.81$ m/s² and $\mu = 2$ N·s/m² we obtain

$$V_i = 0.0207 \text{ m/s} \approx 2.1 \text{ cm/s}.$$

Flow between Rotating Concentric Cylinders

In this case we again consider the annular region between two cylinders, as in the previous case. This time, however, the cylinders rotate with the angular velocities, Ω_i and Ω_o, and the boundary conditions are

$$q_r = q_z = 0, \qquad q_\theta = R_i\Omega_i \qquad \text{at} \quad r = R_i,$$
$$q_r = q_z = 0, \qquad q_\theta = R_o\Omega_o \qquad \text{at} \quad r = R_o. \tag{6.68}$$

We consider the cylinders being long and the flow steady, resulting in $\partial/\partial z = 0$ and $\partial/\partial t = 0$. We assume $q_r = q_z = 0$ everywhere, and obtain from the continuity equation

$$\frac{\partial}{\partial \theta} q_\theta = 0. \tag{6.69}$$

Hence, q_θ can depend on r only.

The Navier–Stokes equations (5.64) – (5.66) with the modified pressure, Eq. (6.13), substituted become for this case

$$\rho\left(\frac{1}{r}q_\theta^2\right) = \frac{dP}{dr}, \tag{6.70}$$

$$\frac{d}{dr}\left[\frac{1}{r}\frac{d}{dr}(rq_\theta)\right] = 0. \tag{6.71}$$

with the solution of Eq. (6.71) for the velocity profile

$$q_\theta = Ar + \frac{B}{r}. \tag{6.72}$$

The integration constants are evaluated with the help of the boundary conditions given in Eq. (6.68), resulting in the velocity profile

$$q_\theta = \frac{-\left[(R_i/R_o)^2\Omega_i - \Omega_o\right]r + (\Omega_i - \Omega_o)R_i^2/r}{1 - (R_i/R_o)^2}. \tag{6.73}$$

This is known as *Couette flow* .

Couette flow is the only example in this chapter where some nonlinear term remained in the equations at the solution stage. It is noted that the part actually solved, Eq. (6.71), is still linear and that the nonlinearity appears only in Eq. (6.70), and even then as a given function, i.e., the solution, Eq. (6.72), is now substituted into Eq. (6.70), which is then solved for the modified pressure P by direct integration.

For the special case where the outer cylinder is stationary ($\Omega_o = 0$) while the inner cylinder is rotating with an angular velocity Ω_i, Eq. (6.73) simplifies to

$$q_\theta = \frac{-\Omega_i(R_i/R_o)^2\left(r - R_i^2/r\right)}{1 - (R_i/R_o)^2}. \tag{6.74}$$

An instrument consisting of an outer stationary and inner rotating cylinder is called a Couette viscometer and is used to measure the viscosity of liquids.

Example 6.10

Two long concentric cylinders have the radii R_i and R_o. The fluid between them has a density ρ and a viscosity μ. Consider the following cases:

a. $\Omega_i = 0$, $\Omega_o = \Omega$,
b. $\Omega_i = \Omega$, $\Omega_o = 0$,
c. $\Omega_i = -\Omega/2$, $\Omega_o = \Omega/2$.

Find the velocity distribution, the pressure distribution and the moments acting on both cylinders in cases a, b and c.

Solution

a.   For $\Omega_i = 0$, $\Omega_o = \Omega$, Eq. (6.73) simplifies to

$$q_\theta = \Omega \frac{r - R_i^2/r}{1-(R_i/R_o)^2} .$$

The pressure distribution is given by Eq. (6.70) as

$$\frac{dP}{dr} = \rho \frac{q_\theta^2}{r} = \frac{\rho\Omega^2\left[r - 2(R_i^2/r) + R_i^4/r^3\right]}{\left[1-(R_i/R_o)^2\right]^2}$$

and

$$P = P_i + \frac{\rho\Omega^2 R_i^2\left[(r^4 - R_i^4)/2R_i^2 r^2 - 2\ln(r/R_i)\right]}{\left[1-(R_i/R_o)^2\right]^2} .$$

The shear stress at, say, the inner surface, is, Eq. (5.51),

$$\tau_{r\theta} = 2\mu\varepsilon_{r\theta} = \mu r \frac{d}{dr}\left(\frac{q_\theta}{r}\right)\bigg|_{r=R_i} = \frac{2\Omega\mu}{1-(R_i/R_o)^2} ,$$

and the shear moment per unit height of cylinder becomes

$$|M| = \left|2\pi R_i^2 \tau_{r\theta}\right| = \frac{4\pi\mu\Omega R_i^2 R_o^2}{R_o^2 - R_i^2} .$$

The same moment acts, of course, on the outer cylinder.

b.   For $\Omega_i = \Omega$, $\Omega_o = 0$, the velocity distribution is given by Eq. (6.74):

$$q_\theta = \frac{-\Omega(R_i/R_o)^2\left(r-R_o^2/r\right)}{1-(R_i/R_o)^2}.$$

The pressure distribution is

$$\frac{dP}{dr} = \rho\frac{q_\theta^2}{r} = \frac{\rho\Omega^2(R_i/R_o)^4\left(r-2R_o^2/r+R_o^4/r^3\right)}{\left[1-(R_i/R_o)^2\right]^2}$$

and

$$P = P_i + \frac{\rho\Omega^2 R_i^2\left(\dfrac{R_i}{R_o}\right)^4\left[\dfrac{1}{2}\left(\dfrac{r}{R_i}\right)^2-\dfrac{1}{2}-\dfrac{1}{2}\left(\dfrac{R_o^4}{R_i^2 r^2}\right)+\dfrac{1}{2}\left(\dfrac{R_o}{R_i}\right)^4-2\left(\dfrac{R_o}{R_i}\right)^2\ln\left(\dfrac{r}{R_i}\right)\right]}{\left[1-(R_i/R_o)^2\right]^2}.$$

The shear stress at, say, the outer surface is, Eq. (5.51),

$$\tau_{r\theta} = 2\mu\varepsilon_{r\theta} = \mu r\frac{d}{dr}\left(\frac{q_\theta}{r}\right)\bigg|_{r=R_o} = \frac{-2\mu\Omega(R_i/R_o)^2}{1-(R_i/R_o)^2},$$

and the shear moment per unit height becomes

$$|M| = \left|2\pi R_o^2\tau_{r\theta}\right| = \frac{4\pi\mu\Omega R_i^2 R_o^2}{R_o^2-R_i^2}$$

for both inner and outer cylinders.

c.   For $\Omega_i = -\Omega/2$, $\Omega_o = \Omega/2$, Eq. (6.73) yields

$$q_\theta = \frac{\Omega\left(r\left[(R_i/R_o)^2+1\right]-R_i^2/r\right)}{1-(R_i/R_o)^2}.$$

The pressure distribution is

$$\frac{dP}{dr} = \rho\frac{q_\theta^2}{r} = \frac{\rho\Omega^2\left(r\left[\left(\dfrac{R_i}{R_o}\right)^2+1\right]^2-\dfrac{2R_i^2}{r}\left[\left(\dfrac{R_i}{R_o}\right)^2+1\right]+\dfrac{R_i^4}{r^3}\right)}{\left[1-(R_i/R_o)^2\right]^2}$$

and

$$P = P_i + \frac{\rho\Omega^2\left\{\left(\frac{r^2-R_i^2}{2}\right)\left[\left(\frac{R_i}{R_o}\right)^2+1\right]^2 - 2R_i^2\left[\left(\frac{R_i}{R_o}\right)^2+1\right]\ln\frac{r}{R_i}+\frac{R_i^4}{2}\left(\frac{1}{R_i^2}-\frac{1}{r^2}\right)\right\}}{1-\left(R_i/R_o\right)^2}.$$

The shear stress at, say, the inner radius is

$$\tau_{r\theta} = \frac{2\mu\Omega}{1-\left(R_i/R_o\right)^2},$$

and the shear moment for unit height becomes

$$|M| = 2\pi R_o^2 \tau_{r\theta} = \frac{4\pi\mu\Omega R_i^2 R_o^2}{R_o^2 - R_i^2}.$$

Note that the moments come out to be the same in all three cases, but not the pressures. An explanation for this is that rotating coordinate systems can be found in which cases b and c are kinematically reduced to case a. However, the dynamics in the rotating system requires the inclusion of appropriate D'Alambert forces which modify the pressure.

References

R.B. Bird, W.E. Stewart and E.N. Lightfoot, "Transport Phenomena," Wiley, New York, 1960.

W.F. Hughes, "An Introduction to Viscous Flow," Hemisphere, Washington, DC, 1979.

H. Schlichting, "Boundary-Layer Theory," 7th ed., McGraw-Hill, New York, 1979, Chapter 5.

Problems

6.1   The space between two long parallel plates (Fig. P6.1) is filled with a fluid of viscosity $\mu = 9\times10^{-3}$ dyn·s/cm^2. The upper plate moves with a velocity of 3 m/s and the lower plate is stationary. What is the shear stress distribution in the fluid?

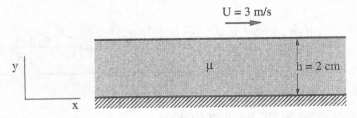

Figure P6.1

6.2   Coating of electric wire with insulating material is done by drawing the wire through a tubular die as shown in Fig. P6.2. The viscosity of the coating material is 100 poise. Simplify the flow equations for this case and calculate the force F required to draw the wire.

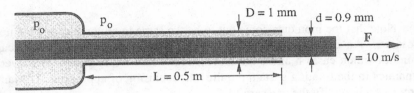

Figure P6.2 Wire coating.

6.3   A laminar layer of glycerin slides down on a semi-infinite vertical wall, Fig. P6.3.
    a.   Calculate the shear stress on the wall and the flowrate of the glycerin.
    b.   What will be the answers to part a if the wall leans at an angle α?

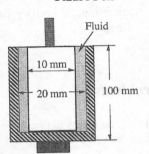

Figure P6.3

6.4   An instrument for measuring viscosity consists of a rotating inner cylinder and stationary outer cylinder as shown in Fig. P6.4. The inner cylinder rotates at 3,600 rpm and the viscosity of the fluid is (1) 10 poise, (2) 100 poise.
    a.   What is the moment acting on the outer cylinder in the two cases?
    b.   What is the efficiency of this instrument as a hydraulic transmission of moment? Find the moment as a function of the rpm of the outer cylinder and the power transmitted.

Figure P6.4 Viscometer.

6.5   A long pipe bend, Fig.P6.5, has the inner diameter of 0.025 m and is 2 m long. A solution of sugar in water, which has the same
density as water but whose viscosity is ten
times that of water, flows through the bend.
Measurements show that the pressure distribution along the pipe is the same as that in a
straight pipe of the same dimensions. At
point B the pressure is the outside pressure,
i.e., atmospheric. The flowrate is 6 kg/s.
Neglect gravity and calculate the forces
transmitted through flange A.

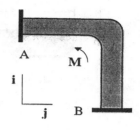

Figure P6.5 Pipe bend.

6.6   A very wide, shallow layer of water is approximately two dimensional. The
bottom is a rigid plate with a small inclination angle α, Fig. P6.6, and the
constant water depth, measured vertically, is h. The flow is assumed fully
developed.

   a. Write the simplified form the Navier–Stokes equations assume in this
   case. Write the boundary conditions at the bottom and the top of the
   layer.
   b. Solve the equation and obtain the velocity profile and the mean velocity.
   c. Find the shear stress at the bottom.

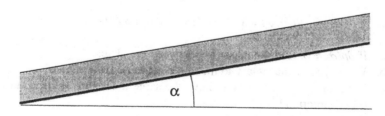

Figure P6.6 Shallow water layer.

6.7   A two-dimensional air bearing consists of an upper plate A that is $2L$ m
wide and 1 m long and a lower plate B of the same size but with a slot in its
middle, Fig. P6.7. Air is forced into the slot through a series of pipes C and
comes out at the edges. As a result the upper plate, which carries a load F,
is raised to the height h. Assuming laminar flow find the air supply pressure
necessary to support F, and find the supply rate necessary to maintain the
height h. Both depend on L, of course.

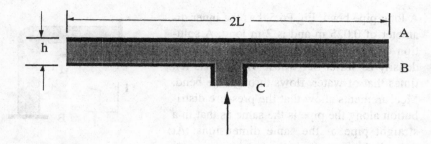

Figure P6.7 Air bearing.

6.8   The distance between two parallel plates in a two-dimensional flow, Fig.
P6.8, is $d = 0.05$ m. The gap between the plates extends to infinity and is
filled with a fluid whose density is $\rho = 1,000$ kg/m³ and whose viscosity is
$\mu = 500$ poise. The lower plate is stationary and the upper one moves in
the x-direction with the velocity of 1 m/s.

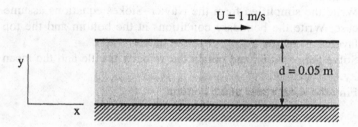

Figure P6.8 Stationary and moving plates.

a. If $\partial p/\partial x = 0$, find the mass flux between the plates.
b. What $\partial p/\partial x$ is necessary to make the net mass flow vanish?
c. What $\partial p/\partial x$ is necessary to have the same mass flux as in a, but in the
    other direction?
d. Find $\partial p/\partial x$ needed to double the mass flux of part a.

6.9   The servomechanism in Fig. P6.9 consists of a piston, A, equipped with
rods, B, moving inside a cylindrical sleeve, C. The servomechanism is filled
with oil and a gear pump maintains a pressure difference Δp between the
two sides of the piston. Once started, the piston motion is approximately at
a constant speed V and it moves until its side is flush with the sleeve edge,
i.e., the total length of its travel from one edge to the other is $L_2 - L_1$. The
time required for this motion is $\Delta t = (L_2 - L_1)/V$ and the force which the
servomechanism must put out is F.

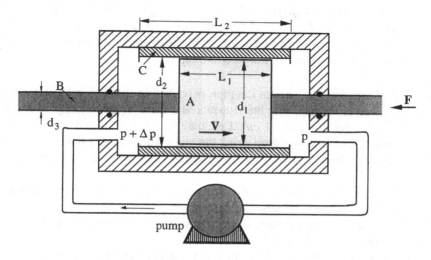

Figure P6.9 Servomechanism.

All the dimensions shown in Fig. P6.9 are given, and so are Δt, F, and the viscosity of the oil, μ. Find the pressure difference Δp the pump must maintain and also Q, the volumetric flux of oil it must supply. You may assume fully developed flow in the gap between the piston and its cylindrical sleeve and also that $(d_2 - d_1)/d_1 \ll 1$, i.e., the gap need not be considered as an annulus but may be taken as that between two plates.

Note that the force applied to the piston by the oil consists of two parts: pressure on the flat surface and shear on the curved surface. The shear force is to the right for small V but to the left for large V.

6.10 Repeat Problem 6.9 without the assumption $(d_2 - d_1)/d_1 \ll 1$. The gap must now be considered an annulus.

6.11 With reference to Problems 6.9 and 6.10:

$d_1 = 60$ mm, $d_2 = 63$ mm, $d_3 = 10$ mm,
$L_1 = 80$ mm, $L_2 = 120$ mm, $F = 2000$ N.
$\mu = 20$ poise, $\Delta t = 5$ s.

Compute Δp [Pa] and Q [m³/s] for the gear pump, using once the simpler analysis of Problem 6.9 and once the more elaborate one of Problem 6.10. Compare the results and decide whether the approximation of Problem 6.9 is sufficient for $(d_2 - d_1)/d_1 = 0.05$.

6.12 The viscosity of a given fluid is $\mu = 2\times10^{-3}$ kg/m·s. The fluid flows in a pipe which has the diameter of 0.025 m and is 20 m long. The average flow velocity is 1 m/s. The pipe opens to the atmosphere. Calculate the flowrate and the pressure at the entrance to the pipe.

6.13 It is suggested to change the pipe described in Problem 6.12 and use two smaller pipes such that the flowrate and the average velocity of the flow remains the same, i.e., 1 m/s. Calculate the required pressure drop in these pipes. For a pressure drop along the pipes which is only that used in Problem 6.12, find the average velocity and the flowrate in the two pipes.

6.14 The same amount of fluid as in Problem 6.12 must now be supplied through a 0.0125-m-diameter pipe.
   a. Calculate the pressure drop in this pipe and compare your result with that of Problem 6.12.
   b. Note that for a constant flowrate, the pressure drop is proportional to the diameter of the pipe raised to the power n. Find n.
   c. Also note that for a constant pressure drop the flowrate is proportional to the diameter of the pipe raised to the power s. Find s.

6.15 A fluid flows in a pipe which has the diameter D. The same fluid flows in the gap between two parallel flat plates. The size of the gap is also D. The same pressure gradient exists in both systems. Calculate the ratio between the flowrate of the fluid in the pipe and the flowrate between the plates, per width of D.

6.16 The pipe and the plates of Problem 6.15 are now used such that the flowrate in the pipe is the same as that between the plates, per width D. Calculate the ratio between the power needed to pump the fluid through the pipe and that needed to pump between the plates.

6.17 The velocity distribution in a fully developed flow through a rectangular square duct with the sides D is, approximately,

$$q = kw = kC\sin\frac{\pi x}{D}\sin\frac{\pi y}{D},$$

where C is a constant.
   a. Calculate the flow through this duct and compare with those in the pipe and the plates as in Problem 6.15, for the same pressure gradients.
   b. Calculate the power needed to pump the fluid and compare with the pipe and the plates as in Problem 6.16.

6.18 A concentric cylinder viscometer is shown in
Fig. P6.18. The height of the inner cylinder is
0.15 m, and its diameter is $D_i = 0.10$ m. The inner
diameter of the outer cylinder is $D_o = 0.11$ m. The
viscometer is used to measure viscosities in the
range of $\mu = 2 \times 10^{-3}$ kg/m·s. The outer cylinder is
stationary and the inner one turns at the rate of
3,000 rpm

  a. Find the range of the moments measured.

  b. It is necessary to measure viscosities in the
range of $\mu = 0.2$ kg/m·s but to have the range
of moments measured unchanged. One possibility is to decrease the number of revolutions of the inner cylinder. Calculate this new
number.

  c. Another possibility is to keep the turning rate
but to decrease the diameter of the inner
cylinder. Calculate this new diameter.

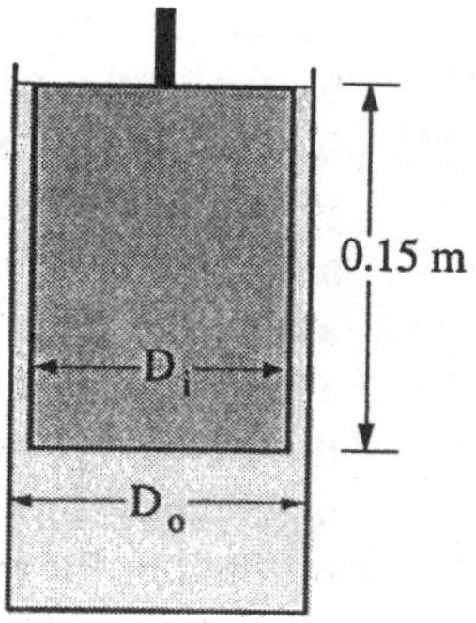

Figure P6.18

6.19 The viscometer described in Problem 6.18, before the suggested modifications are made, is used to measure the viscosity of water at room temperature. The inner cylinder of the viscometer is suddenly set to rotate at 3,000
rpm Assume that even at the transient stage of the flow there exist velocities in the angular direction only and that for very short times the flow field
between the two cylinders resembles the solution of the Rayleigh flow.
Find the time at which the velocity at $r = 0.0501$ m is 90% of that of the
surface of the inner cylinder. Now, without using again the error function,
but rather using the similarity properties of the Rayleigh flow, find times at
which this velocity appears at $r = 0.0502$ m, at $r = 0.0503$ m and at
$r = 0.0504$ m.

6.20 The pressure gradient in a laminar flow in a pipe is given by Eq. (6.58). An
engineer wants to express the modified pressure difference between two
points along a pipe by an equation of the form

$$P_1 - P_2 = C_f \frac{L}{D}\left(\frac{\rho \cdot \overline{w}^2}{2}\right),$$

where L is the distance between the two points, D is the pipe diameter, $\overline{w}$ is
the mean velocity and C_f is a "friction coefficient." Using Eq. (6.58), find an
expression for C_f.

6.21 Two flat plates are set as shown in
Fig. P6.21. The lower plate is stationary and the upper plate moves
to the right at the velocity of
2 m/s. The pressure at point 1 is
$p_1 = 100,000$ Pa, and the pressure
at point 2, which is 1 m further
along the plates, is p_2, as shown.
The viscosity of the fluid between
the plates is $\mu = 2\times10^{-2}$ kg/m·s.

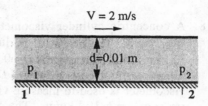

Figure P6.21

a. Find p_2 for which the net flow between the plates is nil.
b. Find p_2 for which the flow per unit width is 0.10 m³/m·s.
c. Find p_2 for which the flow per unit width is (-0.10) m³/m·s.

6.22 A long cylinder with its lower side closed is
shown in Fig. P6.22. The cylinder contains a
fluid with the viscosity of $\mu = 3\times10^{-3}$ kg/m·s. A
piston in the form of a long cylinder moves into
the fluid at the rate of 1 m/s. The flow between
the cylinder and the piston is assumed fully developed.
When flow starts, the length of the inserted part
of the piston is 0.60 m, and the length it can still
travel is 0.40 m.

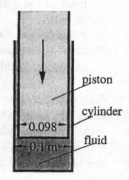

Figure P6.22

a. Find the velocity profile in the gap between
the piston and cylinder.
b. Find all the points where the velocity of the
fluid is zero.
c. Find the shear stress and shear rate at the
cylinder wall.
d. Find the force pushing the piston.

6.23 A viscometer is designed along the general form of the configuration
shown in Fig. P6.22, and is used to measure viscosities in the range
0.001 – 0.02 kg/m·s.

a. As a first step in the design use the dimensions of Fig. P6.22 and calculate the time needed for the piston to travel 0.1 m, provided it is pushed
down with a force of 10.0 N.
b. It is desired that the times to be measured are of the order of 10.0 s.
Suggest convenient weights to be used to push the piston down for
the range of viscosities measured.

6.24 A shock absorber is shown in Fig. P6.24. The viscosity of
the fluid is 1×10^{-3} kg/m·s. Assume the flow is laminar and
fully developed. Find the characteristic behavior of this
shock absorber, i.e., the force which resists the motion of
the piston as a function of the speed of the piston.

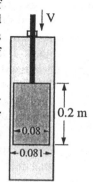

6.25 Water flows at the rate of 0.05 m³/s through the annular
gap with an inner radius of 0.10 m and an outer radius of
0.12 m. The length of the annulus is 100 m. Assume the
flow laminar and fully developed.
a. Find the pressure drop along the annulus.
b. Find the magnitude and location of the maximum
velocity in the annular gap.

Figure P6.24

6.26 Careful measurements show that there is a 0.003-m eccentricity in the
annulus described in Problem 6.25, i.e., there is a distance of 0.003 m
between the centers of the inner cylinder and the outer one. The flow in
this eccentric annulus is still 0.05 m/s and is still assumed fully developed.
An experienced engineer suggests that calculations for a concentric annulus having an inner radius of 0.1 + 0.003 m and an outer radius of 0.12 m or
an inner radius of 0.10 m and an outer radius of 0.12 − 0.003 m, both yield
upper bounds to the pressure drop. He also suggests that calculations for a
concentric annulus having an inner radius of 0.1 − 0.003 m and an outer
radius of 0.12 m or an inner radius of 0.1 m and an outer radius of 0.12 +
0.003 m both yield lower bounds to the pressure drop. Find the better
upper and lower bounds to the pressure drop.

6.27 The vertical concentric annulus in Fig. P6.27 has
an inner radius of 0.100 m and an outer radius of
0.141 m. The outer cylinder is stationary and the
flow in it is upward. The wall of the inner cylinder is very thin and there is a downward flow
inside the inner cylinder. The same fluid flows in
the inner cylinder and in the annulus. Neglect
the weight of the thin-walled inner cylinder, and
find the ratio between the two flow rates necessary to keep the inner cylinder floating, i.e., such
that the shear force on the inside of its walls is
just balanced by that on the outside.

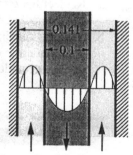

Figure P6.27

6.28 A gas turbine can operate using several different fuels. It is desirable to measure the density of the fuel while it is flowing in a pipe to the combustion chamber. A scheme to do this is shown in Fig. P6.28. The fuel has the viscosity μ and the density ρ. Find expressions for the pressure drop between points 1 and 2 and between points 3 and 4. Show that indeed

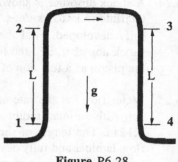

Figure P6.28

$$(p_1 - p_2) + (p_4 - p_3) = 2\rho g L,$$

i.e., the density ρ may be thus obtained from pressure drop measurements, though the viscosity and the velocity of the fluid are not known.

6.29 The pipe shown in Fig. P6.28 has a diameter of 0.025 m; the lengths between points 1 and 2 and between points 3 and 4 are 2 m each. The pipe carries fuel with a viscosity of 2×10^{-3} kg/m·s and a density of 900 kg/m³ at a velocity of 3 m/s. Assuming fully developed laminar flow, find the pressure drops between points 1 and 2 and between points 3 and 4.

6.30 A very long pencil of 1 cm in diameter is centrally located in a pipe, Fig. P6.30. Water flows in the pipe at the rate of 0.005 m³/s. The pencil density is the same as that of water. The pencil is at first held in place, and then it is released.
   a. Neglect the effects of the ends of the pencil and find its acceleration at the moment of release.
   b. Find the velocity of the pencil once it has reached steady motion. Estimate the error made by not taking into account the effect of the edges in the calculation of the acceleration.

0.03 m

Figure P6.30

6.31   Figure P6.31 shows three flat plates. The upper and lower ones are stationary, and the middle one can move in its own plane. The x-wise pressure gradient in the fluid above the moving plate is +300 Pa/m, and that in the fluid below is −300 Pa/m.

Find the velocity of the midplate.

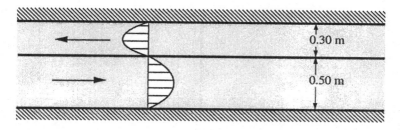

Figure P6.31

7. ENERGY EQUATIONS

Bernoulli's Equation

To obtain a clear physical appreciation of the various terms which appear in the energy equation, we begin our discussion by considering an idealized frictionless case.

It seems that the notion of conservation of mechanical energy was first inferred from the integration of the equations of motion of solids, idealized for no-friction situations. Newton's equation $md^2x/dt^2 = F_x$ can be integrated by multiplication with $dx/dt = V$ to become

$$m\frac{dx}{dt}\frac{d^2x}{dt^2} = \frac{1}{2}m\frac{d}{dt}\left(\frac{dx}{dt}\right)^2 = F_x\frac{dx}{dt}$$

or

$$m\,d\left(\frac{V^2}{2}\right) = F_x\,dx,$$

which can be interpreted as the change in the *kinetic energy* being equal to the work of the external force. The extension to three-dimensional motions and to forces that correspond to potentials, hence the appearance of *potential energy*, is not difficult. The inclusion of rotational motions is slightly more complicated, but still leads to the same result, i.e., that the sum of the kinetic and potential energies of a rigid body changes by exactly the amount of work performed on the body.

We use this result of classical mechanics as a guideline and begin by rewriting the Euler equation, i.e., the equations of motion for frictionless flows. To simplify the analysis we consider only time-independent flows, and Eq. (5.70) with its left-hand side expanded using Eq. (5.11) becomes

$$\frac{1}{2}\nabla q^2 - \mathbf{q}\times\nabla\times\mathbf{q} = -\frac{1}{\rho}\nabla p + \mathbf{g}. \tag{7.1}$$

We further restrict the flow to be incompressible. Compressible flows are considered later.

Let the body force **g** be expressible as

$$\mathbf{g} = -g\,\nabla h. \tag{6.11}$$

Hence, h is the elevation potential, and the negative sign simply means that we want objects to fall "down." Equation (7.1) may now be written as

$$-\mathbf{q} \times \nabla \times \mathbf{q} = -\frac{1}{\rho}\nabla p - g\nabla h - \frac{1}{2}\nabla q^2$$

or

$$\mathbf{q} \times \nabla \times \mathbf{q} = \nabla\left[\frac{1}{2}q^2 + \frac{p}{\rho} + gh\right]. \tag{7.2}$$

The expression in the square brackets is known as the Bernoulli polynomial. We first seek the rate of change of this expression along a streamline: Let **s** be the unit vector tangent to a streamline. The scalar product of **s** with Eq. (7.2) is

$$\mathbf{s} \cdot (\mathbf{q} \times \nabla \times \mathbf{q}) = \mathbf{s} \cdot \nabla\left[\frac{1}{2}q^2 + \frac{p}{\rho} + gh\right]$$
$$= \frac{\partial}{\partial s}\left[\frac{1}{2}q^2 + \frac{p}{\rho} + gh\right], \tag{7.3}$$

where $\partial/\partial s$ means, of course, the directional derivative, i.e., the rate of change of the expression in the square brackets in the direction of **s**.

The vector **s** is parallel to **q** and therefore perpendicular to $\mathbf{q} \times (\nabla \times \mathbf{q})$. Hence

$$\mathbf{s} \cdot (\mathbf{q} \times \nabla \times \mathbf{q}) = 0,$$

and therefore

$$\frac{\partial}{\partial s}\left[\frac{1}{2}q^2 + \frac{p}{\rho} + gh\right] = 0, \tag{7.4}$$

or $q^2/2 + p/\rho + gh = B$ does not change along **s**. Since **s** lies on a streamline, B is conserved along streamlines, i.e., for two points on the same streamline,

$$\frac{1}{2}q_1^2 + \frac{p_1}{\rho} + gh_1 = \frac{1}{2}q_2^2 + \frac{p_2}{\rho} + gh_2. \tag{7.5}$$

In engineering practice the equation is sometimes divided by g, resulting in

$$\frac{p_1}{\gamma} + h_1 + \frac{q_1^2}{2g} = \frac{p_2}{\gamma} + h_2 + \frac{q_2^2}{2g}, \tag{7.6}$$

where all terms have the dimensions of length. Engineers use the expressions *pressure head*, *static head* and *velocity head* to denote these various terms, respectively. Equation (7.6) is known as Bernoulli's equation.

Returning to Eq. (7.3), we now let **s** have the direction of $\nabla \times \mathbf{q} = \boldsymbol{\xi}$, the so-called *vorticity vector*. Because $\mathbf{q} \times (\nabla \times \mathbf{q}) = \mathbf{q} \times \boldsymbol{\xi}$ is also perpendicular to $\boldsymbol{\xi}$, the same argument that led to the conservation of B on streamlines now leads to its conservation along vorticity lines, i.e., along the field lines tangent to $\boldsymbol{\xi}$ everywhere. Streamlines and vorticity lines intersect and form surfaces, called *Bernoulli surfaces*. Each point on a Bernoulli surface can be reached from any other point on the same surface by going along streamlines and vorticity lines. Because Bernoulli's polynomial is conserved along both these families of lines, it is conserved on Bernoulli surfaces.

So far Bernoulli's equation (7.6) has been shown to hold on Bernoulli surfaces only. For such cases the equation is referred to as the *weak Bernoulli equation*. There are flow fields in which Bernoulli's polynomial has the same value everywhere, i.e., not only on Bernoulli's surfaces. This situation is said to satisfy the *strong Bernoulli equation*. The two most important flows satisfying the strong Bernoulli equation are those for which:

a. $\nabla \times \mathbf{q} = 0$ in the whole domain of flow. Equation (7.2) then becomes

$$\nabla B = \nabla \left[\frac{q^2}{2} + \frac{p}{\rho} + gh \right] = 0 \tag{7.7}$$

and, obviously, B is conserved in the whole domain.

b.   Every point in the domain of interest can be traced back along a streamline to another domain, where the strong Bernoulli equation holds. The flow field in this other domain may be trivially simple in many cases, e.g., at a great distance in front of an approaching airplane. Because of the weak Bernoulli equation B values are conserved along streamlines; and since all the streamlines in the domain of interest originate in the other domain, where B values are one and the same, the streamlines carry with them their B values into the new domain.

Mathematically, Bernoulli's equation is an *intermediate integral* of the equations of frictionless motion, i.e., it is an algebraic relation between the dependent variables. It is a useful tool in calculations of flows, and even when the flow in general is not frictionless, the equations can still serve as a good approximation in many engineering situations.

Example 7.1

Figure 7.1 shows an airplane wing, which is to be analyzed for a constant velocity flight. Show that once the velocity field around the wing is known, the strong Bernoulli equation holds, which may then be used to calculate the pressure on the wing and hence the lift force of the wing.

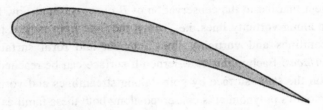

Figure 7.1 Wing in a flow field.

Solution

All streamlines in the vicinity of the wing started very far in front of it. A stationary observer located very far in front of the wing sees its environment at rest, with its pressure the atmospheric one. For him $B = p = p_o =$ const, and the strong Bernoulli equation holds. A moving observer, sitting on the wing, sees the vicinity of the first observer moving toward him with the velocity of the wing, V. For the second observer the Bernoulli polynomial on all streamlines very far in front is the same, being $B = p_o/\rho + V^2/2$. Thus all these streamlines carry with them their Bernoulli polynomials all over the vicinity of the wing, which therefore satisfies the strong Bernoulli equation. The pressure near the wing, where the velocity is $\mathbf{q}$, can therefore be obtained as

$$p = p_o + \frac{\rho}{2}\left(V^2 - q^2\right).$$

Example 7.2

Figure 7.2 shows a device called a *venturi tube*, which consists of a converging pipe section followed by a diverging section. The device is used to measure the flowrate of a fluid in a pipe. The velocity of the fluid in section 1 is lower than that in section 2, which results in a pressure difference between the sections, according to the Bernoulli equation. This pressure difference is measured and used to calculate the flow rate. Derive expressions for velocity and mass flowrate to be used for the venturi tube.

Solution

Neglecting friction, we write Bernoulli's equation (7.6):

$$\frac{p_1}{\gamma}+\frac{v_1^2}{2g}=\frac{p_2}{\gamma}+\frac{v_2^2}{2g},$$

where, for simplicity, we have assumed

$$h_1=h_2.$$

Because of continuity,

$$v_2=v_1\frac{A_1}{A_2},$$

and hence

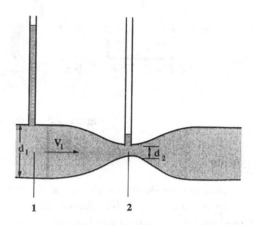

Figure 7.2 Venturi tube.

$$p_1-p_2=\frac{\rho}{2}\left(v_2^2-v_1^2\right)=\frac{\rho}{2}v_2^2\left[1-\left(\frac{A_2}{A_1}\right)^2\right]=\frac{\rho}{2}v_2^2\left[1-\left(\frac{d_2}{d_1}\right)^4\right],$$

or

$$v_2=\sqrt{\frac{\frac{2}{\rho}\left(p_1-p_2\right)}{1-\left(d_2/d_1\right)^4}}$$

and

$$\dot{m}=\frac{\pi}{4}d_2^2\rho v_2=\frac{\pi}{4}d_2^2\sqrt{\frac{2\rho\left(p_1-p_2\right)}{1-\left(d_2/d_1\right)^4}}.$$

Example 7.3

Old train engines operated without condensers and had to replenish the water in the boiler rather frequently. One way to do this without having to stop the train was to lower a water scoop, as shown in Fig. 7.3, into an open water tank laid beside the railway.

For a train moving to the left with the speed V_1 and a water tank which permits a 50 m run with the scoop lowered, find how much water can be taken into the train and what resistance force is applied to the train.

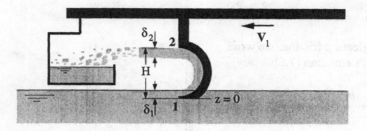

Figure 7.3 Water scoop.

Solution

An observer sitting on the scoop observes a sheet of water of thickness δ_1 entering the scoop with the relative velocity V_1 to the right. The average height of the layer is $\delta_1/2$ from the bottom of the scoop, and its average pressure is $p_o + \gamma\delta_1/2$, p_o being the atmospheric pressure. The Bernoulli equation, (7.6), may now be written for a streamline connecting points 1 and 2:

$$\frac{p_o}{\gamma} + \delta_1 + \frac{V_1^2}{2g} = \frac{p_o}{\gamma} + \left(H - \frac{\delta_2}{2}\right) + \frac{V_2^2}{2g} .$$

Hence the exit velocity from the scoop V_2, relative to the observer, is given by

$$V_2^2 = V_1^2 - \left(2H - \delta_2 - 2\delta_1\right)g ,$$

which is less than V_1^2. Thus, because of continuity,

$$\delta_2 = \delta_1 \frac{V_1}{V_2} > \delta_1 ,$$

and the water layer leaving the scoop is thicker. We assume, however, that δ is much smaller than H and that V_2 may be approximated by

$$V_2^2 = V_1^2 - 2gH .$$

For a scoop b meters wide, the amount of water taken is $50\delta_1 b$, which takes the train $t = 50/V_1$ seconds to fill.

To find the force transferred by the scoop to the train, we may use the integral form of the momentum theorem, Eq. (4.45),

$$\mathbf{F}_s = i\rho b\left[\delta_1 V_1^2 + \delta_2 V_2^2\right] .$$

The pressure terms cancel out because we assume the water to follow the wet back side of the scoop. This is not quite correct, but the error involved is not more than the mean hydrostatic pressure there, i.e., $\gamma b \delta_1^2$.

Still, this is not the resistance force applied to the train. The water layer leaving the scoop, Fig. 7.3, enters a tank where it settles to the train velocity. Thus the momentum $\mathbf{i} b \rho \delta_2 V_2^2$ is destroyed in this tank, and the tank transfers to the train a force

$$\mathbf{F}_T = -\mathbf{i} \rho b \, \delta_2 V_2^2 .$$

The resistance force is therefore

$$\mathbf{F}_R = \mathbf{F}_S + \mathbf{F}_T = -\mathbf{i} \rho b \, \delta_1 V_1^2 .$$

A man who has not studied fluid mechanics reasons as follows: As the train moves ahead 1 m it scoops $b \delta_1 \rho$ kg water. This water has been at rest, but once scooped, it has the kinetic energy $b \delta_1 \rho V_1^2 / 2$. This kinetic energy is imparted while the train moves 1 m, hence the resistance to its motion must be $b \delta_1 \rho V_1^2 / 2$, i.e., half of the result obtained above!

A student of fluid mechanics tells him that he has forgotten the kinetic energy lost in the tank and also the higher water level in the train, by about H.

Static, Dynamic and Total Pressures

Consider a streamline in a flow along which the Bernoulli polynomial

$$B = \frac{1}{2}q^2 + \frac{p}{\rho} + gh$$

is conserved, i.e., for points 1 and 2 on the streamline Eq. (7.5) holds,

$$\frac{1}{2}q_1^2 + \frac{p_1}{\rho} + gh_1 = \frac{1}{2}q_2^2 + \frac{p_2}{\rho} + gh_2 . \qquad (7.5)$$

Let the contributions of gh_1 and gh_2 be so small that they may be ignored. Now suppose a point 2 is found with $q_2 = 0$. The Bernoulli equation now reads

$$p_2 = p_1 + \tfrac{1}{2}\rho q_1^2 .$$

Relating to point 1 we see that by stopping the flow the pressure there, p_1, may be increased to p_2 by the amount $\rho q_1^2 / 2$. We now define the following nomenclature:

$\tfrac{1}{2}\rho q_1^2$ is the *dynamic pressure* at point 1;

p_1 is the *static pressure* there;

p_2 is the *total pressure*, or the *stagnation pressure*, at point 1.

All these terms describe point 1, and given q_1 and p_1, their magnitude may be computed. Their values do not depend on whether point 2 is physically realized, i.e., on whether there really exists a stagnation point on the streamline. Point 2 has been used just to indicate the reason for choosing the names *dynamic*, *static* and *total* pressures. In many engineering applications one finds these terms useful, and it is important to note that they arise from the Bernoulli equation; hence they imply the Bernoulli equation approximation.

Example 7.4

A Pitot tube, Fig. 7.4, is used to measure the speed of a fluid moving relative to the tube. It consists of an inner tube with its open end directed against the flow located inside a sleeve with holes, H, Fig. 7.4. The sleeve generators are parallel to the flow and the holes' axes are thus perpendicular to the flow direction. The pipe and the sleeve are connected to a manometer. Find the relation between the pressure difference measured by the manometer and the speed of the fluid. The flow is incompressible.

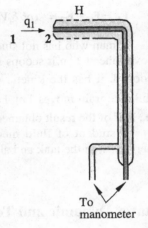

Figure 7.4 Pitot tube.

Solution

We choose point 1 far upstream from the Pitot tube, on the streamline which eventually hits the center of the hole in the inner tube, point 2. This streamline must have a stagnation point there. The Bernoulli equation, (7.5), reads for this case

$$\frac{p_1}{\rho} + \frac{q_1^2}{2} = \frac{p_2}{\rho}$$

and indeed p_2 is the *total pressure* at the stagnation point.

Now at point H, close to point 2 but on a streamline which bypasses the sleeve, the pressure p is still p_1, because this streamline is not affected by the presence of the tube. The strong Bernoulli equation is assumed to hold, and the fluid on this second streamline does not change its velocity as it moves parallel to the sleeve. The holes in the sleeve thus measure p_1, the *static pressure*. Therefore,

$$q_1 = \sqrt{\frac{2}{\rho}(p_2 - p_1)}.$$

Let the pressure difference be measured by a manometer in which a manometric fluid of density ρ_m is used, and let the manometer show a reading Δh_m. Then from Example 3.1

$$\frac{p_2 - p_1}{\rho} = \frac{\rho_m - \rho}{\rho} g \Delta h_m,$$

and therefore, the velocity at point 1, q_1, is related to the manometer reading, Δh_m, by

$$q_1 = \sqrt{2g\,\Delta h_m \frac{\rho_m - \rho}{\rho}}.$$

Extension to Compressible Flows

Let Bernoulli's equation (7.4) be now extended to apply to compressible flows. This is a mathematical step and requires no additional physical laws. To obtain Eq. (7.2), we have used

$$\frac{1}{\rho}\nabla p = \nabla\left(\frac{p}{\rho}\right),$$ (7.8)

which holds for incompressible flows only. In order to retain the gradient form for variable density, we define a *pressure function, F*, which we expect to satisfy

$$\frac{1}{\rho}\nabla p = \nabla F.$$ (7.9)

With this function the Bernoulli polynomial becomes just

$$B = \tfrac{1}{2}q^2 + gh + F$$ (7.10)

with all the previous results intact.

We now recall the Leibnitz rule for differentiation of an integral,

$$\frac{\partial}{\partial x}\int_c^{g(x)} f(\xi)d\xi = f(g)\frac{\partial g}{\partial x}$$

and also

$$\nabla \int_c^{g(x)} f(\xi)d\xi = f(g)\nabla g.$$

Comparison of this rule with Eq. (7.9) suggests at once

$$F = \int_{p_o}^{p}\frac{d\xi}{\rho(\xi)} = \int_{p_o}^{p}\frac{dp}{\rho(p)} = \int_{p_o}^{p}\frac{dp}{\rho},$$ (7.11)

where in the last term on the right ρ is considered a function of p only.

Generally for compressible fluids the density is a function of both the pressure and the temperature. However, there are some important flows where the density may be expressed as a function of the pressure only. These satisfy the form

$$\rho = \rho(p). \tag{7.12}$$

Examples for such flows are the isentropic flow (p/ρ^k = const) and the isothermal flow (p/ρ = const). Flows in which the density depends on the pressure only are called *barotropic flows*. Obviously, Eq. (7.11) and with it Eq. (7.10) hold for barotropic flows only.

The most important family of barotropic flows is that of the isentropic flows, and for this case the pressure function has a direct interpretation. The first and second laws of thermodynamics, combined, state that

$$T\,ds = di - \frac{1}{\rho}dp, \tag{7.13}$$

which for isentropic ($ds = 0$) processes becomes

$$di = \frac{1}{\rho}dp. \tag{7.14}$$

The differential of the enthalpy function, di in Eq. (7.14), is a complete differential and therefore dp/ρ is also a complete one. Isentropic flows are therefore barotropic, with no restriction to ideal gas behavior. Furthermore, from Eqs. (7.11) and (7.14),

$$F = \int_{p_o}^{p} \frac{dp}{\rho} = \int_{i_o}^{i} di = i - i_o. \tag{7.15}$$

Hence the pressure function is just the thermodynamic enthalpy, i.

Isentropic flows serve as very good approximations for many real flows, and Eq. (7.15) combined with Eq. (7.10) give the form

$$B = i + \tfrac{1}{2}q^2 + gh = i_T, \tag{7.16}$$

where the term *total enthalpy*, i_T, is used instead of B.

Returning to Eq. (7.11) it is emphasized that the identification of the pressure function with enthalpy is a particular case, although a most important one, restricted to isentropic flows. In general

$$dp = \left(\frac{dp}{d\rho}\right)d\rho = \left(\frac{d\rho}{dp}\right)^{-1} d\rho \tag{7.17}$$

and $(d\rho/dp)$ is just the compressibility of the fluid multiplied by its density, as given by Eq. (1.28). The more general interpretation of F for compressible flows is, therefore, the elastic energy of the fluid, which for isentropic flows equals the change in its enthalpy.

Example 7.5

A tank filled with fluid has a hole in its bottom. The pressure on the inside, just near the hole, is 1.5×10^5 Pa. The outside pressure is 10^5 Pa. Assuming the fluid flow out of the tank to be isentropic, find its average speed
a.   for a fluid which behaves like water,
b.   for a fluid which behaves like air.

Solution

Let subscript i denote in and o denote out; then:

a.   For water, from Eq. (7.5)

$$\tfrac{1}{2} V_o^2 + \frac{p_o}{\rho} = \frac{p_i}{\rho},$$

$$V_o = \sqrt{2 \left(\frac{p_i - p_o}{\rho} \right)} = \sqrt{\frac{2 \times 10^5 \times 0.5}{1,000}} = 10 \text{ m/s}.$$

b.   For air, from Eq. (7.16)

$$\tfrac{1}{2} V_o^2 + i_o = i_i,$$

$$V_o = \sqrt{2 \left(i_i - i_o \right)}.$$

Now

$$i_i - i_o = \int_{i_o}^{i_i} di = \int_{p_o}^{p_i} \frac{dp}{\rho},$$

$$\frac{p_i}{\rho_i^k} = \frac{p_o}{\rho_o^k} = \frac{p}{\rho^k}$$

and

$$\frac{dp}{\rho} = \frac{p_i}{\rho_i^k} k \rho^{k-2} dp.$$

Thus

$$\int_{p_o}^{p_i} \frac{dp}{\rho} = \int_{\rho_o}^{\rho_i} \frac{p_i}{\rho_1^k} k\rho^{k-2} d\rho = \frac{k}{k-1}\frac{p_i}{\rho_1^k}\left[\rho_i^{k-1} - \rho_o^{k-1}\right]$$

$$= \frac{k}{k-1}\left[\frac{p_i}{\rho_i}\frac{p_o}{\rho_o}\right] = \frac{k}{k-1}RT_i\left[1 - \left(\frac{p_o}{p_i}\right)^{\frac{k-1}{k}}\right].$$

Let $T_i = 300$ K. Then

$$V_o = \sqrt{\frac{2k}{k-1}RT_i\left[1 - \left(\frac{p_o}{p_i}\right)^{\frac{k-1}{k}}\right]} = \sqrt{\frac{2 \times 1.4}{0.4} \times 287 \times 300\left[1 - \left(\frac{1}{1.5}\right)^{\frac{0.4}{1.4}}\right]} = 257\,\text{m/s}.$$

Example 7.6

The hole at the bottom of the tank in Example 7.5 is fitted with an ideal engine, through which the fluid expands isentropically from the inside pressure to the environment. Find the work obtained from the transfer of 1 kg of fluid through the engine.

Solution

From thermodynamics we know that for an isentropic process:

$$w_{io} = i_i - i_o .$$

a. For water:

$$i_i = e_i + p_i / \rho_i, \qquad i_o = e_o + p_o / \rho_o,$$

where e is the internal thermal energy of the water. The engine is ideal and we do not expect e to change. Also ρ does not change. Thus,

$$W_{io} = i_1 - i_o = \frac{p_i}{\rho_i} - \frac{p_o}{\rho_o} = \frac{1}{\rho}(p_i - p_o) = \frac{1}{1,000} \times 0.5 \times 10^5 = 50\,\text{J}.$$

b. For air: From Example 7.5,

$$W_{io} = i_i - i_o = \frac{k}{k-1}RT_i\left[1 - \left(\frac{p_o}{p_i}\right)^{\frac{k-1}{k}}\right].$$

Thus

$$W_{io} = \frac{1.4}{0.4} \times 287 \times 300 \left[1 - \left(\frac{1}{1.5} \right)^{\frac{0.4}{1.4}} \right] = 32,963 \text{ J.}$$

In Example 7.5 the same energies were manifested as the kinetic energies of the emerging fluids. We may check this by

$$\frac{W_{iob}}{W_{ioa}} = \frac{32,963}{50} = 659.3 = \left(\frac{V_{ob}}{V_{oa}} \right)^2 = \left(\frac{257}{10} \right)^2 = 659.3,$$

which indeed comes out correct.

Modifications of the Bernoulli Equation for Conduit Flow

a. Average Velocities

The results obtained so far for the Bernoulli equation are strictly applicable to streamlines only. In conduit flow, where the velocity generally varies between streamlines, it becomes more convenient to use the average velocity for the whole cross section. The situation is analogous to that of using the average velocity in the application of the integral momentum theorem, in Chapter 4, and indeed the difficulty is resolved in the same manner as in Chapter 4, i.e., by the introduction of an *energy correction factor* β_E.

Let a conduit have the velocity distribution $\mathbf{q}$ and let the total flow rate of mechanical energy through its cross section, A, be computed:

$$\int_A \left(\frac{p}{\gamma} + \frac{q^2}{2g} + h \right) \rho \mathbf{q} \cdot \mathbf{n} \, dA = \text{const.} \tag{7.18}$$

Because of the Bernoulli equation, the same value is obtained for all cross sections, A. The mass flowrate through A is just

$$\int_A \rho \mathbf{q} \cdot \mathbf{n} \, dA = \rho \overline{V} A = \text{const.} \tag{7.19}$$

The average mechanical energy per unit mass is obtained by dividing Eq. (7.18) by Eq. (7.19). This average is also conserved along the conduit.

Let all streamlines be approximately parallel to the conduit axis. For this case $p/\gamma + h$ is the same over any given cross section and therefore is not modified by the averaging process. The kinetic energy term, however, is modified, and the resulting form is

$$\frac{p}{\gamma} + h + \beta_E \frac{\overline{V}^2}{2g} = \text{const.} \tag{7.20}$$

The kinetic energy correction factor β_E is obtained from

$$\beta_E = \frac{1}{A\overline{V}^3} \int\limits_A q^3 dA. \tag{7.21}$$

For a uniform velocity profile $\beta_E = 1$, while for a parabolic profile, as in laminar pipe flow, Eq. (6.53), $\beta_E = 2$. Thus all applications of Bernoulli's equation to streamlines may be extended to conduits, provided the correction factor, β_E, is used. In many cases β_E may be taken as 1, e.g., in turbulent flows.

Example 7.7

Find how would the results in Example 7.2 be modified when the flow in the pipe is the viscous Poiseuille flow, i.e., has a velocity distribution as given by Eq. (6.51).

Solution

In Example 7.2, V_1 and V_2 are the average velocities. From Eqs. (6.53) and (6.57) with q the velocity,

$$q = 2\overline{q}\left[1 - \left(\frac{r}{R}\right)^2\right].$$

and Eq. (7.21) reads

$$\beta_E = \frac{1}{\pi R^2 \overline{q}^3} \int\limits_0^R 2\pi 8\overline{q}^3 \left[1 - \left(\frac{r}{R}\right)^2\right]^3 r\, dr$$

$$= \frac{16}{R^2} \int\limits_0^R \left[r - 3\frac{r^3}{R^2} + 3\frac{r^5}{R^4} - \frac{r^7}{R^6}\right] dr = 16\left[\frac{1}{2} - \frac{3}{4} + \frac{3}{6} - \frac{1}{8}\right] = 2.$$

Bernoulli's equation now reads

$$\frac{p_1}{\gamma} + \beta_E \frac{V_1^2}{2g} = \frac{p_2}{\gamma} + \beta_E \frac{V_2^2}{2g}.$$

Hence

$$V_2 = \sqrt{\frac{\frac{1}{\rho}(p_1 - p_2)}{1 - (d_2/d_1)^4}}$$

and

$$\dot{m} = A_2 \rho V_2 = A_2 \sqrt{\frac{\rho(p_1 - p_2)}{1 - (d_2/d_1)^4}},$$

i.e., $\sqrt{2}$ times less than the mass flowrate in the original example.

b. Friction Head and Work Head

Another correction required in conduit flows is for friction. Friction is defined in thermodynamics, and one of its pronounced effects is to reduce the mechanical energy of the system. Since Bernoulli's equation is a balance of mechanical energies, scaled to have the dimensions of length ("head"), careful measurements always show the downstream total head to be deficient with respect to the upstream total one. This deficiency is exactly the amount of mechanical energy dissipated by friction, per unit mass, i.e., scaled to the dimension of length, and denoted *friction head loss* h_f.

To balance the equation, h_f must be included:

$$\frac{p_1}{\gamma} + \beta_E \frac{q_1^2}{2g} + h_1 = \frac{p_2}{\gamma} + \beta_E \frac{q_2^2}{2g} + h_2 + h_f. \tag{7.22}$$

The numerical evaluation of h_f is considered in a later chapter. When h_f is small, Eq. (7.22) is approximated by Eq. (7.6), the unmodified Bernoulli equation.

Suppose the flow passes through a pump. A pump increases the mechanical energy of the fluid, and this increase per unit mass, i.e., scaled to the dimension of length, is called the *work head* h_w. The relation for two points separated by a pump thus becomes

$$\frac{p_1}{\gamma} + \beta_E \frac{q_1^2}{2g} + h_1 + h_w = \frac{p_2}{\gamma} + \beta_E \frac{q_2^2}{2g} + h_2 + h_f, \tag{7.23}$$

where for negligible friction the h_f term may be dropped.

It is noted that the h_w term is not the power required to run the pump, but rather that part of the power which actually reaches the fluid as work. The ratio between the two is denoted the *hydraulic efficiency* of the pump.

Example 7.8

Find an expression for the friction head, h_f, for Poiseuille flow in a circular pipe.

Solution

From Eq. (6.56),

$$\Delta p = \frac{32}{D^2}\overline{V}L\mu .$$

Hence,

$$h_f = \frac{\Delta p}{\gamma} = \frac{32\mu}{D^2}L\frac{\overline{V}}{\rho g}.$$

Example 7.9

Water is being discharged from a large tank open to the atmosphere through a vertical tube, as shown in Fig. 7.5. The tube is 10 m long, 1 cm in diameter, and its inlet is 1 m below the level of the water in the tank. Find the velocity and the volumetric flowrate in the pipe, assuming:

a. Frictionless flow.
b. Laminar viscous flow.
c. A turbine is connected at the tube out-
    let, point 2. Find the maximum power
    obtainable under assumptions a and b,
    above.

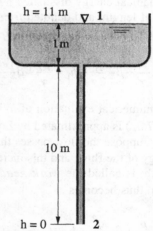

Solution

a. We select a streamline connecting the
    top surface, point 1, with the discharge,
    point 2. These two points have atmo-
    spheric pressure, and Eq. (7.6) simplifies
    to

Figure 7.5 Flow from a water tank
through a vertical tube.

$$\frac{v_1^2}{2g} + h_1 = \frac{v_2^2}{2g} + h_2 .$$

Neglecting the velocity v_1 at the tank surface, we obtain

$$v_2 = \sqrt{2gh_1} = \sqrt{2 \times 9.8 \times 11} = 14.7 \, \text{m/s},$$

and the flowrate is

$$Q = Av = \frac{\pi}{4} \times 0.01^2 \times v_2 \times 3{,}600 = 0.2827 v_2 \ \text{m}^3/\text{hr} = 4.16 \ \text{m}^3/\text{h}.$$

b. Bernoulli's equation between points 1 and 2, including friction and velocity variation across the stream tube, Eq. (7.22), is

$$h_1 = \beta_E \frac{v_2^2}{2g} + h_f.$$

Assuming laminar flow ($\beta_E = 2$) we substitute h_f for the Poiseuille flow, Eq. (6.56) (see Example 7.8),

$$h_f = \frac{\Delta p}{\gamma} = \frac{32 v_2 L \mu}{D^2 \rho g}$$

and obtain

$$gh_1 = v_2^2 + \frac{32 vL}{D^2} v_2.$$

For water at 20°C, $v = 10^{-6}$ m²/s; hence

$$9.8 \times 11 = v_2^2 + \frac{32 \times 10^{-6} \times 10}{0.01^2} v_2$$

or

$$v_2^2 + 3.2 v_2 - 107.8 = 0,$$

resulting in

$$v_2 = 8.9 \, \text{m/s}$$

and

$$Q = 0.2827 \times 8.9 = 2.516 \, \text{m/s}.$$

c/a. For this case the modified Bernoulli equation becomes

$$h_1 = \beta_E \frac{v_2^2}{2g} + h_f + h_w.$$

For negligible friction and $\beta_E = 1$

$$h_w = h_1 - \frac{v_2^2}{2g}$$

and the power of the turbine is

$$P = \dot{m}gh_w = \gamma Q \times h_w = \gamma h_w \times \frac{\pi D^2}{4} v_2.$$

Substitution of h_w gives P as a function of v_2:

$$P = \frac{\pi D^2 \gamma}{4} v \left(h_1 - \frac{v^2}{2g} \right).$$

To find the maximum power, we differentiate P with respect to v and equate the derivative to zero,

$$\frac{\partial P}{\partial v} = \frac{\pi D^2 \gamma}{4}\left(h_1 - \frac{3}{2g}v^2\right) = 0,$$

resulting in

$$v = \sqrt{\frac{2}{3}gh_1} = \sqrt{\frac{2}{3} \times 9.8 \times 11} = 8.48 \text{ m/s}$$

and

$$h_w = h - \frac{v^2}{2g} = 11 - \frac{71.9}{2 \times 9.8} = 7.33 \text{ m}.$$

Hence, the maximum power is

$$P_{max} = 9.8 \times 10^3 \times 7.33 \frac{\pi}{4} \times 0.01^2 \times 8.48 = 47.8 \text{ W}.$$

c/b. The work head including frictional effects is

$$h_w = h_1 - \beta_E \frac{v_2^2}{2g} - \frac{32 vL}{gD^2}v$$

and the power is

$$P = \frac{\pi D^2 \gamma}{4}v\left(h_1 - \beta_E \frac{v_2^2}{2g} - \frac{32 vL}{gD^2}v\right).$$

Now the power is maximal when

$$\frac{\partial P}{\partial v} = \frac{\pi D^2 \gamma}{4}\left(h_1 - \frac{3\beta_E v_2^2}{2g} - \frac{64 vL}{gD^2}v\right) = 0,$$

$$v^2 + \frac{64 vL}{3D^2}v - \frac{gh_1}{3} = 0,$$

$$v^2 + \frac{64 \times 10^{-6} \times 10}{3 \times 10^{-4}}v - \frac{9.8 \times 10}{3} = 0,$$

$$v^2 + 2.133v - 32.7 = 0,$$

$$v = 4.75 \text{ m/s}.$$

Hence

$$h_w = 11 - \frac{4.75^2}{9.8} - \frac{32 \times 10^{-6} \times 10}{9.8 \times 0.01^2} \times 4.75 = 7.15 \text{ m}$$

and

$$P_{max} = 9.8 \times 10^3 \times 8.37 \times \frac{\pi}{4}0.01^2 \times 4.75 = 26.14 \text{ W}.$$

Example 7.10

Equation (7.23) is the Bernoulli equation for conduit flow modified to include friction and mechanical work. Compare this equation with the first law of thermodynamics written for a control volume applicable to this case and evaluate the friction head h_f in terms of the appropriate thermodynamic variables.

Solution

We write the first law of thermodynamics for a control volume at steady state having heat and work interactions with the surroundings and a fluid entering the control volume at point 1 and leaving at point 2 as

$$\dot{Q} - \dot{W} = \dot{m}\left[(i_2 - i_1) + \tfrac{1}{2}\left(q_2^2 - q_1^2\right) + g\left(h_2 - h_1\right)\right],$$

where $\dot{m}$ is the mass flowrate of the fluid, $\dot{Q}$ the heat input and $\dot{W}$ the work output. Rewriting this equation per unit mass flow and rearranging yields

$$i_1 + \frac{q_1^2}{2} + gh_1 - \frac{\dot{W}}{\dot{m}} = i_2 + \frac{q_2^2}{2} + gh_2 + \frac{\dot{Q}}{\dot{m}}.$$

The work output, $\dot{W}$, may be rewritten in terms of the work head *input*, h_w, as

$$\dot{W} = -\dot{m}gh_w.$$

Also, the enthalpy, i, is given by

$$i = e + p/\rho,$$

where e is the internal energy of the fluid. Hence, the first law of thermodynamics takes on the form

$$\frac{p_1}{\rho} + \frac{q_1^2}{2} + gh_1 + gh_w = \frac{p_2}{\rho} + \frac{q_2^2}{2} + gh_2 + \left[(e_2 - e_1) - \frac{\dot{Q}}{\dot{m}}\right].$$

Dividing this equation by g renders it in terms of "heads,"

$$\frac{p_1}{\gamma} + \frac{q_1^2}{2g} + h_1 + h_w = \frac{p_2}{\gamma} + \frac{q_2^2}{2g} + h_2 + \left[\frac{e_2 - e_1}{g} - \frac{\dot{Q}}{\dot{m}g}\right].$$

Comparing this with Eq. (7.23), we obtain an expression for the friction head h_f,

$$h_f = \frac{1}{g}\left[e_2 - e_1 - \frac{\dot{Q}}{\dot{m}}\right].$$

Hence, the mechanical energy of the fluid "lost" as friction is "found" in the internal energy of the fluid and in the heat transferred to the surroundings.

Low Pressures Predicted by Bernoulli's Equation

Consider the pipe and the reservoir shown in Fig. 7.6. We assume the height of the reservoir above the pipe, H, to be large and neglect friction in the pipe, i.e., $h_f \approx 0$.

Bernoulli's equation applied between points 1 and 3 yields

$$\frac{p_o}{\rho} + \frac{v_1^2}{2} + gH = \frac{p_o}{\rho} + \frac{v_3^2}{2}.$$

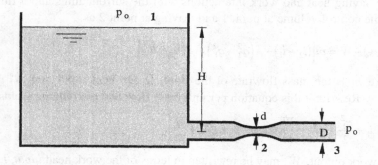

Figure 7.6 Contraction in a pipe.

We assume a very wide reservoir, hence $v_1 \approx 0$, and obtain

$$v_3^2 = 2gH,$$

which is not new (see Example 7.9). Now we apply Bernoulli's equation again, but this time between points 2 and 3:

$$\frac{p_2}{\rho} = \frac{p_o}{\rho} + \frac{v_3^2}{2} - \frac{v_2^2}{2}.$$

Because of continuity, $v_2^2 = v_3^2 (D/d)^4$, and therefore,

$$\frac{p_2}{\rho} = \frac{p_o}{\rho} - \frac{v_3^2}{2}\left[\left(\frac{D}{d}\right)^4 - 1\right] = \frac{p_o}{\rho} - gH\left[\left(\frac{D}{d}\right)^4 - 1\right]. \tag{7.24}$$

Equation (7.24) indicates that the pressure at point 2 may be forced to arbitrarily low values, including negative values, by the manipulation of either D/d or H.

From physics we know that gases cannot have negative pressures and that fluids boil once the pressure becomes lower than their vapor pressures. We conclude, therefore, that Eq. (7.24), and with it Bernoulli's equation, may not be applied once they predict negative or even saturation pressures. Such flow conditions require considerations which are too elaborate to be included in a basic text such as this.

Example 7.11

A dentist uses a suction device as shown in Fig. 7.7. Water from the main line flows in the larger pipe, which changes its diameter from D to d and back to D. Eventually the water flows out to the sink, at point 3. The pressure in the main water line, which feeds the device, is induced by a water tower 5 m high. The atmospheric pressure is $p_o = 10^5$ Pa. The vapor pressure of water at 30°C is 4246 Pa, and to ensure smooth operation, the pressure must never drop below 5,000 Pa. Find the highest D/d ratio.

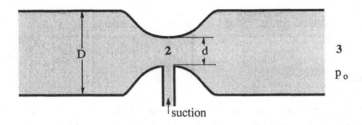

Figure 7.7 Dentist's suction device.

Solution

Point 1 is chosen at the top of the water tower. Repeating the analysis which has led to Eq. (7.24), we obtain

$$gh + \frac{p_o}{\rho} = \frac{p_o}{\rho} + \frac{v_3^2}{2}, \qquad v_3^2 = 2gh,$$

$$p_2 = p_o - \rho gh \left[\left(\frac{D}{d} \right)^4 - 1 \right],$$

and for the lowest value of $p_2 = 5,000$ Pa

$$5,000 = 100,000 - 1,000 \times 9.81 \times 5 \left[\left(\frac{D}{d} \right)^4 - 1 \right],$$

leading to

$$\left(\frac{D}{d}\right)_{max} = 1.309 .$$

References

R.B. Bird, W.E. Stewart and E.N. Lightfoot, "Transport Phenomena," Wiley, New York, 1960, Chapter 10.

W.F. Hughes, "An Introduction to Viscous Flow," Hemisphere Publishing, Washington, DC, 1979, Chapter 4.

R.H. Sabersky, A.J. Acosta and E.G. Hauptman, "A First Course in Fluid Mechanics," 2nd ed., Macmillan, New York, 1971, Chapter 3.

Problems

7.1   A dentist uses a suction device as shown in Fig. P7.1. Water from the line flows in the larger pipe, which changes its diameter from D to d and back to D; eventually the water flows out to the sink. The pressure in the main water line, before the water enters the device, is $p_1 = 1.5 \times 10^5$ Pa. Find the suction pressure for $d/D = 0.8$.

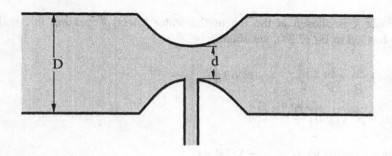

Figure P7.1 Dentist's suction device.

7.2 Before installing the water pipes in his new house, a man measured the pressure in the main water line near the house and found it to be $p_L = 10^6$ Pa. He then installed pipes with an inner diameter of 2 cm. Find the maximum rate of water supply he may expect.

7.3 A pump is used to raise water into a reservoir, Fig. P7.3. Both the water source and the reservoir have free surfaces, i.e., are open to the atmosphere. Water is raised at the rate $Q = 3$ m³/min to the height $h = 10$ m. The average velocity of the water in the pipe is 8 m/s. Find the power required to run the pump and the value of the various terms of the Bernoulli polynomial at the entrance and the exit of the pipe and also at points in the water source and in the reservoir far away from the pipe.

7.4 A Pitot tube, Fig. 7.4, is used to measure the speed relative to a moving fluid. The details of the Pitot tube are described in Example 7.4. For $p_2 = 1.5 \times 10^5$ Pa and $p_1 = 10^5$ Pa, find the flow velocity when the moving fluid is
a. water, b. air.
Note that for air it is not enough to know the dynamic pressure
$p_D = p_2 - p_1$.

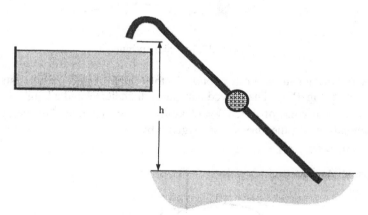

Figure P7.3 Pump raising water.

7.5 A water turbine consists of an outer shell, propeller and diffuser as shown in Fig. P7.5. At point 1 the diameter is 1 m, the velocity of the water is 30 m/s and the pressure is 250 kPa. At point 2 the diameter is 2 m, while at point 3 it is 3 m. The flow losses in the turbine are about 25% of the power of an ideal turbine.
a. What is the power of an ideal turbine?
b. What is the actual power?

c. What is the ideal and actual power without the diffuser?
d. If the propeller rotates at 900 rpm, what is the moment acting on the outer cover and what is the direction of this moment relative to that of the propeller rotation?

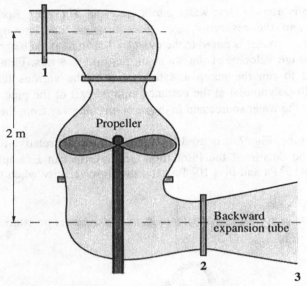

Figure P7.5 Water turbine.

7.6 A two-dimensional hollow body in the shape of an "igloo" rests on the ground, Fig. P7.6. The curved part has a diameter D and a length $L = \pi D/2$. There is a wind of speed $V = 40 \, \text{km/h}$. Assuming that the average wind velocity along the curved part is given by

$$V_L = VL / D = V \pi / 2,$$

calculate the lift on the igloo.

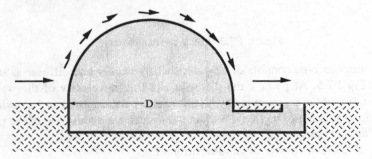

Figure P7.6 Igloo in cross wind.

7.7   A large tank of water has a hole at its side, $h = 4$ m below the water level. The hole diameter is $d = 0.05$ m, Fig. P7.7. Neglect friction and find the rate of the water flow out of the tank.

Note: The answer is not $A\sqrt{2gh}$; the shape of the stream tube coming out of the tank is not cylindrical, but rather curved. This constriction of the cross section is called vena contracta.

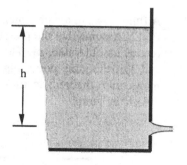

h

Figure P7.7 Vena contracta.

7.8   Find the correction factors,

β_M for the use of the average velocity in the momentum theorem,
β_E for the use of the average velocity in the Bernoulli equation,

for the flow in a rectangular duct of sides a and b. Assume the velocity to be approximated by

$$w = w_o \sin\left(\frac{\pi x}{a}\right)\sin\left(\frac{\pi y}{b}\right).$$

7.9   Water at 20°C is flowing through a 10-cm pipe at 1.5 m/s. At one point, the static pressure is 135 kPa gauge, and the friction head between this point and a second point 10 m below the first is 3.5 m. Find the static pressure at the second point.

Now suppose the second point is 10 m above the first point; can the flow take place?

7.10   Oil at 20°C (viscosity 900 cp, specific gravity 0.95) is to flow by gravity through a 1-mile-long pipeline from a large reservoir to a lower station. The difference in levels between the tanks is 14 m. The discharge rate required is 1.6 m³/min. Calculate the diameter of the steel pipe necessary for these conditions.

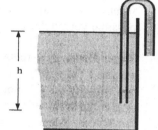

h

7.11   A tank 1.25 m deep contains water, Fig. P7.11. The inlet end of the siphon is $h = 1$ m below the surface. Friction losses amount to 10 kPa. How far below the sur-

Figure P7.11
Tank with siphon.

face must the outlet end be placed to give a velocity of 0.75 m / s?

7.12 A fluid flows in a pipe which has a sudden increase in its cross section, Fig. P7.12. Assume that the pressure in the wider section right after the jump p_2 retains its old value, p_1. Then choose a control volume containing sides 1 and 3 of the jump point, Fig. P7.12, and use conservation of mass and the momentum theorem to find the new average velocity. Find the head loss across the jump.

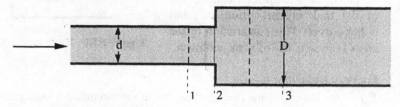

Figure P7.12 Sudden increase in cross section and control volume.

7.13 An inventor who did not study fluid mechanics suggests to put a conical funnel in a fast-flowing river and to place a turbine rotor at the funnel's apex, Fig. P7.13. Because of continuity, he claims, the ratio of the river flow speed to the speed at the rotor is the inverse of the funnel cross sectional areas. Thus, in principle, any speed can be attained at the

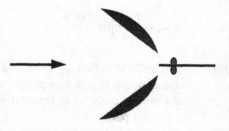

Figure P7.13 Turbine with guide funnel.

apex. "Not so fast," says a student of fluid mechanics, "the maximum attainable speed is quite finite." Find this finite speed and explain what happens when the area ratio becomes larger than that corresponding to the highest possible velocity ratio.

7.14 Water flows in a rectangular channel open to air at p_o, Fig. P7.14. At section 1 the width of the channel is b_1, the height of water h_1 and its average velocity q_1. At another section the channel width is b_2, while the pressure at the surface is still p_o. Find q_2 and h_2. Find conditions for $h_2 > h_1$ and for $h_2 < h_1$. Identify situations when the Bernoulli equation cannot hold between the two sections.
Hint: $h_2 < 0$.

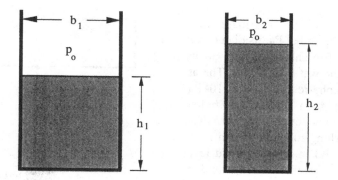

Figure P7.14 Sections of a rectangular channel.

7.15 A stationary nozzle ejects a jet of water with the mass-flowrate of 200 kg/s and with the velocity $u_1 = 20$ m/s, Fig. P7.15. The water jet hits a plane vane with the angle $\alpha = 30°$ between the jet axis and the plane surface. The vane recedes in the direction of its normal with the speed $v = 6$ m/s. The environment pressure is constant and gravitation and viscosity effects are negligible.

   a. Choose a control volume and define it carefully.
   b. Find the division of the jet mass-flowrate on the vane: the flux to the left and the flux to the right, i.e., how thick are δ_2 and δ_3.
   c. Find the force applied by the jet to the vane.

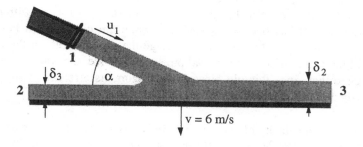

Figure 7.15

7.16 A water turbine operates between two reservoirs open to the atmosphere, as shown in Fig. P7.16. All friction losses are negligible, except at the pipe leading to the lower reservoir which is terminated with a sudden enlargement of the cross section; and the total velocity head there, i.e., $V_2^2/2$, is lost. Find the maximum power obtainable by this turbine.

7.17 The pressure in a main water line is 4×10^5 Pa, its diameter is 0.150 m and the mean velocity of the water is 10 m/s. The atmospheric pressure is 10^5 Pa. The main line is 1 m below ground level. An industrial washing machine which is located 3 m above ground level has a peak water demand of 0.020 m³/s.
Find the diameter of the pipe leading from the main line to the machine.

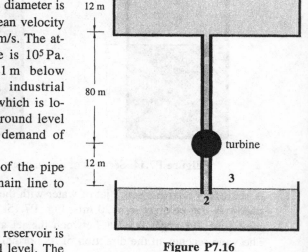

Figure P7.16

7.18 The water level in a reservoir is 20 m above ground level. The top of the tank is open to the atmosphere. Water is supplied to a field at ground level using a pipe with a diameter of 0.012 m. Neglecting friction, find the maximum flowrate that can be supplied. To increase the supply the pipe is changed to another having a diameter of 0.019 m. Find the new flowrate.

7.19 A small length of the old pipe, that of 0.012 m diameter in Problem 7.18, is laid under a wall. In changing the old pipe for the new one with the larger diameter it has been suggested to change the pipe section between the reservoir and up to the wall, to change the pipe starting on the other side of the wall, but to leave the old pipe section under the wall. It has been argued that since this old piece is very short, its effect is negligible. Find whether this effect is indeed negligible.

7.20 The mean velocity of water in a main line is 10 m/s, and the water pressure there is 3×10^5 Pa. The water then passes through a booster pump and is distributed into houses located 100 m above the main line. When a tap is opened in a house, the water comes out at the speed of 20 m/s. The atmospheric pressure is 10^5 Pa. Friction head losses between the main line and the houses are estimated as 5 times the velocity head at the houses. The booster pump overall efficiency is 0.8. Find the power needed to run the booster pump.

7.21 A two-dimensional fluid jet hits a curved vane, which recedes from the jet, Fig. P7.21. Show that the thickness of the water layer, as it follows the contour of the vane surface, is constant.

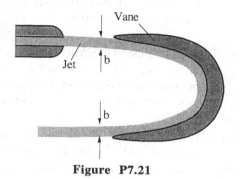

7.22 Water comes out of a 25-m-high water tower, open at the top, and flows through a 0.15-m-diameter, 600-m-long pipe. The water is then dis-

Figure P7.21

tributed through 0.025-m-diameter pipes, one of which is 100 m long, Fig. P7.22. Measurements of pressure drops along the horizontal pipes yield $P_A - P_B = 1,000$ Pa, and $P_D - P_E = 2,000$ Pa. The mean velocity of the flow in the 0.15-m-diameter pipe is $V = 0.15$ m/s.

Assume that at a given velocity the friction losses in a pipe are proportional to the pipe's length, and find the pressure at point F.

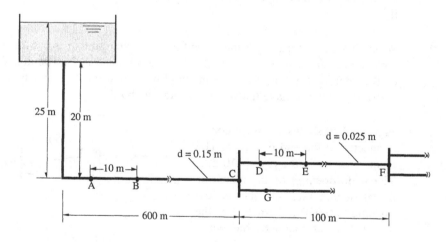

Figure P7.22

7.23 Measurements show that friction losses in straight pipes are proportional to L/d, where L is the pipe's length and d is its diameter, and approximately proportional to ρV^2, where V is the mean velocity in the pipe. In the system of Fig. P7.22 all pipes are closed except the 0.15-m-diameter pipe and the 0.025-m-diameter pipe designated by the letters DEF. Thus the whole flow is now through these two pipes, along the path ABCDEF. At point F the pipe is open to the atmosphere. Find the flowrate.

7.24   Another pipe of 0.025 m diameter, identical to that along DEF in Fig. P7.22, is opened, such that the water coming out at point C is equally divided between the two parallel 0.025-m-diameter pipes, both of which are open to the atmosphere. Find the flowrate.

7.25   It is necessary to double the flowrate of the water under the conditions of Problem 7.23. To do this, a booster pump is installed at point B in Fig. P7.22. The booster pump overall efficiency is 80%. Find the power needed to run the booster pump.

7.26   The booster pump designed in Problem 7.25 is used under the conditions of Problem 7.24. Its motor draws the same power as obtained in Problem 7.25. Find the flowrate under these operating conditions.

7.27   Given a laminar viscous flow in a circular pipe. Find the friction head, expressed as length, as a function of the fluid properties and of the velocity.

7.28   Water is poured form a jar into a bottle. The water stream is assumed to have a circular cross section, and its flowrate is 0.0005 m³/s. The diameter of the neck of the bottle is 0.02 m. Find how high must the jar be held above the bottle to have the water jet enter the bottle.

7.29   Figure P7.29 shows a conical nozzle connected at the end of a pipe. The mean velocity in the pipe is 10 m/s. The wide opening of the nozzle is 0.150 m diameter and the narrow opening is 0.050 m diameter. At the exit to the atmosphere the water pressure is atmospheric.

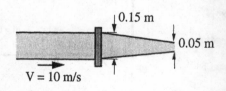

Figure P7.29

   a. Find the mean velocity of the water at the exit. Find the distribution of the pressure inside the cone.
   b. Find the distribution of the axial force along the walls of the cone.

7.30   A production process consists of lowering a ceramic form into a bath of hot molten plastic, and then taking the form, now coated with a plastic layer, out to cool. The plastic solidifies as it cools. Finally the plastic is cut along predetermined lines, while still on the form, using a high-speed thin water jet. A surface stress, i.e., a pressure of 1,000 kPa, is required in order to cut the plastic, while a pressure of 10,000 kPa results in pitting on the ceramic surface.

a. Find the necessary velocity of the water jet.
b. Find the pressure needed to produce the jet.
c. Find the maximal pressure and jet velocity which still do not harm the ceramic form.

8. SIMILITUDE AND ORDER OF MAGNITUDE

Dimensionless Equations

All the equations derived until now are dimensional, i.e., their various terms have physical dimensions. These terms represent numerical values, which thus depend on the system of physical units adopted. While there are some arguments why a particular system of units should be preferred, e.g., the SI system, still this choice is just a convention and is highly arbitrary. It seems, therefore, that by an orderly and preplanned removal of this arbitrariness one could gain some additional information. Such a process would result in Dimensionless Equations.

There are several goals for using dimensionless equations; two of the most important ones are *similitude* and *order of magnitude*. While these two goals are quite different from one another, both require an intermediate step consisting of essentially the same algebraic procedure which leads to dimensionless equations. We choose to show this procedure first.

Let us rewrite the continuity equation, (5.21), and the momentum equations, (5.27) - (5.29), for some incompressible flow, adding "stars" to all dependent and independent variables. The role of these stars is to denote dimensional variables, such that when these same variables appear later without stars they are considered dimensionless:

$$\frac{\partial u^*}{\partial x^*} + \frac{\partial v^*}{\partial y^*} + \frac{\partial w^*}{\partial z^*} = 0 \tag{8.1}$$

$$\rho\left(\frac{\partial u^*}{\partial t^*} + u^*\frac{\partial u^*}{\partial x^*} + v^*\frac{\partial u^*}{\partial y^*} + w^*\frac{\partial u^*}{\partial z^*}\right) = \rho g_x^* - \frac{\partial p^*}{\partial x^*} + \mu\left(\frac{\partial^2 u^*}{\partial x^{*2}} + \frac{\partial^2 u^*}{\partial y^{*2}} + \frac{\partial^2 u^*}{\partial z^{*2}}\right),$$
$$\tag{8.2}$$

$$\rho\left(\frac{\partial v *}{\partial t *}+u *\frac{\partial v *}{\partial x *}+v *\frac{\partial v *}{\partial y *}+w *\frac{\partial v *}{\partial z *}\right)=\rho g_y^* -\frac{\partial p *}{\partial y *}+\mu\left(\frac{\partial^2 v *}{\partial x *^2}+\frac{\partial^2 v *}{\partial y *^2}+\frac{\partial^2 v *}{\partial z *^2}\right),$$

(8.3)

$$\rho\left(\frac{\partial w *}{\partial t *}+u *\frac{\partial w *}{\partial x *}+v *\frac{\partial w *}{\partial y *}+w *\frac{\partial w *}{\partial z *}\right)=\rho g_z^* -\frac{\partial p *}{\partial z *}+\mu\left(\frac{\partial^2 w *}{\partial x *^2}+\frac{\partial^2 w *}{\partial y *^2}+\frac{\partial^2 w *}{\partial z *^2}\right).$$

(8.4)

We now assume that some characteristic physical quantities have been defined and attempt to express the dimensional starred variables in terms of these characteristic quantities. The problem of how to define the characteristic quantities will be considered later, because their definitions are quite different for similitude and for order of magnitude. Here it suffices to assume that such characteristic quantities exist and that they are dimensional. Thus if τ is the characteristic time, which is measured in, say, seconds, then

$$t=\frac{t *}{\tau}, \qquad\qquad t* = t\tau.$$

(8.5)

The first of these equations is really a definition of the dimensionless time t.

Similarly, a characteristic length L, which may be measured in meters, satisfies

$$x* = Lx, \qquad y* = Ly, \qquad z* = Lz,$$

(8.6)

meaning that we choose to measure length in units of L meters. A characteristic velocity V, measured in m/s, satisfies

$$u* = Vu, \qquad v* = Vv, \qquad w* = Vw$$

(8.7)

and a characteristic pressure π satisfies

$$p* = \pi p.$$

(8.8)

We also choose the constant of gravity g as the characteristic body force, i.e., $g_x^* = gg_x$. The starred variables, Eqs. (8.5) - (8.8), are now substituted in Eqs. (8.1) - (8.4), resulting in the disappearance of stars from these. Equation (8.1) thus becomes

$$\frac{V}{L}\left(\frac{\partial u}{\partial x}+\frac{\partial v}{\partial y}+\frac{\partial w}{\partial z}\right)=0,$$

(8.9)

and Eq. (8.2)

$$\rho\left[\frac{V}{\tau}\frac{\partial u}{\partial t}+\frac{V^2}{L}\left(u\frac{\partial u}{\partial x}+v\frac{\partial u}{\partial y}+w\frac{\partial u}{\partial z}\right)\right]=\rho g g_x -\frac{\pi}{L}\frac{\partial p}{\partial x}+\frac{\mu V}{L^2}\left(\frac{\partial^2 u}{\partial x^2}+\frac{\partial^2 u}{\partial y^2}+\frac{\partial^2 u}{\partial z^2}\right).$$

(8.10)

We now select some term in each equation and divide the whole equation by the group of physical constants and characteristic quantities that precedes it.

The term now stands alone and is dimensionless. Because the equations we use are always dimensionally homogeneous, all other terms in the equation have also become dimensionless, and so must be the groups composed of the characteristic quantities which appear as multipliers in the equation. Thus Eq. (8.9) divided by V/L becomes

$$\frac{\partial u}{\partial x} + \frac{\partial v}{\partial y} + \frac{\partial w}{\partial z} = 0. \tag{8.11}$$

Equation (8.10) divided by $\rho V^2/L$ becomes

$$\left(\frac{L}{\tau V}\right)\frac{\partial u}{\partial t} + \left[u\frac{\partial u}{\partial x} + v\frac{\partial u}{\partial y} + w\frac{\partial u}{\partial z}\right]$$
$$= \left(\frac{Lg}{V^2}\right)g_x - \left(\frac{\pi}{\rho V^2}\right)\frac{\partial p}{\partial x} + \left(\frac{\mu}{\rho VL}\right)\left(\frac{\partial^2 u}{\partial x^2} + \frac{\partial^2 u}{\partial y^2} + \frac{\partial^2 u}{\partial z^2}\right). \tag{8.12}$$

The dimensionless groups which have emerged in the equations are recognized as

$$\frac{L}{\tau V} = \Omega \quad - \quad \text{the time number,}$$

$$\frac{\pi}{\rho V^2} = \text{Eu} \quad - \quad \text{the Euler number,}$$

$$\frac{\rho VL}{\mu} = \frac{VL}{\nu} = \text{Re} \quad - \quad \text{the Reynolds number,}$$

and

$$\frac{V^2}{gL} = \text{Fr}^2, \quad \text{Fr is the Froude number.}$$

Using these dimensionless numbers, Eqs. (8.11)-(8.12) are rewritten in their dimensionless forms as

$$\frac{\partial u}{\partial x} + \frac{\partial v}{\partial y} + \frac{\partial w}{\partial z} = 0 \tag{8.13}$$

$$\Omega\frac{\partial u}{\partial t} + u\frac{\partial u}{\partial x} + v\frac{\partial u}{\partial y} + w\frac{\partial u}{\partial z} = \frac{1}{\text{Fr}^2}g_x - \text{Eu}\frac{\partial p}{\partial x} + \frac{1}{\text{Re}}\left(\frac{\partial^2 u}{\partial x^2} + \frac{\partial^2 u}{\partial y^2} + \frac{\partial^2 u}{\partial z^2}\right). \tag{8.14}$$

The characteristic physical quantities are now hidden in the dimensionless numbers, and Eqs. (8.13) and (8.14) may be used for both similitude and order of magnitude analysis. The selection of the characteristic quantities is considered now separately for each objective.

Similitude

We begin by considering a particular example of steady unidirectional vertical film flow shown in Fig. 8.1a, the x-component of the dimensionless Navier–Stokes equations. Equation (8.14) becomes in this case

$$0 = \frac{1}{\text{Fr}^2} g_x + \frac{1}{\text{Re}} \frac{d^2u}{dy^2} .$$

With the coordinates used in Fig. 8.1a, $g_x = 1$, and the equation takes the form

$$\frac{d^2u}{dy^2} + N = 0, \tag{8.15}$$

with

$$N = \frac{\text{Re}}{\text{Fr}^2} = \frac{L^2 g}{V v} .$$

The expression for N contains two physical constants, the constant of gravity, g [m/s^2], and the kinematic viscosity of the fluid, v [m^2/s]. It also contains the characteristic length, L [m], and the characteristic velocity, V [m/s]. These must be selected now, and we choose them to be (see Fig. 8.1a)

$$L = \delta,$$

the thickness of the film;

$$V = \frac{1}{\delta} \int_{o}^{\delta} u \, dy,$$

the average fluid velocity.

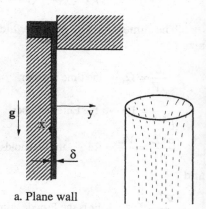

a. Plane wall

b. Elliptical conduit

Figure 8.1 Steady film flow on vertical wall.

The boundary conditions that the velocity, u, must satisfy are

$$u = 0 \quad \text{at} \quad y = 0,$$
$$\frac{du}{dy} = 0 \quad \text{at} \quad y = 1. \tag{8.16}$$

The first boundary condition is recognized as the no-slip viscous boundary condition, Eq. (5.62). The second boundary condition expresses the fact that there is zero shear at the gas–liquid interface.

The solution of Eq. (8.15) with the boundary conditions of Eq. (8.16) is

$$u = N\left(-\frac{y^2}{2} + y\right),$$

which relates the dimensionless velocity to the dimensionless y-coordinate. Thus, for example, at the free edge of the film, i.e., at $y=1$, $u = N/2$; the velocity there is $N/2$ times the average velocity.

Let two flows have the same N value. These flows need not be identical, yet they have the same velocity distribution. We express this result by stating that a condition for the *similarity* of the velocity distributions in two such flows is that they have the same *similarity parameter N* . This similarity permits one to experiment with a *model* and apply the results to *real life* flows.

Equations (8.15) and (8.16) have one similarity parameter. We note that when an equation is written in its dimensionless form, the similarity parameters can be identified by simple inspection, i.e., by listing the various dimensionless groups that have appeared.

When looking for *similitude,* i.e., for similarity conditions, it is necessary to first have the equations set in their dimensionless form. As already seen, one must start with the definition of the characteristic quantities. At this point it seems helpful to consider an additional example.

Consider Eq. (8.17), which is recognized as the same problem with the flow taking place on the inside surface of an elliptical conduit, Fig. 8.1b:

$$\frac{\partial^2 u}{\partial y^2} + \frac{\partial^2 u}{\partial z^2} = -N = -\left(\frac{\text{Re}}{\text{Fr}^2}\right). \tag{8.17}$$

The boundary conditions are of the form $u = 0$ on the solid wall, and $\partial u/\partial n = 0$ at the free surface. Although Eq. (8.17) is also one-parametric, conditions for similarity here must be formulated more carefully. Here *similarity of geometry* is also a necessary condition, which has its implication when the characteristic quantities are chosen. In this example the cross section of the conduit is elliptical. For two cases to be similar, they must both have elliptical cross sections, with similar ellipses, i.e., they must have *geometrical similarity,* with all corresponding lengths having the same ratio. They must have *similar boundary conditions,* i.e., the same boundary conditions applied on corresponding surfaces on both boundaries. And also they must have the *same similarity parameters.*

At this point it is clear that the characteristic length may be chosen as the small axis of the ellipse, or as the large one, *provided it is chosen in the same way for all cases.*

This conclusion is now generalized: The similarity of two problems requires similarity of geometry, similarity of boundary conditions, and corresponding equality of all similarity parameters. The characteristic physical quantities used to construct the similarity parameters can be *chosen arbitrarily,* provided they are *well defined* and correspond to the *same geometrical locations* in both cases.

Another way to state this is that two problems, A and B, are similar, if their dimensionless formulations contain no clue as to whether it is problem A or problem B that has led to the formulation.

Example 8.1

Steamships use sea water to cool their condensers. The intake of the sea water may be designed as shown in Fig. 8.2. Sea water streams into the intake nozzle, accelerates there into the pump, and proceeds into the condenser. The designer idea is that at the cruising speed of the ship the pump would run idly, with the required pumping power supplied by the main motion of the ship, i.e., the relative kinetic energy of the incoming water. On the other hand, at other speeds the pump must do some pumping. A model is tested in a water tunnel. Find conditions for similarity.

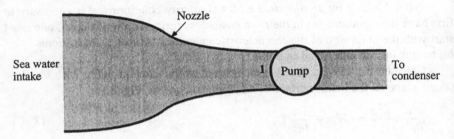

Figure 8.2 Cooling water intake.

Solution

Obviously, the model must be geometrically similar to the ship's system. Using the subscript s for the ship's system and M for the model, similarity requires that

$$\text{Re}_s = \text{Re}_M, \qquad \text{Eu}_s = \text{Eu}_M.$$

Therefore: $V_s L_s / v_s = V_M L_M / v_M$, and because both run in water, $V_s L_s = V_M L_M$, or

$$V_M = V_S \frac{L_S}{L_M}.$$

The characteristic pressure π is chosen as the pressure difference between a point just before the pump, point 1 in the figure, and the hydrostatic pressure at the equivalent depth outside the ship. Thus $\pi = p_1 - p_o$, and

$$\frac{\pi_S}{\rho V_S^2} = \frac{\pi_M}{\rho V_M^2},$$

or

$$\pi_M = \pi_S \frac{V_M^2}{V_S^2} = \pi_S \left(\frac{L_S}{L_M}\right)^2.$$

During the test the pump of the model must be run such that this condition on the pressure is satisfied.

Hidden Characteristic Quantities

There are cases where several similarity parameters do appear in the dimensionless equations but some of the characteristic physical quantities do not occur in the formulation of the problem. These missing characteristic quantities must therefore be defined using other quantities. Under such circumstances the actual number of parameters that must be kept similar may be smaller than the apparent one. An example of such a case is the incompressible flow in a cylindrical conduit.

Figure 8.3 Cylindrical conduit.

Consider a cylindrical conduit of arbitrary cross section, with the cylinder generators parallel to the z-axis, Fig. 8.3. Let the cross-sectional area of the cylinder be A_c and its circumference be C_c. A characteristic length may be defined as* $D = 4A_c/C_c$, and a characteristic velocity may be chosen as the average velocity in the cylinder,

$$V = \frac{1}{A_c}\int_{A_c} w\, dA.$$

The dimensionless z-component of the time-independent momentum equation (8.14) is

$$u\frac{\partial w}{\partial x} + v\frac{\partial w}{\partial y} + w\frac{\partial w}{\partial z} = -\text{Eu}\frac{\partial P}{\partial z} + \frac{1}{\text{Re}}\left(\frac{\partial^2 w}{\partial x^2} + \frac{\partial^2 w}{\partial y^2} + \frac{\partial^2 w}{\partial z^2}\right) \qquad (8.18)$$

where P is the modified pressure. There are two additional equations, for the x- and y-components, which are not written. The important thing to notice here is

---

* This characteristic length is called the hydraulic diameter. The number 4 appears just to make it the regular diameter for a circular cylinder.

the appearance of the two similarity parameters:

$$\mathrm{Re} = \frac{DV}{\nu}, \qquad \mathrm{Eu} = \frac{\pi}{\rho V^2}.$$

It seems that for two cylinders of similar cross section the flows would come out to be similar when

$$\mathrm{Re}_1 = \mathrm{Re}_2, \qquad \mathrm{Eu}_1 = \mathrm{Eu}_2.$$

We notice, however, that while the characteristic length and the characteristic velocity emerged rather naturally, the statement of the problem does not suggest a particular characteristic pressure. We know that there is going to be a pressure gradient along the pipe, but its magnitude is not known a priori. We therefore define the characteristic pressure π in terms of the characteristic velocity $\pi = \frac{1}{2}\rho V^2$. (The $\frac{1}{2}$ is put there just for consistency with the Bernoulli equation. Everything works out without the $\frac{1}{2}$ too). Now for similarity,

$$\mathrm{Re}_1 = \mathrm{Re}_2, \qquad \frac{D_1 V_1}{\nu_1} = \frac{D_2 V_2}{\nu_2}. \tag{8.19}$$

The Euler numbers, however, come out now to be

$$\mathrm{Eu}_1 = \frac{\pi_1}{\rho V_1^2} = \frac{1}{2}, \qquad \mathrm{Eu}_2 = \frac{\pi_2}{\rho V_2^2} = \frac{1}{2} \tag{8.20}$$

and they are automatically the same, i.e., we do not have to do anything physical to make them the same.

Because the dimensionless continuity equation (8.11) contains no additional similarity parameters and the x and y momentum equations contain the same parameters as does the z-component equation, we have now complete similarity between the two flows, provided Eq. (8.19) holds.

The knowledge of the details of the flow in one cylinder gives us the details of the flow in the second one. Solving for or measuring the pressure drop along one cylinder, say, C_o times the characteristic pressure (the only unit to measure dimensionless pressures), supplies at once the same information for the second cylinder. Far from the zone of entrance to the cylinder we expect this pressure drop to become constant per unit length (i.e., per characteristic length measured along the cylinder axis*), and indeed that same constant, say C_f, must apply to both flows. Thus, for a section of length $L^* = DL$ we expect the dimensionless pressure drop to be $C_f L$ and the dimensional one to be

---

* Nothing important changes if the characteristic length is taken as the largest diameter of the cross-section and the characteristic velocity as the largest velocity. One finds a variety of values to choose from, all of which are well defined.

$$\Delta P = C_f L \pi = C_f \frac{L^*}{D} \frac{\rho V^2}{2},$$

(8.21)

for all similar cylinders. Since Eu = 1/2 holds for all these cylinders C_f depends on Re only, i.e., $C_f = f(\text{Re})$, and can be obtained, for instance, by a set of experiments using one particular cylinder. The results apply to all similar cylinders provided C_f is taken for the same Reynolds numbers as in the tested cylinder.

Example 8.2

A duct has a rectangular cross section, with the sides 1 m by 0.5 m. It is to be used to transfer water at the rate of $Q = 2$ m³/s. It is proposed to construct a smaller model first and to find experimentally what pump is required for the full-scale duct. The pump available for the experiment with the model in the laboratory delivers $Q_M = 0.4$ m³/s water. Find the dimensions of the model. Experiments with the model just found yield a pressure gradient of $\Delta P_M = 30,000$ Pa/m. Find the pressure gradient in the full-scale duct. Note that D and V in Eq. (8.21) are dimensional.

Solution

Let the characteristic length for the duct be chosen as its hydraulic diameter,

$$D = \frac{4A}{C} = \frac{4 \times (1 \times 0.5)}{2 \times 1 + 2 \times 0.5} = 0.667\,\text{m}.$$

The characteristic velocity is chosen as the mean velocity,

$$V = \frac{Q}{A} = \frac{2}{1 \times 0.5} = 4 \text{ m/s}.$$

The Reynolds number for the flow in the duct is then

$$\text{Re} = \frac{V \cdot D}{\nu} = \frac{2.667}{\nu}.$$

Now for the model

$$D_M = \frac{4A_M}{C_M}, \qquad V_M = \frac{Q_M}{A_M},$$

$$\text{Re}_M = \frac{V_M \cdot D_M}{\nu} = \frac{4Q_M / C_M}{\nu},$$

and for similitude,

$$\text{Re}_M = \text{Re}, \qquad \frac{4Q_M}{C_M} = \frac{4 \times 0.4}{C_M} = 2.667,$$

or

$C_M = 0.6$ m.

Similitude also requires similarity of geometry. Let the larger side of the model be b; then the smaller side is $b/2$, and $C_M = 2(b + b/2) = 3b = 0.6$. The model is, therefore, rectangular, with the sides 0.2 m and 0.1 m. Also

$$V_M = \frac{Q_M}{A_M} = \frac{0.4}{0.2 \times 0.1} = 20 \text{ m / s,}$$

and

$$D_M = \frac{4A_M}{C_M} = \frac{4 \times (0.2 \times 0.1)}{0.6} = 0.133 \text{ m.}$$

To evaluate the pressure drop, we conveniently use Eq. (8.21). For the model, with $L^* = 1$ m, we find

$$\Delta P_M = 30,000 = C_f \times \frac{1}{0.133} \times \frac{1,000}{2} \times 20^2;$$

hence $C_f = 0.02$.

For the full-scale duct C_f is the same. With $L^* = 1$ m we find

$$\Delta P = 0.02 \times \frac{1}{0.667} \times \frac{1,000}{2} \times 4^2 = 240 \text{ Pa .}$$

The pressure gradient in the main duct is 240 Pa/m.

Example 8.3

An important step in the preparation of frozen chickens is just after cleaning, when the chicken is moved through a blast tunnel where very cold air is blown on it to chill it very fast. Larger chickens stay in the tunnel longer than smaller ones.

Assuming all chickens similar, find similarity rules and deduce how long should a chicken of each size stay in the tunnel. The differential equation governing the cooling process is given as

$$\frac{\partial T^*}{\partial t^*} = \alpha \nabla^{*2} T^*,$$

where T^* is temperature, t^* is time and α [m²/s] is the thermal diffusivity.

Solution

Let the characteristic length, L, be some well-defined length of the chicken (e.g., its width), and the characteristic time, τ, be the time it has to stay in the tunnel. Then

$$t^* = t\tau, \qquad x^*, y^*, z^* = (x, y, z)L,$$
$$\nabla^{*2} = \nabla^2 / L^2,$$

and the equation becomes

$$\frac{\partial T}{\partial t} = \text{Fo}\,\nabla^2 T, \qquad \text{where} \quad \text{Fo} = \frac{\alpha\tau}{L^2} \text{ is the Fourier number.}$$

Thus conditions for similarity for two chickens are

$$\text{Fo}_1 = \text{Fo}_2, \qquad \frac{\alpha_1 \tau_1}{L_1^2} = \frac{\alpha_2 \tau_2}{L_2^2}.$$

Let the mass of a chicken be m and its density be ρ. Then, for geometrically similar chickens,

$$\frac{m_1}{m_2} = \frac{\rho_1}{\rho_2}\left(\frac{L_1}{L_2}\right)^3.$$

Assuming $\alpha_1 = \alpha_2$ and $\rho_1 = \rho_2$, the condition for the equality of the Fourier numbers becomes

$$\frac{\tau_1}{\tau_2} = \left(\frac{L_1}{L_2}\right)^2 = \left(\frac{m_1}{m_2}\right)^{2/3}.$$

A chicken which is twice the mass of another should stay $2^{2/3} = 1.587$ times the time of stay of the other.

Example 8.4

The force of air resistance to a train moving at the speeds of 3, 10 and 25 m/s is to be obtained by the use of a model. The linear dimension of the model is 1/5 of that of the train. The model is run in a wind tunnel, in air. Find the corresponding air speeds in the wind tunnel. Find the relations between the forces measured on the model and those which will act on the full-size train.

Solution

The governing equations are the continuity equation and the Navier–Stokes equations. The similarity dimensionless numbers are, therefore, the Reynolds number and the Euler number.

For the Reynolds number,

$$\text{Re}_T = \text{Re}_M, \qquad \frac{V_T L_T}{\nu_T} = \frac{V_M L_M}{\nu_M}.$$

Let the characteristic length be the height, H, and assume $\nu_T = \nu_M$. Then $L_T = H_T$, $L_M = H_M$ and

$$V_M = V_T \frac{H_T}{H_M} = 5V_T.$$

The corresponding speeds in the wind tunnel are

$$V_{M1} = 3 \times 5 = 15 \text{ m / s}; \quad V_{M2} = 10 \times 5 = 50 \text{ m / s}; \quad V_{M3} = 25 \times 5 = 185 \text{ m / s}.$$

For the Euler number

$$\text{Eu}_T = \text{Eu}_M, \qquad \frac{\pi_1}{\rho_1 V_1^2} = \frac{\pi_2}{\rho_2 V_2^2}.$$

We assume $\rho_1 = \rho_2$, and since the problem has no typical characteristic pressure, we choose

$$\pi = \tfrac{1}{2}\rho V^2, \qquad \text{i.e.,} \quad \pi_T = \tfrac{1}{2}\rho_T V_T^2, \qquad \pi_M = \tfrac{1}{2}\rho_M V_M^2.$$

With this the condition for the equality of the Euler numbers is satisfied automatically, because

$$\text{Eu}_T = \tfrac{1}{2}, \qquad \text{Eu}_M = \tfrac{1}{2}.$$

To find relations between the forces, we need the dimensionless forces, which are equal in similar problems. We define a characteristic force by

$$\varphi = \pi \cdot A = \tfrac{1}{2}\rho V^2 A.$$

Thus the dimensionless forces, F, relate to the dimensional ones, F^*,

$$F^* = F\varphi = F \times \tfrac{1}{2}\rho V^2 A,$$

i.e.,

$$F_T^* = F_T \times \tfrac{1}{2}\rho_T V_T^2 A_T, \qquad F_M^* = F_M \times \tfrac{1}{2}\rho_M V_M^2 A_M.$$

We assume $\rho_T = \rho_M$. Also $A_T / A_M = L_T^2 / L_M^2$. At similarity $F_T = F_M$. Equality of the Reynolds numbers gives

$$V_M / V_T = L_T / L_M.$$

Hence, substitution yields

$$F_T^* = F_M^*.$$

The force acting on the model is the same as that acting on the train. Note the high speeds of the air flowing past the model.

Order of Magnitude

The previous section deals with similitude and shows how solutions of particular cases can be extended and applied to whole families of similar flows. However, there are many varieties of flows that a fluid dynamicist is compelled to consider. For many of them no exact solutions are known, experiments are not practical and approximate methods must be used.

A way to affect an approximate solution is to identify in the equations some terms which are quite smaller than the other terms. One may try then to drop the smaller terms and to solve the truncated equation, obtaining, hopefully, an approximation. Two distinct questions arise in this procedure. The first one is the correct estimate of the order of magnitude of the various terms in the equations. The second question is whether dropping a small term in the equation results in a small error in the solution.

The first question is considered under the heading of the proper choice of the characteristic quantities; the second comes under perturbations but is still closely connected with the characteristic quantities chosen.

Here, again, the equations are written in their dimensionless form, but now we make an effort to choose the characteristic quantities such that *all terms containing dependent variables are of order 1*. The order of magnitude of each term in the dimensional equation is reproduced in the dimensionless equation by the dimensionless parameters associated with the term.

The dimensionless groups, i.e., the similarity parameters, become now the order of magnitude parameters. The following example further illustrates this point. Let Eq. (8.2) be rewritten, with the order of magnitude of each term indicated by a number written below it; for simplicity we assume $v = w = 0$ and consider flows in which body forces are negligible:

$$\frac{\partial u^*}{\partial t^*} + u^* \frac{\partial u^*}{\partial x^*} = -\frac{1}{\rho} \frac{\partial p^*}{\partial x^*} + v\left(\frac{\partial^2 u^*}{\partial x^{*2}} + \frac{\partial^2 u^*}{\partial y^{*2}} + \frac{\partial^2 u^*}{\partial z^{*2}}\right). \tag{8.22}$$
$\quad$ O(1) $\quad$ O(100) $\qquad$ O(1,000) $\qquad\qquad$ O(1,000)

In its dimensionless form, as in Eq. (8.14), Eq. (8.22) becomes

$$\Omega \frac{\partial u}{\partial t} + u \frac{\partial u}{\partial x} = -\text{Eu}\frac{\partial p}{\partial x} + \frac{1}{\text{Re}}\left(\frac{\partial^2 u}{\partial x^2} + \frac{\partial^2 u}{\partial y^2} + \frac{\partial^2 u}{\partial z^2}\right). \tag{8.23}$$

Assuming success in our effort to enforce order 1 on all terms containing dependent variables, we have

$$\frac{\partial u}{\partial t} = O(1), \qquad\qquad u\frac{\partial u}{\partial x} = O(1),$$

$$\frac{\partial p}{\partial x} = O(1), \qquad \left(\frac{\partial^2 u}{\partial x^2} + \frac{\partial^2 u}{\partial y^2} + \frac{\partial^2 u}{\partial z^2}\right) = O(1). \tag{8.24}$$

and since the $u\,\partial u/\partial x$ term has no coefficient, the whole equation must have been rescaled with respect to it, i.e., divided by 100. Hence, now,

$$\Omega = O(1/100) = O(0.01),$$
$$\mathrm{Eu} = O(1,000/100) = O(10), \tag{8.25}$$
$$1/\mathrm{Re} = O(1,000/100) = O(10).$$

Equation (8.22) describes a certain physical phenomenon i.e., it is a *physical truth*. This *physical truth* cannot be changed by nondimensional formulation, and therefore the relative magnitude of the four terms in Eq. (8.23) must be the same as in Eq. (8.22). In trying to make all dependent variable groups of order 1, the relative correct order of magnitude is transferred to the dimensionless parameters, Eqs. (8.25). At this point it seems that an approximate solution to either Eq. (8.22) or Eq. (8.23) may be obtained neglecting the $\partial u/\partial t$ term on the left-hand side.

Estimates of the Characteristic Quantities

Until now we have taken the order of magnitude of the various terms in the equation as known. We must remember, however, that the object of the whole procedure is to simplify the equations, and therefore we must have a method to estimate these orders of magnitude a priori.

The major difficulty here is the selection of quantities that are really characteristic of the problem and thus have the correct order of magnitude. This is by no means an automatic procedure, and clearly, the criterion of "well defined but otherwise arbitrary," which applies in our similarity considerations, completely fails here. The proper selection depends on the understanding and the information the selector has on the physics of the problem, and there are examples of disagreement on this selection. There are, however, a few helpful indicators for a reasonable selection, which can be presented rather generally.

i.   The investigated phenomena must be sensitive to variations in the selected characteristic quantity. Consider, for instance, the flow field at point A which is at a distance L from a submerged body which is an ellipsoid of axes $2a$, $2b$, $2c$, Fig. 8.4. Under certain circumstances, e.g., for large L, the flow at the point of consideration is quite insensitive to variations in a, b or c. Under these conditions a, b or c are not characteristic lengths and should not be chosen as such. A possible reasonable choice for this case is the distance L between the ellipsoid and the considered point; we assume here that the flow at least "feels" the presence of the body, through the $\mathbf{q} = 0$ boundary condition on the rigid body surface.

ii.   When scaling a derivative, say $\partial u/\partial y$, a fair scaling is $\Delta u/\Delta y$ in the region of interest. Although $\partial u/\partial y$ may vary in this region, there exists at least one point where $\partial u/\partial y = \Delta u/\Delta y$, and since only one scaling is permitted, it might as well be this one. The characteristic velocity in such a case is $u_2 - u_1$ and the characteristic length $y_2 - y_1$, 2 and 1 being subscripts for the limits of the region.

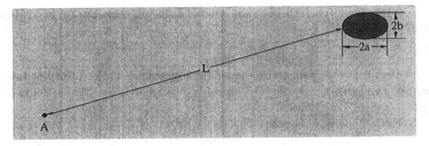

Figure 8.4 Submerged ellipsoid.

iii.   A second derivative is scaled as a first derivative of a derivative, and so on to higher derivatives.

iv.   When no characteristic quantity is apparent, such a quantity may be constructed using other characteristic quantities and physical coefficients, provided the constructed combination has the right dimensions and satisfies the sensitivity test.

An example of such a situation has already appeared in incompressible pipe flow, where the characteristic pressure had to be constructed in terms of ρV^2.

Example 8.5

The flow between a stationary inner cylinder and an external rotating one is given by the differential equation (6.71):

$$\frac{d^2 q_\theta^*}{dr^{*2}} + \frac{1}{r^*}\frac{dq_\theta^*}{dr^*} - \frac{1}{r^{*2}}q_\theta^* = 0 \qquad\qquad (6.71)$$

with the boundary conditions,

$$q_\theta^* = 0 \quad \text{at} \quad r^* = R_i, \quad \text{the inner cylinder;}$$

$$q_\theta^* = \omega R_o \quad \text{at} \quad r^* = R_o, \quad \text{the outer cylinder.}$$

Using order of magnitude considerations, find an approximation to this flow for small dimensionless gaps, ε, between the cylinders, where

$$\varepsilon = \frac{R_o - R_i}{R_i}.$$

Solution

With rules (ii) and (iii) in mind we denote the gap between the two cylinders by δ,

$$\delta = R_o - R_i,$$

and select it as the characteristic length; the difference of the velocities between the two cylinders,

$$V = q_\theta^* \Big|_{R_o} - q_\theta^* \Big|_{R_i} = \omega R_o,$$

is chosen as the characteristic velocity. Now rules (ii) and (iii) yield

$$\frac{dq_\theta^*}{dr^*} = \frac{\omega R_o}{\delta} \frac{dq_\theta}{dr}, \quad \frac{d^2 q_\theta^*}{dr^{*2}} = \frac{\omega R_o}{\delta^2} \frac{d^2 q_\theta}{dr^2},$$

where we keep our convention that starred *variables* are dimensional and nonstarred ones are not. The scaling of r^* itself is done by R_i, because $r^* \approx R_i$. Hence,

$$r^* = R_i r = \frac{(R_o - R_i)}{\varepsilon} r = \left(\frac{\delta}{\varepsilon}\right) r.$$

Substitution of all the dimensional scaled quantities into the differential equation (6.71) yields

$$\left(\frac{\omega R_o}{\delta^2}\right) \frac{d^2 q_\theta}{dr^2} + \left(\frac{\varepsilon}{\delta}\right) \frac{1}{r} \left(\frac{\omega R_o}{\delta}\right) \frac{dq_\theta}{dr} - \left(\frac{\varepsilon}{\delta}\right)^2 \frac{1}{r^2} (\omega R_o) q_\theta = 0,$$

or through division by $\omega R_o / \delta^2$,

$$\frac{d^2 q_\theta}{dr^2} + \varepsilon \frac{1}{r} \frac{dq_\theta}{dr} - \varepsilon^2 \frac{1}{r^2} q_\theta = 0.$$

Now for $\varepsilon \ll 1$ the approximation becomes

$$\frac{d^2 q_\theta}{dr^2} = 0,$$

with the *dimensional* solution

$$q_\theta = \omega R_o \left(\frac{r - R_i}{\delta} \right);$$

and when ε is not very small compared to 1, but still $\varepsilon^2 \ll 1$, the approximation becomes

$$\frac{d^2 q_\theta}{dr^2} + \frac{\varepsilon}{r} \frac{dq_\theta}{dr} = 0,$$

with the *dimensional* solution

$$q_\theta = \omega R_o \frac{\ln(r / R_i)}{\ln(R_o / R_i)} \approx \omega R_o \left(\frac{r - R_i}{\delta} \right) \left(\frac{3R_i - r}{3R_i - R_o} \right).$$

Both approximations may now be compared with the exact solution

$$q_\theta = \omega R_o \frac{r - R_i}{\delta} \left(\frac{R_o}{r} \right) \left(\frac{r + R_i}{R_o + R_i} \right),$$

obtained from Eq. (6.73) for $\Omega_i = 0$, $\Omega_o = \omega$.

*The Concept of Perturbations

Let a differential equation have the form

$$F(y, x_i, \alpha) = 0, \tag{8.26}$$

where α is some parameter, x_i are independent variables and y is the dependent variable. Equation (8.26) may also mean that the partial derivatives $\partial y / \partial x_i$ and higher ones do appear. The equation is assumed to be accompanied by sufficient boundary conditions, and the uniqueness and existence of the solution are also implied.

Let the solution of this equation for a given α, say $\alpha = 0$, be known at a particular point, i.e., at a particular set of x_i values. The dependent variable y at this point is, of course, also known. The whole problem may be reformulated and solved anew for different values of the parameter α, and y may now be considered a function of α at that particular point.

Let this $y(\alpha)$ have a series expansion in α, around the given value $\alpha = 0$. Let this expansion hold for a whole region of points, i.e., for a continuous region of x_i values. Thus

$$y(x_i, \alpha) = y_o(x_i, 0) + \alpha y_1(x_i, 0) + \alpha^2 y_2(x_i, 0) + \cdots \qquad (8.27)$$

and $y_o(x_i, 0)$ must satisfy Eq. (8.26) with $\alpha = 0$:

$$F(y_o, x_i, 0) = 0. \qquad (8.28)$$

Equation (8.27) is a series solution of Eq. (8.26) around the point $\alpha = 0$. The solution at that point, Eq. (8.28), is obtained by a "perturbation" in the original equation (8.26), i.e., by setting $\alpha = 0$. A solution in the form of Eq. (8.27) is therefore called a perturbation solution.

In some difficult cases the series is terminated after $y_o(x_i, 0)$, simply because the evaluation of additional terms is not practical. Then $y_o(x_i, 0)$ is used as an approximation to $y(x_i, \alpha)$ provided α is small. This approximation is said to have been obtained by *perturbation* in the original equation (8.26). The equation is considered "perturbed" by the substitution of zeros for small order of magnitude terms, and the smaller this perturbation is, the better the approximation.

*Singular Perturbations

The solution of a differential equation satisfies, by definition, both the equation and the boundary conditions. The number of the boundary conditions to be satisfied depends in general on the order of the equation.

The process of perturbation as described in the previous section has the effect of deleting some terms from the equation. It may happen that the deleted term is the highest derivative and the equation left after the perturbation is of a lower order. The solution of this new equation cannot satisfy all the boundary conditions. Therefore, near a boundary it is simply wrong and cannot serve even as an approximation. Far from a boundary it may still be an approximation. However, near the boundary the perturbation process must be modified and performed very carefully, so as to result in an equation of the *same differential order* as the original one. This new process is called *Singular Perturbation*.

Singular perturbation, i.e., perturbation in the highest derivative of the equation, is even more complicated than the regular perturbation considered before. It will therefore be used here only in the particular cases necessary to obtain the boundary layer equations. Still, the concept is quite general and is used in the solution of differential equations.

The Boundary Layer Equations

Consider the dimensionless Navier–Stokes equations written out for two-dimensional time-independent flows:

$$u\frac{\partial u}{\partial x} + v\frac{\partial u}{\partial y} = -\mathrm{Eu}\frac{\partial p}{\partial x} + \frac{1}{\mathrm{Re}}\left(\frac{\partial^2 u}{\partial x^2} + \frac{\partial^2 u}{\partial y^2}\right),$$ (8.29)

$$u\frac{\partial v}{\partial x} + v\frac{\partial v}{\partial y} = -\mathrm{Eu}\frac{\partial p}{\partial y} + \frac{1}{\mathrm{Re}}\left(\frac{\partial^2 v}{\partial x^2} + \frac{\partial^2 v}{\partial y^2}\right)$$ (8.30)

together with the continuity equation

$$\frac{\partial u}{\partial x} + \frac{\partial v}{\partial y} = 0.$$ (8.31)

In many practical cases these equations cannot be solved analytically, and approximate formulations become very useful.

Inspection of Eqs. (8.29) - (8.31) reveals that extreme values of the Reynolds number may indeed lead to such approximations:

Very small Re enhance the relative importance of the second derivative terms, making them dominant. Then the nonlinear terms may be neglected. This situation is considered in Chapter 9.

Very large Re tend to make the second derivative terms negligible. This changes the Navier–Stokes equations into the Euler equations, already mentioned as Eq. (5.70). This direction is followed in Chapter 10. However, the Euler equation is of the first order and cannot satisfy the two boundary conditions

$$q_n = 0, \qquad q_t = 0$$ (5.62)

on rigid boundaries. The Euler equations, therefore, cannot be the correct approximation near a rigid boundary. Thus close to a boundary the perturbation associated with Re $\rightarrow \infty$ is a singular perturbation. The resulting approximation, called the boundary layer equations, are further treated in Chapter 11. Their derivation, however, being a singular perturbation, follows here.

We consider again Eqs. (8.29) - (8.31). We look for the perturbed form of the equations for Re $\rightarrow \infty$, and our region of interest is in the vicinity of rigid boundaries. It is impossible to use here a regular perturbation where Re $\rightarrow \infty$ is simply substituted in the equations, which leads to the Euler equations. The two boundary conditions do not permit this. A singular perturbation must therefore be performed.

We assume the rigid surface to be generally flat, e.g., a flat plate in parallel flow. Let a coordinate system be chosen as in Fig. 8.5, with y normal to the rigid surface. The x-coordinate points along the surface and coincides with the direction of the velocity, U, at some distance far from the rigid surface.

It seems reasonable to choose U as the characteristic velocity and because ν, the kinematic viscosity of the fluid, is known, the Reynolds number is almost determined; we still need only to select the characteristic length.

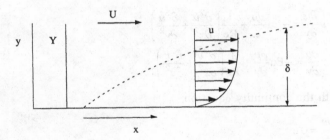

Figure 8.5 Boundary layer flow.

The problem we are faced with is what relations must exist between 1/Re and $\left(\partial^2 u/\partial x^2 + \partial^2 u/\partial y^2\right)$ such that for Re $\rightarrow \infty$ and at very small y this second derivative term is not completely lost.

No help comes from the $\partial^2 u/\partial x^2 = 0$ term. On the rigid surface $u = 0$ because of the no-slip boundary condition. For flat surfaces $\partial u/\partial x = \partial^2 u/\partial x^2 = 0$ because u is constant, i.e., $u = 0$, for all x. Thus this term vanishes even without Re $\rightarrow \infty$. We must therefore direct our efforts at

$$\frac{1}{\text{Re}} \frac{\partial^2 u}{\partial y^2} = 0.$$

We note that for a fixed $\partial^2 u/\partial y^2$,

$$\lim_{\text{Re} \rightarrow \infty} \left[\frac{1}{\text{Re}} \left(\frac{\partial^2 u}{\partial y^2} \right) \right] = 0 .$$

Hence, to keep $1/\text{Re}\left(\partial^2 u/\partial y^2\right)$ finite as Re $\rightarrow \infty$, $\partial^2 u/\partial y^2$ cannot be held constant but must also become infinitely large. In other words, when we consider a series of flow fields, quite similar to one another except that each flow has a Reynolds number larger than the flow which has just preceded it, the flow must also have larger $\partial^2 u/\partial y^2$ values; as Re $\rightarrow \infty$, say because $U \rightarrow \infty$, then $\partial^2 u/\partial y^2$ must also diverge.

To find the rate at which this second derivative must diverge, we refer to the sections on order of magnitude and estimates of the characteristic quantities and find, under rules (ii) and (iii),

$$\frac{\partial^2 u}{\partial y^2} = O\left(\frac{U}{\delta^2} \right) ,$$

where δ is the thickness of the layer near the boundary over which u changes from U_o to zero. We expect now that with increasing Re this δ-thick layer becomes thinner and thinner, so as to make

$$\lim_{\text{Re}\to\infty}\left[\frac{1}{\text{Re}}\left(\frac{\partial^2 u}{\partial y^2}\right)\right] = \lim_{\text{Re}\to\infty}\left[\frac{1}{\text{Re}}\frac{U}{\delta^2}\right]$$

finite, i.e., of order 1. Inspection of this relation reveals that $\text{Re}\cdot\delta^2$ or $\delta\cdot\text{Re}^{1/2}$ must remain finite. Because δ does not appear explicitly in the equations we proceed as follows: Let a new y-coordinate, say Y, be defined:

$$Y = y\cdot\text{Re}^{\frac{1}{2}}, \qquad y = Y\cdot\text{Re}^{-\frac{1}{2}}. \tag{8.32}$$

This makes

$$\frac{1}{\text{Re}}\frac{\partial^2 u}{\partial y^2} = \frac{\partial^2 u}{\partial Y^2},$$

which no longer depends on the Reynolds number, and thus does not vanish. The layer δ over which u decreases from U_o to 0 is very thin, i.e., of order $\text{Re}^{-1/2}$. The whole change in the velocity takes place in this narrow layer near the boundary, called the *boundary layer*.

To find the order of magnitude of v inside the boundary layer, we substitute $y = Y\text{Re}^{1/2}$ in the continuity equation (8.31) to obtain

$$\frac{\partial u}{\partial x} + \text{Re}^{1/2}\frac{\partial v}{\partial Y} = 0,$$

and to keep this equation meaningful, i.e., to have both terms of the same order, we must define

$$V = v\text{Re}^{1/2}, \qquad v = V\text{Re}^{-1/2} \tag{8.33}$$

to make it

$$\frac{\partial u}{\partial x} + \frac{\partial V}{\partial Y} = 0 \tag{8.34}$$

The y-component of the velocity, v, is of order $\text{Re}^{-1/2}$.

We now proceed to substitute Y and V in Eq. (8.30), which becomes

$$\text{Re}^{-1}\left(u\frac{\partial V}{\partial x} + V\frac{\partial V}{\partial Y}\right) = -\text{Eu}\frac{\partial p}{\partial Y} + \text{Re}^{-1}\left(\text{Re}^{-1}\frac{\partial^2 V}{\partial x^2} + \frac{\partial^2 V}{\partial Y^2}\right). \tag{8.35}$$

For large Re, Eq. (8.35) yields the order of magnitude of $\partial p/\partial Y$ as

$$\text{Eu}\frac{\partial p}{\partial Y} = O\!\left(\text{Re}^{-1}\right), \tag{8.36}$$

and for $\text{Re}\to\infty$ this term vanishes. Hence, the pressure does not depend on y and therefore does not vary across the boundary layer. Equation (8.30) reduces to

$$\frac{\partial p}{\partial y} = 0. \tag{8.37}$$

The same substitution in Eq. (8.29) yields

$$u\frac{\partial u}{\partial x} + V\frac{\partial u}{\partial Y} = -\text{Eu}\frac{\partial p}{\partial x} + \text{Re}^{-1}\left(\frac{\partial^2 u}{\partial x^2} + \text{Re}\frac{\partial^2 u}{\partial Y^2}\right). \tag{8.38}$$

For $\text{Re} \to \infty$, $\text{Re}^{-1}(\partial^2 u/\partial x^2)$ vanishes, and because $\partial p/\partial y$ also vanishes, $\partial p/\partial x = dp/dx$. Equation (8.38) thus becomes

$$u\frac{\partial u}{\partial x} + v\frac{\partial u}{\partial y} = -\text{Eu}\frac{dp}{dx} + \frac{1}{\text{Re}}\frac{\partial^2 u}{\partial y^2} \tag{8.39}$$

which together with Eqs. (8.29) and (8.37) constitute the *boundary layer equations*. In dimensional form the boundary layer equations are

$$u\frac{\partial u}{\partial x} + v\frac{\partial u}{\partial y} = -\frac{1}{\rho}\frac{dp}{dx} + v\frac{\partial^2 u}{\partial y^2}, \tag{8.40}$$

$$\frac{\partial p}{\partial y} = 0, \tag{8.41}$$

$$\frac{\partial u}{\partial x} + \frac{\partial v}{\partial y} = 0 \tag{8.42}$$

together with the boundary conditions

$$u = v = 0 \quad \text{at} \quad y = 0, \tag{8.43}$$

$$u = U \quad \text{at} \quad y\text{Re}^{1/2} \to \infty, \tag{8.44}$$

where $y\text{Re}^{1/2} \to \infty$ means the outer edge of the boundary layer.

Buckingham's π Theorem

In a previous section similitude has been motivated by the need to apply results obtained for a model to a similar real-life situation. Similitude has more applications, and as an example consider the determination of the drag force acting on a sphere moving in a liquid. At this stage we do not know how to solve this case, so we may turn to experiments. We need data for several combinations of sphere diameters, fluid viscosities and sphere velocities, and therefore a certain range of the Reynolds number is chosen and k experiments are performed, say $k = 10$. Had we not known that the Reynolds number was a similarity parameter for this case, we would have had to choose a range of velocities, a range of viscosities and a range of diameters; and to come out with the same coverage we would have had to perform k^3 experiments, 1,000 of them. The difference between 10 and 1,000 is very significant, and similarity parameters have now gained the additional role of information concentrators; in this example the concentration is a hundred-fold. An engineer therefore views similarity parameters as quite desirable. The engineer knows how to get them from the governing equations, as already shown. But is there a way to hunt for them when even the governing equations are not completely available? The Buckingham's π theorem indicates such a procedure.

Imagine there exist two sheets of paper. On one an experienced engineer has written down the correct set of governing equations, in their dimensionless form. The engineer therefore sees right away the various similarity parameters, which are, of course, dimensionless numbers. Unfortunately we cannot read the paper. On the second sheet *we* have listed down *all* the physical quantities,

$$Q_j, \quad j = 1, 2, \ldots m,$$

that might affect the phenomena. Now we start a procedure: each physical quantity Q_j is raised to some power n_j, and a product is formed:

$$\pi = Q_1^{n_1} \cdot Q_2^{n_2} \cdots Q_j^{n_j} \cdots = \prod_{j=1}^{m} Q_j^{n_j} \tag{8.45}$$

If all n_j are adjusted such that π comes out to be dimensionless, we consider the procedure successful. There may be many combinations of n_j values, including some $n_i = 0$ values, that satisfy the rules of this game. But note: *all* the similarity parameters on the experienced engineer's paper are reproduced by the procedure. This is so because *all* relevant physical quantities are presumed listed. This game may also produce some dimensionless combinations which are not on the engineer's list. These extraneous parameters are bothersome but not dangerous because the experiments would prove them irrelevant. What should worry one is the chance of missing some important similarity parameter. To avoid such a mishap, it is important to know the largest number of dimensionless parameters that the procedure can produce.

Suppose some of the physical quantities Q_j, which are dimensional, include the dimension of time, in the form t^{a_i}. Then for Eq. (8.45) to be dimensionless, i.e., for t to cancel out of the equation, we require that

$$a_1 n_1 + a_2 n_2 + \cdots = \sum_{j=1}^{m} a_j n_j = 0. \tag{8.46}$$

Suppose some of the Q_j contain the dimension of length, in the form L^{b_i}; then again

$$b_1 n_1 + b_2 n_2 + \cdots = \sum_{j=1}^{m} b_j n_j = 0. \tag{8.47}$$

Thus each additional dimension contained in some of the Q_j produces an additional linear equation similar to (8.46) or (8.47).

Let the total number of physical quantities, and therefore of n_j, be m, and the total number of independent dimensions be r; then there are r linear equation for the m variables n_j. Now a set of r linear equations for m unknowns has in general $m - r + 1$ sets of linearly independent solutions. Inspection of Eqs. (8.46) and (8.47) reveals that all the equations are homogeneous, and therefore

$$n_1 = n_2 = \cdots = n_j = \cdots = n_m = 0$$

is also a solution. This trivial solution yields no similarity parameter and therefore may be ignored. The number of the remaining solutions is therefore $m - r$, which is the π theorem. Thus, when we find $m - r$ linearly independent solutions, and therefore have a list of $m - r$ dimensionless groups, we may be assured that all the similarity parameters from the wise engineer's paper are included in our list.

It is noted that products, quotients and powers of dimensionless groups are not counted as new groups, and this is a direct result of the requirement that the solutions for the n_j combinations be linearly independent.

Thus Buckingham's π theorem is indeed a powerful tool which facilitates the establishment of similitude even when the governing equations are not known. Like most powerful tools it must be handled with care. The main pitfall in its application is the omission, due to ignorance, of some physical quantities which affect the phenomenon. It is noted that in a situation where even the governing equations cannot be formulated, ignorance is not a harsh word but rather a call for caution. An experienced engineer does use the π theorem when necessary, but whenever the governing equations are available they would be preferred as the source of the similarity parameters.

Example 8.6

A sphere submerged in a liquid is released and sinks down, or floats up. Find the similarity parameters of the problem.

Solution

The following quantities are presumed to affect the phenomenon:

D[m]   the diameter of the sphere;
ρ_s [kg / m^3]   the density of the sphere material;
ρ_f[kg / m^3]   the density of the fluid;
V[m / s]   the instantaneous velocity of the sphere;
μ [kg / m·s]   the viscosity of the fluid;

$g\,[\,\mathrm{m/s^2}\,]$ the acceleration of gravity.

The number of physical quantities is $m = 6$, and the number of involved dimensions is $r = 3$. We may therefore expect three dimensionless parameters.

Rather than plow through the steps leading to the formal solution of the set of equations, let us try to guess these parameters. This approach permits one to use some physical intuition, and the Buckingham π theorem reveals when one may stop guessing because enough parameters have been found.

Let our guess be

$$\left(\frac{DV\rho_f}{\mu}\right)\;,\;\left(\frac{\left(\rho_f-\rho_s\right)D^2g}{\mu V}\right)\;,\;\left(\frac{\rho_f-\rho_s}{\rho_s}\right)$$

and the reader may check that indeed all three are dimensionless.

A clue to the selection of the list of physical quantities is also a clue to the guess of the similarity parameters:

As the sphere moves, the fluid undergoes acceleration and viscous effects, and the ratio between the relevant terms is the Reynolds number, i.e., the first parameter. The second parameter presents the ratio between buoyancy and viscous forces, and the third one gives the ratio between buoyancy and the sphere's own acceleration.

The sphere does accelerate, and when the time, $\tau\,[\,\mathrm{s}\,]$, during which acceleration is important is within the range of interest another dimensionless group emerges

$$\left(\frac{\rho g D\tau}{\mu}\right);$$

and when a glass container of diameter B is used in the experiments, the dimensionless group D/B must be included.

References

G. Birkhoff, "Hydrodynamics, a Study in Logic, Fact and Similitude," Dover, New York, 1955.

J.D. Cole, "Perturbation Methods in Applied Mathematics," Blaisdell, Waltham, MA, 1968.

W.J. Duncan, "Physical Similarity and Dimensional Analysis," E. Arnold & Co., London, 1953.

M. Holt, "Dimensional Analysis," in V.L. Streeter (ed.), "Handbook of Fluid Dynamics," McGraw-Hill, New York, 1961.

S.J. Kline, "Similitude and Approximation Theory," McGraw-Hill, New York, 1965.

H.L. Langhaar, "Dimensional Analysis and Theory of Models," Wiley, New York, 1951.

L.I. Sedov, "Similarity and Dimensional Methods in Mechanics," Academic Press, New York, 1959.

M.D. Van Dyke, "Perturbation Methods in Fluid Mechanics," Parabolic Press, Stanford, CA, 1975.

Problems

8.1 A model of a car is 1/4 of the length of the car itself. The car runs between 50 and 100 km/h. It is suggested to test air resistance to the motion of the car by running the model (a) in an air tunnel, (b) in a water tunnel. Find the running speed in each tunnel and the interpretation of the measured resistance forces.

8.2 Complete similarity between a ship model and a full-size ship requires the same Reynolds and Froude numbers. Suppose that

Re $= LV/\nu = 11 \times 10^8$ and Fr$^{-1} = gL/V^2 = 33.25$.

Assuming a model 1/100 the size of the ship, can you design an experiment having full similarity? If not, why? Could you design a full-similarity experiment for a submarine?

8.3 Find conditions for similarity for time-independent flows between parallel plates: shear flows, plane Poiseuille flows and combinations of the two. Find rules for the interpretation of the results.

8.4 Find conditions for similarity between Rayleigh flows. Find rules for the interpretation of the results.

8.5 Find conditions for similarity between annulus flows where the inside cylinder does not have the same center as the outer one. Find rules for the interpretation of the results.

8.6   Find conditions for similarity between Couette flows for two rotating cylinders. Find rules for the interpretation of the results. Repeat for non-concentric cylinders.

8.7   For the flow between parallel plates, using the differential equation (6.21) and the boundary conditions, Eq. (6.23), find conditions under which the shear flow part can be neglected and conditions for which the Poiseuille flow part can be neglected. Check your results using Eq. (6.24).

8.8   A plane shear flow has been demonstrated by the instructor setting the upper plate suddenly into motion at its full velocity. Find times for which the flow field is approximated by the Rayleigh flow and times for which the plane shear flow is a good approximation.

8.9   For the Couette flow between concentric rotating cylinders find conditions for which the rotation of the inner cylinder may be ignored. Do the same for the outer cylinder.

8.10   A cylindrical drum full of fluid is suddenly set into rotation at the angular velocity ω. Find times and regions of flow where the phenomenon is approximated by the Rayleigh flow. Find times for which rigid body rotation is a good approximation.

8.11   It has been suggested to construct a cylinder with a bullet head on one side and with a thin soft plastic fish-tail on the other. A top view of the cylinder is shown in Fig. P8.11. The cylinder is made of soft iron, and when placed in a periodic magnetic field, it wiggles its tail and moves forward. By making it hollow, it can have neutral buoyancy. The idea was to let it swim inside blood vessels, for medical purposes. Find the similarity parameters for this magnetic fish.

Figure P8.11 Cylinder with tail fin.

8.12   Assume you have not seen the solution of the Rayleigh problem: the suddenly accelerated flat plate. Find its similarity parameters using Buckingham's π theorem. Check the similarity parameters with those obtained from the differential equation, and note the clue they give you toward the exact analytical solution.

8.13 In Example 8.3 the physical quantities that affect the cooling rate of the chicken are

D [m]   its size,
ρ [kg/m³]   its density,
c [kJ/kg°C]   its specific heat,
k [w/m°C]   its thermal conductivity,
h [w/m² °C]   the heat convection coefficient,
                  (assumed infinite in Example 8.3 but not here),
τ [s]   the time it stays in the tunnel.

Find the similarity parameters of the problem using Buckingham's π theorem.

8.14 A submerged body is designed to move in oil at the speed of 2 m/s. A model 1/8 the size of the body is run in water and yields a drag force of F_d = 300 N.

For oil ρ_1 = 880 kg/m³, μ_1 = 0.082 Pa·s.
For water ρ_2 = 998 kg/m³, μ_2 = 0.082 Pa·s.

a. At what speed must the model be run?
b. What is the drag force on the submerged body?

8.15 In fighting forest fires, specially equipped airplanes release a body of water above the fire. As it falls down, the water is broken into smaller and smaller drops. Once these drops reach a certain size, they do not break any more. The details of this phenomenon are to be investigated experimentally. Using Buckingham's π theorem, find the similarity parameters of this phenomenon.

8.16 An airplane wing is tested in a wind tunnel, Fig. P8.16. The purpose of the test is to obtain drag forces and lift forces acting on the wing at various flight speeds. Drag forces are those acting on the wing in the direction of the velocity of the oncoming air,

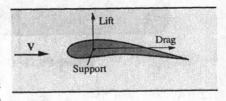

Figure P8.16

while lift forces act in the direction perpendicular to that velocity. These forces are measured on the model. Translate them into the forces which will act on the full-size wing.

8.17   One way to drill holes in sand, in order to set posts, is to push a water hose, with the water running, into the sand, Fig. P8.17. Find the similarity parameters of this method.

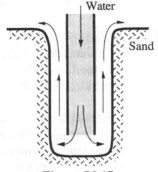

Figure P8.17

8.18   While a fuel tank in a ship is used, it must have a free surface. As the ship rolls, the fuel motion inside the tank is rather complex. The fuel applies forces to the ship and these forces are to be measured experimentally, using a model. Find the similarity parameters for these experimental forces and the way to translate them into the full-size ship forces.

8.19   A model of a ship's propeller is first tested in a water tunnel, and then it is assembled on a ship's model and tested again. Find the similarity parameters for the propeller alone and for the assembled propeller. Formulate the translation of the model experimental results into the real propeller performance.

8.20   A weather balloon has a diameter of 2 m and is to be used in air at 20°C. To find the drag force on the balloon, an experiment was conducted in which a 0.02-m-diameter sphere was held in water moving with the velocity of 10 m/s. The drag force on the sphere was measured as 6.5 N. Find the speed of the wind past the balloon which corresponds to the experiment. Find the drag force on the balloon at this wind velocity.

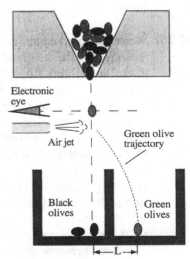

8.21   A sorting machine for olives is shown in Fig. P8.21. Olives drop down in front of an electronic eye. Black olives just proceed down. When a green olive passes the eye, air is released from a nozzle deflecting the green olive to

Figure P8.21 A machine for sorting olives.

another bin. It is necessary to construct a similar system for olives which are twice the linear size of those treated now. Find relations between the radii of the old and new air jets, between their velocities and between the distances the rejected green olives travel.

8.22 A glider model is 1/10 of the full-size glider. The model is connected to a car by a wire, and at a speed of 140 km/h the force in the wire is 1,000 N. Find the speed of the full-size glider relative to that of the model and the power expended by an airplane pulling the glider at that speed.

8.23 Electrical wires are subject to stresses caused by winds. The effect of the wind may look like a force distributed along the wire, pulling it in the down-wind direction, or it may take the form of oscillations. A 1/50 model is to be tested in a wind tunnel. Find the necessary similarity parameters.

8.24 A new hull design for a ship is to be 100 m long. A model is 3 m long. The model is tested in a tow tank, which can accommodate towing speeds of up to 2 m/s. Only the Froude number is conserved for similarity, because it is impractical to conserve the Reynolds number too. A correction is later made for the different Reynolds numbers. Find to what ship's speed corresponds the 2-m/s speed of the model and what are the ship and the model Reynolds numbers at these speeds.

8.25 The size of raindrops depends on their speed, on the trajectory they have traveled and on the surface tension of water. Find the similarity parameters for this size.

8.26 Some crude oil comes out of wells mixed with pebbles, gravel and sand. It is suggested to let the oil flow through settling tanks, in which most of these solids sink down. The experimental investigation of these tanks utilizes water instead of oil, because water is clear and the settling process may be viewed, and because crude oil entails continuous cleaning of instrumentation. Find similarity parameters. Assume that satisfactory settling has been achieved in water and give the major design parameters for the oil in terms of those for the water.

8.27 A turbo-pump which runs at 2000 rpm takes in water at atmospheric pressure at the rate of 0.010 m³/s and discharges it at 200,000 Pa above atmospheric pressure.
   a. If the pump rpm is increased to 3,000, find the flow rate and the discharge pressure at which the new operating conditions are completely similar to the first ones. Neglect viscous dissipation

b. Find the ratio of the powers necessary to run the pump in the two cases.

8.28 A cylindrical drum completely filled with water is set on a turntable. The turntable is suddenly set into rotation, at 2,000 rpm. After 30 seconds the water inside the drum turns with the drum in a solid body rotation.

a. Find how long it will take the water to reach solid body rotation if the drum is suddenly set to 3,000 rpm.

b. Find how long it will take the water in a drum twice the linear size of the old one, set to 2,000 rpm, and set to 3,000 rpm, to rotate like a solid body.

8.29 The water in Problem 8.28 is changed to glycerin. Find the times for the old and the new drums, with 2,000 and 3,000 rpm.

8.30 Two fluids flow into a cylindrical mixing chamber where they are mixed at steady state by a paddle wheel that rotates at 50 revolutions per second, Fig. P8.30. It is suggested to scale up the process to twice the amounts of liquids mixed. Because the original mixing is quite satisfactory, complete similarity is favorable. One design suggestion was to increase all piping to twice their cross sections and to keep the cylindrical mixing chamber height but to increase its diameter by $\sqrt{2}$. Another suggestion was to keep the old structure but to have the flowrates in the piping increased by 2. Find if these suggestions can satisfy similarity, and if they can, find the size and the number of revolutions of the paddle wheel.

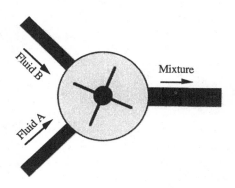

Figure P8.30 Paddle-wheel mixer.

8.31 Find the similarity parameters for the water hammer described in Example 4.10.

9. FLOWS WITH NEGLIGIBLE ACCELERATION

The nonlinear terms on the left-hand side of the Navier–Stokes equations result from the acceleration of the fluid. We already know that these terms contribute much to the difficulties one encounters in the solution of the equations. Indeed, we have noted the relative ease with which exact solution have been obtained in Chapter 6 for fully developed flows, where those nonlinear terms vanish. As we now look for approximations, obtained by the solutions of approximate forms of the Navier–Stokes equations, the first idea that comes to mind is to remove the terms which cause the greatest difficulty, i.e., the nonlinear acceleration terms.

Our subject here is fluid mechanics, and in this context we must ask if there exist real flows in which the acceleration terms are negligible; and if they exist, how do we identify them? The answer to this enquiry is that there are at least two such families of flows: *flows in narrow gaps* and *creeping flows.*

Flow in Narrow Gaps

Consider the two-dimensional flow of an incompressible fluid in the narrow gap between the two plates shown in Fig. 9.1. For simplicity we assume the plates to be flat. The results obtained in this analysis may be extended to cases where the plates are curved, as long as the gap is much smaller than the radius of curvature. Another possible extension is to three-dimensional flows, where a w-component of the velocity also exists. Here we present the simplest examples of gap flows and limit the analysis to two-dimensional flows between flat plates.

The lower plate is set stationary along the x-axis and the upper plate may be inclined to it with the small angle α. If the upper plate moves, its motion is restricted to its own plane We call such a flow *a flow in a narrow gap* when

$$\frac{\delta}{L} \ll 1, \qquad \alpha < \frac{\delta}{L}. \tag{9.1}$$

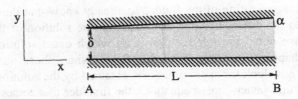

Figure 9.1 Flow in a narrow gap.

Let $\bar{u}$, the mean velocity in the x-direction, be defined,

$$\bar{u} = \frac{1}{\delta}\int_0^\delta u\, dy.$$

Conservation of mass requires

$$\bar{u}_A \cdot \delta = \bar{u}_B(\delta + L\alpha).$$

Hence

$$\left|\frac{\partial u}{\partial x}\right| \approx \left|\frac{\partial \bar{u}}{\partial x}\right| \approx \frac{\bar{u}_B - \bar{u}_A}{L} = \bar{u}\frac{\alpha}{\delta}. \tag{9.2}$$

The equation of continuity, Eq. (5.21), states

$$\frac{\partial u}{\partial x} = -\frac{\partial v}{\partial y}.$$

Hence with Eq. (9.2),

$$\left|\frac{\partial v}{\partial y}\right| \approx \left|\frac{\bar{u}\alpha}{\delta}\right|. \tag{9.3}$$

The boundary conditions for v on both plates are

$$v = 0 \quad \text{at } y = 0 \quad \text{and} \quad \text{at } y = \delta.$$

The largest value that $|v|$ may attain, in the vicinity of the middle of the gap, is approximately

$$|v|_{\max} \approx \frac{\delta}{2}\left|\frac{\partial v}{\partial y}\right| \approx \left|\frac{\bar{u}\alpha}{2}\right| \ll |\bar{u}|. \tag{9.4}$$

The y-component of the velocity may be neglected in comparison with the x-component.

Furthermore, for a moving or stationary upper plate

$$\left|\frac{\partial u}{\partial y}\right| \approx \frac{\bar{u}}{\delta},$$

(9.5)

and with Eq. (9.2),

$$\left|\frac{\partial u}{\partial x}\right| \Big/ \left|\frac{\partial u}{\partial y}\right| \approx \alpha < \frac{\delta}{L} \ll 1.$$

(9.6)

The x-derivative of u is negligible compared with the y-derivative.

The approximate form of the Navier–Stokes equations for two-dimensional gap flows thus becomes

(9.7)

$$0 = -\frac{dp}{dx} + \mu \frac{d^2 u}{dy^2}.$$

Reynolds Lubrication Theory

An important application of flows in narrow gaps, which are flows with their acceleration terms neglected, is in Reynolds lubrication theory. This theory yields the forces which appear in bearings and other lubricated sliding surfaces, provided that indeed acceleration forces may be neglected.

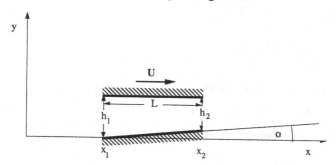

Figure 9.2 Flow in a bearing.

Consider the two-dimensional steady flow between the two plates shown in Fig. 9.2. The x-wise momentum equation for this narrow gap flow is just Eq. (9.7), which looks exactly like the one for a fully developed flow.* For an upper plate

---

* It is the solution, Eq. (9.8), with $h\,(x)$ varying with x very slowly, that justifies $\partial^2 u / \partial x^2 = 0$.

moving with the velocity U, Eq. (9.7) has the solution [see shear flow with pressure gradient, Eq. (6.24)]

$$u = \frac{1}{2\mu}\left(\frac{dP}{dx}\right)y[y - h(x)] + U\frac{y}{h(x)}. \tag{9.8}$$

Assuming $w = 0$ (a very wide bearing), the mass flow between the plates is conserved

$$\dot{m} = \int_{o}^{h(x)} \rho u\, dy = \tfrac{1}{2}\rho U h(x) - \frac{\rho}{12\mu}\left(\frac{dP}{dx}\right)h^3(x) = \text{const} = \tfrac{1}{2}\rho U h_o \tag{9.9}$$

or

$$\frac{dP}{dx} = 6\mu U\frac{h - h_o}{h^3}. \tag{9.10}$$

The lower plate, Fig. 9.2, is inclined to the horizontal by the angle

$$\alpha \cong \frac{dh}{dx}.$$

Now

$$\frac{dP}{dx} = \frac{dP}{dh}\cdot\frac{dh}{dx} = \alpha\frac{dP}{dh},$$

which may be substituted in Eq. (9.10) to yield

$$\frac{dP}{dh} = 6\mu\frac{U}{\alpha}\cdot\frac{h - h_o}{h^3}. \tag{9.11}$$

Integration yields

$$P = 6\mu\frac{U}{\alpha}\left[-\frac{1}{h} + \frac{h_o}{2h^2}\right] + B$$

and h_o and B have to satisfy the two boundary conditions:

$$P = P_o \quad \text{at } h_1 \text{ and } \text{ at } h_2.$$

This is achieved by setting $2/h_o = 1/h_1 + 1/h_2$. Hence

$$P - P_o = 6\mu\frac{U}{\alpha}\frac{(h - h_1)(h - h_2)}{(h_1 + h_2)h^2}. \tag{9.12}$$

The mass flow rate is obtained from Eq. (9.9) as

$$\dot{m} = \tfrac{1}{2}\rho U h_o = \tfrac{1}{2}\rho U\frac{2h_1 h_2}{h_1 + h_2} \qquad (h_1 > h_o > h_2). \tag{9.13}$$

The *lift force* **L** per unit width acting on the upper plate is

$$\mathbf{L} = \int\limits_{x_1}^{x_2} (P - P_o) dx = \frac{1}{\alpha} \int\limits_{h_1}^{h_2} (P - P_o) dh$$

$$= 6\mu \frac{U}{\alpha^2} \left[\frac{2(h_2 - h_1)}{h_1 + h_2} - \ln \frac{h_2}{h_1} \right] = \frac{6\mu UL^2}{(k-1)^2 h_2^2} \left[\ln k - \frac{2(k-1)}{k+1} \right], \tag{9.14}$$

where $k = h_1/h_2 > 1$, $L = x_2 - x_1 = \frac{1}{\alpha}(h_2 - h_1)$. The *drag force* **D** per unit width acting on the upper plate is

$$\mathbf{D} = \int\limits_{x_1}^{x_2} \mu \left(\frac{du}{dy} \right)_{y=h} dx = \int\limits_{h_1}^{h_2} \left[\frac{1}{2} h \frac{dP}{dh} + \mu \frac{U}{\alpha} \frac{1}{h} \right] dh = \mu \frac{U}{\alpha} \int\limits_{h_1}^{h_2} \left[3 \frac{h - h_o}{h^2} + \frac{1}{h} \right] dh.$$

Hence,

$$\mathbf{D} = \frac{2\mu UL}{(k-1)h_2} \left[2 \ln k - \frac{3(k-1)}{k+1} \right]. \tag{9.15}$$

To maximize the lift, let $dL/dk = 0$. This yields

$$k = 2.2 ,$$

which corresponds to

$$\mathbf{L} = 0.4 \mu UR, \qquad \mathbf{D} = 1.2 \mu UR, \qquad \mathbf{D}/\mathbf{L} = 3/R, \tag{9.16}$$

where

$$R = \frac{2L}{h_1 + h_2} .$$

Example 9.1

Figure 9.3 shows a rotating disk with a floating magnetic reader. The linear velocity of the disk just below the reader is 60 m/s. The width of the reader plate is $L = 0.02$ m and the weight it must support is 5 g/cm in the radial direction. The reader's arm is set to an angle α which yields the maximum lift but is otherwise free to float, i.e., it finds its own height h_1 and h_2. The whole system is placed in nitrogen at 300 K. Find h_1, h_2 and α. Find the drag force acting on the reader.

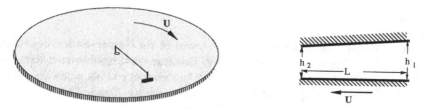

Figure 9.3 Rotating disk and magnetic reader.

Solution

The lift force needed to support 5 g/cm is

$$L = \frac{5 \times 10^{-3} \times 9.81}{10^{-2}} = 4.905 \text{ N / m}.$$

For maximum lift

$$k = h_1/h_2 = 2.2$$

$$\frac{h_1 - h_2}{L} = \frac{h_2(k-1)}{L} = 1.2\frac{h_2}{L} = \tan\alpha.$$

For nitrogen at 300 K, $\mu = 17.84 \times 10^{-6}$ kg/m·s. Equation (9.14) for the lift force now yields h_2,

$$L = \frac{1}{h_2^2}\frac{6\mu UL^2}{k-1}\left[\ln k - 2\frac{k-1}{k+1}\right]$$

or

$$4.905 = \frac{1}{h_2^2}\frac{6 \times 17.84 \times 10^{-6} \times 60 \times 0.02^2}{1.2}\left[\ln 2.2 - \frac{2 \times 1.2}{3.2}\right] = \frac{8.23 \times 10^{-8}}{h_2^2}.$$

Hence,

$$h_2 = 1.3 \times 10^{-4} \text{ m} = 0.13 \text{ mm},$$

$$h_1 = kh_2 = 0.286 \text{ mm},$$

$$\alpha = 0°26.8' = 0.0078 \text{ radians}.$$

The drag force is obtained from Eq. (9.15) as

$$D = \frac{2\mu UL}{(k-1)h_2}\left[2\ln k - \frac{3(k-1)}{k+1}\right]$$

$$= \frac{2 \times 17.84 \times 10^{-6} \times 60 \times 0.02}{1.2 \times (1.3 \times 10^{-4})}\left[2\ln 2.2 - \frac{3 \times 1.2}{3.2}\right] = 0.124 \text{ N/m}.$$

Creeping Flows

We are still looking for approximate forms of the Navier–Stokes equations, in which the nonlinear terms are neglected. Because these troublesome terms are nonlinear in the velocities, they are forced to become very small when the velocities are smaller and smaller. Hence the name *creeping flows*. This idea is expressed more exactly by the use of the Reynolds number.

The Reynolds number, $Re = VL/\nu$, can become small because V, the characteristic velocity, is small or because L, the characteristic length, is small or because ν, the kinematic viscosity, is large. Of course, it is the combination VL/ν which determines the magnitude of Re. However, observing the time number $\Omega = L/\tau V$ and the Euler number $Eu = \pi/\rho V^2$, we choose to say that Re is small because V is small, implying that Ω and Eu are not necessarily small. Equation (8.14) multiplied by Re becomes

$$Re \cdot \Omega \frac{\partial u}{\partial t} + Re\left[u\frac{\partial u}{\partial x} + v\frac{\partial u}{\partial y} + w\frac{\partial u}{\partial z}\right] = -Eu \cdot Re\frac{\partial p}{\partial x} + \left(\frac{\partial^2 u}{\partial x^2} + \frac{\partial^2 u}{\partial y^2} + \frac{\partial^2 u}{\partial z^2}\right). \quad (9.17)$$

For $Re \cdot \Omega = L^2/\nu\tau$ and $Eu \cdot Re = \pi L/\mu V$ which are not very small, the inertia terms in the square brackets are neglected and the equation attains the form

$$\Omega \frac{\partial u}{\partial t} = -Eu\frac{\partial p}{\partial x} + \frac{1}{Re}\left(\frac{\partial^2 u}{\partial x^2} + \frac{\partial^2 u}{\partial y^2} + \frac{\partial^2 u}{\partial z^2}\right), \quad (9.18)$$

with two similar equations for the y- and z-components of the momentum equation. Assuming incompressible flow, the continuity equation (8.11) remains the same. Both equations are written in their vectorial form as

$$\nabla \cdot \mathbf{q} = 0, \quad (9.19)$$

$$\Omega \frac{\partial \mathbf{q}}{\partial t} = -Eu \nabla p - \frac{1}{Re}\nabla \times \nabla \times \mathbf{q}. \quad (9.20)$$

The flows described by Eqs. (9.19) and (9.20) are called *Stokes creeping flows*. The equations are linear and possess some properties which are generally useful in their solution.

Let the divergence of Eq. (9.20) be taken. The term on the left-hand side becomes

$$\nabla \cdot \left(\frac{\partial \mathbf{q}}{\partial t}\right) = \frac{\partial}{\partial t}(\nabla \cdot \mathbf{q}) = \frac{\partial}{\partial t}(0) = 0,$$

and this is because $\nabla \cdot \mathbf{q} = 0$, Eq. (9.19).

On the right-hand side,

$$\nabla \cdot (\nabla \times \nabla \times \mathbf{q}) = \nabla \cdot [\nabla \times (\nabla \times \mathbf{q})] = 0,$$

and this term vanishes because the divergence of a curl of any vector is identically zero. The only term left of Eq. (9.20) is

$$\nabla^2 p = 0. \quad (9.21)$$

Thus the pressure in creeping flows is a harmonic function, i.e., it satisfies the Laplace equation.

Equation (9.21) is not as useful as it looks because the boundary conditions in terms of p are not always known; still this equation is used in many cases.

For time-independent flows an alternative general form can be obtained by taking the curl of Eq. (9.20). Since $\nabla \times (\nabla p) = 0$ for any scalar p, this results in

$$\nabla \times \nabla \times \nabla \times \mathbf{q} = 0. \tag{9.22}$$

Equation (9.22) is particularly useful in flows for which the direction of the $\nabla \times \mathbf{q}$ vector is a priori known, e.g., two-dimensional and axisymmetric flows.

Two-dimensional Flows

In two-dimensional flows with the velocity vector $\mathbf{q} = \mathbf{i}u + \mathbf{j}v$,

$$\nabla \times \mathbf{q} = \left(\mathbf{i}\frac{\partial}{\partial x} + \mathbf{j}\frac{\partial}{\partial y} \right) \times (\mathbf{i}u + \mathbf{j}v) = \mathbf{k}\left(\frac{\partial v}{\partial x} - \frac{\partial u}{\partial y} \right).$$

Taking the curl of $\nabla \times \mathbf{q}$ results in

$$\nabla \times (\nabla \times \mathbf{q}) = \mathbf{i}\frac{\partial}{\partial y}\left(\frac{\partial v}{\partial x} - \frac{\partial u}{\partial y} \right) - \mathbf{j}\frac{\partial}{\partial x}\left(\frac{\partial v}{\partial x} - \frac{\partial u}{\partial y} \right),$$

and the final curl,

$$\nabla \times [\nabla \times (\nabla \times \mathbf{q})] = \mathbf{k}\left(\frac{\partial^2}{\partial y^2} + \frac{\partial^2}{\partial x^2} \right)\left(\frac{\partial v}{\partial x} - \frac{\partial u}{\partial y} \right).$$

Thus, for two-dimensional flows in rectangular coordinates Eq. (9.22) becomes

$$\left(\frac{\partial^2}{\partial y^2} + \frac{\partial^2}{\partial x^2} \right)\left(\frac{\partial u}{\partial y} - \frac{\partial v}{\partial x} \right) = 0. \tag{9.23}$$

With the two-dimensional stream function as defined by Eq. (5.57)

$$u = \frac{\partial \psi}{\partial y}, \qquad v = -\frac{\partial \psi}{\partial x}. \tag{5.57}$$

Equation (9.23) may be written in terms of ψ as

$$\left(\frac{\partial^2}{\partial x^2} + \frac{\partial^2}{\partial y^2} \right)\left(\frac{\partial^2 \psi}{\partial x^2} + \frac{\partial^2 \psi}{\partial y^2} \right) = 0 \tag{9.24}$$

or

$$\nabla^2 (\nabla^2 \psi) = 0. \tag{9.25}$$

Equation (9.25) is sometimes written as

$$\nabla^4 \psi = \left(\frac{\partial^2}{\partial x^2} + \frac{\partial^2}{\partial y^2} \right)^2 \psi = 0, \tag{9.26}$$

which means that the operator inside the parentheses has to be applied twice in succession. The stream function ψ is thus biharmonic.

Axisymmetric Flows

One of the most important creeping flows is that of the flow around a sphere. To treat this case efficiently, we must extend the concept of the stream function, which has been introduced in Chapter 5 for two-dimensional flows, to flows which are axisymmetric.

A flow is defined axisymmetric when its description in polar coordinates (z, r, θ) shows no q_θ velocity and no dependence on the θ-coordinate. Such a flow can be described in the $\theta = 0$ plane, with the velocity vector and the streamlines lying in this plane, Fig. 9.4. Furthermore, $\theta = C$ planes, which are obtained by rotating the $\theta = 0$ plane around the z-axis, show a flow pattern identical to that in the $\theta = 0$ plane. The $\theta = 0$ plane is therefore called the *representative plane* of the axisymmetric flow. Figure 9.4 shows such a representative plane, with four streamlines, denoted by A,B,C,D. The streamlines in this representative plane are also the lines of intersections of stream sheets with the representative plane, in a fashion similar to that in two-dimensional flows. In this case, however, these stream sheets are really stream tubes with circular cross sections, obtained by the rotation of the whole figure around the z-axis. Each streamline may now have a number attached to it signifying the mass flow inside its corresponding stream tube.

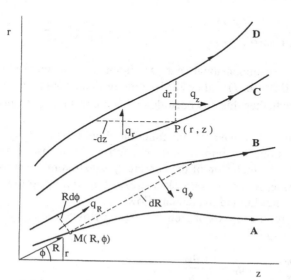

Figure 9.4 Axisymmetric representative plane and streamlines.

The number zero flow is naturally assigned here to the z-axis which, in axisymmetric flows, is always a streamline. These numbers, measuring the mass flow inside the stream tubes, are called *Stokes stream functions* or, where no confusion with the two-dimensional stream function is expected, just *stream functions*.

As in the treatment of the two-dimensional case, let $\psi_D = \psi_C + d\psi$. The cross section available for q_z in this case is $2\pi r\, dr$, and that for q_r is $-2\pi r\, dz$. Hence

$$d\psi = (2\pi r\, dr)(q_z \rho) = (-2\pi r\, dz)(q_r \rho),$$

from which follows

$$q_z \rho = \frac{1}{2\pi r}\frac{\partial \psi}{\partial r}, \qquad q_r \rho = -\frac{1}{2\pi r}\frac{\partial \psi}{\partial z}. \tag{9.27}$$

The spherical coordinates $(R,\ \phi,\ \theta)$ also appear in the representative plane, with no dependence on θ, of course. Using these with

$$\psi_B = \psi_A + d\psi,$$

one obtains

$$d\psi = (2\pi r R\, d\phi)(q_R \rho) = (2\pi R^2 \sin\phi\, d\phi)(q_R \rho)$$
$$= (2\pi r\, dR)(-q_\phi \rho) = (2\pi R \sin\phi\, dR)(-q_\phi \rho),$$

from which follows

$$q_R \rho = \frac{1}{2\pi R^2 \sin\phi}\left(\frac{\partial \psi}{\partial \phi}\right)_R, \qquad q_\phi \rho = -\frac{1}{2\pi R \sin\phi}\left(\frac{\partial \psi}{\partial R}\right)_\phi. \tag{9.28}$$

As in the two-dimensional case, ρ is ignored for incompressible flows. The $1/2\pi$ factor in Eqs. (9.27) and (9.28) is sometimes also suppressed, and one must always check whether this has been done when considering numerical computations.

Obviously the Stokes stream function should also satisfy identically the continuity equation. It is important to remember that the use of stream functions is equivalent to the inclusion of the continuity equation in the considerations.

The considerations of the Stokes stream function up to this point are quite general and are not limited to creeping flows.

We now return to axisymmetric creeping flows. Using cylindrical coordinates, in which $q_\theta = 0$,

$$q_z = \frac{1}{r}\frac{\partial \psi}{\partial r}, \qquad q_r = -\frac{1}{r}\frac{\partial \psi}{\partial z}. \tag{9.29}$$

Substitution of these in Eq. (9.28) results in the axisymmetric equivalent of Eq. (9.26):

$$\left(\frac{\partial^2}{\partial r^2} - \frac{1}{r}\frac{\partial}{\partial r} + \frac{\partial^2}{\partial z^2}\right)^2 \psi = 0,$$ (9.30)

which, however, is not the biharmonic equation.

Similarly, for axisymmetric flows using spherical coordinates

$$q_R = \frac{1}{R^2 \sin\phi}\frac{\partial\psi}{\partial\phi}, \qquad q_\phi = -\frac{1}{R\sin\phi}\frac{\partial\psi}{\partial R},$$ (9.31)

one obtains

$$\left[\frac{\partial^2}{\partial R^2} + \frac{\sin\phi}{R^2}\frac{\partial}{\partial\phi}\left(\frac{1}{\sin\phi}\frac{\partial}{\partial\phi}\right)\right]^2 \psi = 0.$$ (9.32)

It is noted again that Eqs. (9.30) and (9.32) are linear and methods for their solutions are fairly well known.

Stokes Flow around a Sphere

Figure 9.5 shows a stationary sphere set in a parallel flow. The radius of the sphere is a and its center is chosen to coincide with the origin of the coordinate system. There exists a creeping flow field around the sphere, with the velocity far from the sphere satisfying

$$q_z = U_\infty = \text{const}, \qquad \mathbf{q} = \mathbf{k}q_z.$$

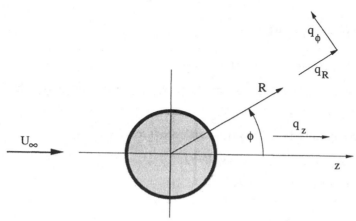

Figure 9.5 Creeping flow around a sphere.

Spherical coordinates are convenient to apply to this configuration, and the Stokes stream function is related to the velocity vector by Eq. (9.31):

$$q_R = \frac{1}{R^2 \sin\phi} \frac{\partial\psi}{\partial\phi}, \qquad q_\phi = -\frac{1}{R\sin\phi}\frac{\partial\psi}{\partial R}. \tag{9.31}$$

The differential equation for the stream function is now Eq. (9.32):

$$\left[\frac{\partial^2}{\partial R^2} + \frac{\sin\phi}{R^2}\frac{\partial}{\partial\phi}\left(\frac{1}{\sin\phi}\frac{\partial}{\partial\phi}\right)\right]^2 \psi = 0. \tag{9.32}$$

This equation is of the fourth order, and there are four boundary conditions that must be satisfied: the two velocity components, q_R and q_ϕ, must vanish on the sphere and they must become the undisturbed parallel flow velocity components far away from the sphere. Formally,

$$q_R = \frac{1}{R^2 \sin\phi}\frac{\partial\psi}{\partial\phi} = 0, \qquad \frac{\partial\psi}{\partial\phi} = 0 \qquad \text{at} \qquad R = a,$$

$$\tag{9.33}$$

$$q_\phi = -\frac{1}{R\sin\phi}\frac{\partial\psi}{\partial R} = 0, \qquad \frac{\partial\psi}{\partial R} = 0 \qquad \text{at} \qquad R = a,$$

and because the undisturbed parallel flow is just

$$\mathbf{q} = \mathbf{k}U_\infty = \mathbf{e}_R q_R + \mathbf{e}_\phi q_\phi = U_\infty\big[\mathbf{e}_R\cos\phi - \mathbf{e}_\phi\sin\phi\big],$$

then

$$q_R = U_\infty\cos\phi = \frac{1}{R^2\sin\phi}\frac{\partial\psi}{\partial\phi} \qquad \text{at} \qquad R\to\infty,$$

$$-q_\phi = U_\infty\sin\phi = \frac{1}{R\sin\phi}\frac{\partial\psi}{\partial R} \qquad \text{at} \qquad R\to\infty,$$

or

$$\frac{\partial\psi}{\partial\phi} = U_\infty R^2\cos\phi\sin\phi \qquad \text{at} \qquad R\to\infty,$$

$$\tag{9.34}$$

$$\frac{\partial\psi}{\partial R} = U_\infty R\sin^2\phi \qquad \text{at} \qquad R\to\infty.$$

Equations (9.34), which hold for very large R, may be integrated to yield

$$\psi = \tfrac{1}{2}U_\infty R^2\sin^2\phi.$$

Now this result must hold for *any* ϕ, and one is tempted to try

$$\psi = f(R)\sin^2\phi \qquad \text{for any } R.$$

Substitution in Eq. (9.32) yields

$$\left(\frac{d^2}{dR^2} - \frac{2}{R^2}\right)^2 f(R) = 0, \tag{9.35}$$

which is homogeneous in R and dR and therefore may have a solution in the form of a polynomial. Indeed, Eq. (9.35) is satisfied by

$$f(R) = \frac{A}{R} + BR + CR^2 + DR^4.$$

Because from the integration of Eq. (9.34), for $R \to \infty$, $f(R) = \frac{1}{2} U_\infty R^2$,

$$D = 0 \quad \text{and} \quad C = \frac{U_\infty}{2},$$

and A and B do not affect the velocity at $R \to \infty$. The stream function thus becomes

$$\psi = \left(\frac{A}{R} + BR + \frac{1}{2} U_\infty R^2 \right) \sin^2 \phi. \tag{9.36}$$

Equation (9.36) must be satisfied for $R = a$. Hence

$$\frac{A}{a} + Ba + \frac{1}{2} U_\infty a^2 = 0,$$

$$\frac{A}{a^2} - B - U_\infty a = 0$$

or

$$A = \frac{1}{4} U_\infty a^3$$

and

$$B = -\frac{3}{4} U_\infty a,$$

and finally

$$\psi = U_\infty a^2 \left[\frac{1}{4} \frac{a}{R} - \frac{3}{4} \frac{R}{a} + \frac{1}{2} \left(\frac{R}{a} \right)^2 \right] \sin^2 \phi, \tag{9.37}$$

$$q_R = U_\infty \left[1 + \frac{1}{2} \left(\frac{a}{R} \right)^3 - \frac{3}{2} \frac{a}{R} \right] \cos \phi, \tag{9.38}$$

$$q_\phi = U_\infty \left[1 - \frac{1}{4} \left(\frac{a}{R} \right)^3 - \frac{3}{4} \frac{a}{R} \right] \sin \phi. \tag{9.39}$$

We may now proceed to calculate the pressure. Equation (9.20) is written again in dimensional form for the present case

$$0 = -\nabla p - \mu \nabla \times \nabla \times \mathbf{q}, \tag{9.40}$$

and the expression for the velocity, Eqs. (9.38) - (9.39), is used to calculate

$$\nabla \times \mathbf{q} = \mathbf{e}_\phi \left(\frac{3}{2} \frac{U_\infty a}{R^2} \sin \phi \right)$$

and

$$\nabla \times \nabla \times \mathbf{q} = \frac{3}{2}\frac{U_\infty a}{R^3}\left(\mathbf{e}_R 2\cos\phi + \mathbf{e}_\phi \sin\phi\right).$$

Substitution into Eq. (9.40) in component form results in

$$\frac{\partial p}{\partial R} = -3\frac{U_\infty a\mu}{R^3}\cos\phi,$$

$$\frac{1}{R}\frac{\partial p}{\partial \phi} = -\frac{3}{2}\frac{U_\infty a\mu}{R^3}\sin\phi \tag{9.41}$$

and integration yields

$$p = \frac{3}{2}\frac{\mu U_\infty a}{R^2}\cos\phi. \tag{9.42}$$

To calculate the drag on the sphere, we note that because of symmetry, the drag force F_D on the sphere acts in the z-direction only. Consider the forces acting on the differential strip shown in Fig. 9.6: S is a tangential shear stress and p is the (normal) pressure. Hence

$$dF_D = (2\pi a\sin\phi)a\,d\phi\,[S\sin\phi - p\cos\phi]_{R=a}, \tag{9.43}$$

where p is given by Eq. (9.42) and S is found from

$$S = -\mu\left(\frac{\partial q_\phi}{\partial R}\right)_{R=a} = -\mu\left(3\frac{A}{R^4} - \frac{B}{R^2}\right)_{R=a}\sin\phi = -\frac{3}{2}\frac{\mu U_\infty}{a}\sin\phi. \tag{9.44}$$

Thus

$$dF_D = -3\mu U_\infty \pi a\left(\sin^2\phi + \cos^2\phi\right)\sin\phi\,d\phi = -3\mu U_\infty a\pi\sin\phi\,d\phi$$

and the expression for the drag force, known as Stokes law, is obtained as

$$\tag{9.45}$$

$$F_D = \int\limits_0^\pi dF_D = 6\pi\mu a U_\infty.$$

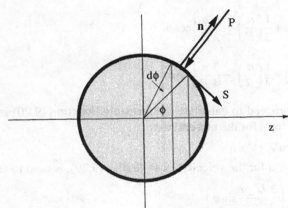

Figure 9.6 Forces on the sphere.

The case just solved has been that of a fluid flowing steadily around a stationary sphere. The transcription of the results to the case where the bulk of the fluid is stationary while the sphere moves with a constant velocity $(-U_\infty)$ is obtained by a simple transformation of coordinates.

The drag force given by Eq. (9.45) is just the hydrodynamic contribution. In the presence of body forces, e.g., gravity, and when the density of the sphere, ρ', is different from the fluid density, ρ, there is an additional hydrostatic force

$$F_B = \tfrac{4}{3}\,\pi a^3 g(\rho - \rho') \tag{9.46}$$

caused by buoyancy. The net force which now acts on the sphere is the sum

$$F_N = F_B + F_D + \tfrac{4}{3}\,\pi a^3 g(\rho - \rho') + 6\pi \mu a U_\infty. \tag{9.47}$$

We now consider a small sphere dropped into a fluid at rest. The sum of the forces acting on the sphere is at first different from zero. Newton's second law of motion requires that the sphere accelerate. The term F_B in Eq. (9.47) remains the same, while F_D, which opposes F_B, increases in magnitude with the increase in the velocity. The net force F_N thus eventually becomes zero. The sphere is then said to have reached its *terminal velocity,* which becomes

$$U_t = -\frac{4}{3}\frac{\pi a^3 g(\rho' - \rho)}{6\pi \mu a} = \frac{2}{9}\frac{a^2 g(\rho - \rho')}{\mu} \tag{9.48}$$

or

$$\mu = \frac{2}{9}\frac{a^2 g}{U_t}(\rho - \rho'). \tag{9.49}$$

The results derived above are verified experimentally for

$$\mathrm{Re}_a = \frac{U_t a \rho}{\mu} < 0.5. \tag{9.50}$$

Equation (9.49) may be used to obtain the viscosity of a fluid from experimental measurements of the terminal velocity of a sphere dropped into it.

Example 9.2

a. The viscosity of printing ink is approximately $\mu = 3.0$ kg/m·s. The exact values of the viscosity must be measured, and it is suggested to drop glass balls into the ink and measure their terminal velocities. The densities of printing ink and glass are 1,000 and 2,000 kg/m³, respectively. Find the size of the balls to be used.

b. Balls of radius $a = 1.0$ cm were ordered, and in a certain experiment a terminal velocity of 12 cm/s was measured. Find the viscosity of the ink in that experiment.

Solution

a.   The equations obtained for Stokes creeping flow are valid for, Eq. (9.50),

$$Re_a = \frac{U_t a \rho}{\mu} < 0.5,$$

i.e., the velocity should not exceed

$$U_t = 0.5\frac{\mu}{\rho} = \frac{0.5 \times 3}{1,000} = 1.5 \times 10^{-3}\,\text{m}/\text{s}.$$

On the other hand, from Eq. (9.48),

$$U_t = \frac{2}{9}\frac{a^2 g}{\mu}\left|(\rho - \rho')\right| = \frac{2}{9}a^2 \times \frac{9.81}{3} \times 1,000 = 726.67a^2.$$

Thus,

$$726.67a^3 = 1.5 \times 10^{-3},$$

and the maximal sphere radius is found as

$$a^3 = 2.06 \times 10^{-6}, \qquad a = 1.273 \times 10^{-2}\,\text{m} = 1.27\,\text{cm}.$$

b.   From Eq. (9.49),

$$\mu = \frac{2}{9}\frac{a^2 g}{U_t}\left|(\rho - \rho')\right| = \frac{2}{9} \times \frac{10^{-4} \times 9.81}{0.12} \times 1,000 = 1.817\,\text{kg/m}\cdot\text{s}.$$

Example 9.3

Small dust particles come out of a chimney of a cement factory. The particles are approximately spherical, with radii in the range of 10^{-5} - 10^{-7} m. The density of the particles is $1,000$ - $2,000$ kg/m³. The viscosity of air is $\mu = 15 \times 10^{-6}$ kg/m·s and its density is 1.2 kg/m³. Find how long it will take the particles to settle on the ground.

Solution

Equation (9.48) states

$$U_t = \frac{2}{9}a^2 g\frac{\left|\rho - \rho'\right|}{\mu} \approx \frac{2}{9}a^2 \times 9.81\frac{\rho}{15 \times 10^{-6}} = 145,333a^2\rho.$$

For $a = 10^{-5}$ m with $\rho = 2,000$ kg/m³

$$U_t = 0.029\,\text{m/s} = 29\,\text{mm/s}.$$

For $a = 10^{-6}$ m with $\rho = 1,000$ kg/m³

$$U_t = 1.45 \times 10^{-3}\,\text{m/s} = 0.145\,\text{mm/s}.$$

For $a = 10^{-7}$ m with $\rho = 1,000$ kg/m³

$$U_t = 1.45 \times 10^{-6}\,\text{m/s} = 1.45 \times 10^{-3}\,\text{mm/s}.$$

For a chimney 30 m high, the $a = 10^{-5}$ m particles will settle in

$$\frac{30}{0.029} = 1034\,s = 35\,\text{min};$$

the $a = 10^{-6}$ m particles will settle in

$$\frac{30}{0.145 \times 10^{-3}} = 2 \times 10^{5}\,s = 58\,\text{h};$$

the $a = 10^{-7}$ m particles will settle in

$$\frac{30}{1.45 \times 10^{-6}} = 2 \times 10^{7}\,s = 240\,\text{days}.$$

In reality the $a < 10^{-6}$ m particles will not settle at all because of gravity, as winds and thermal convection air movements are quite faster than their gravity settling speeds.

Stokes Flow around a Cylinder

Figure 9.7 shows a stationary cylinder set in a two-dimensional parallel flow. The radius of the cylinder is a and its center is chosen to coincide with the origin of the coordinate system. This flow seems to be the two-dimensional analog of the flow around a sphere, and we want to find out whether indeed this is the case.

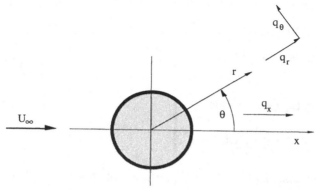

Figure 9.7 Two-dimensional creeping flow around a cylinder.

The convenient coordinate system to describe this flow is the cylindrical one. The relations between velocity and stream function are given by Eq. (5.58):

$$q_r = \frac{1}{r}\frac{\partial \psi}{\partial \theta}, \qquad q_\theta = -\frac{\partial \psi}{\partial r}. \tag{5.58}$$

The differential equation satisfied by this stream function is Eq. (9.26), which in cylindrical coordinates becomes

$$\nabla^4 \psi = \left[\frac{1}{r}\frac{\partial}{\partial r}\left(r\frac{\partial}{\partial r}\right) + \frac{1}{r^2}\frac{\partial^2}{\partial \theta^2}\right]^2 \psi = 0. \tag{9.51}$$

The boundary conditions on the cylinder are

$$q_r = \frac{1}{r}\frac{\partial \psi}{\partial \theta} = 0, \qquad q_\theta = -\frac{\partial \psi}{\partial r} = 0 \qquad \text{at} \qquad r = a. \tag{9.52}$$

The undisturbed parallel flow is

$$\mathbf{q} = \mathbf{e}_r q_r + \mathbf{e}_\theta q_\theta = U_\infty[\mathbf{e}_r \cos\theta - \mathbf{e}_\theta \sin\theta]$$

and the other two boundary conditions are adjusted such that for very large r

$$\mathbf{q} \rightarrow \mathbf{i} U_\infty.$$

Hence

$$\frac{\partial \psi}{\partial \theta} = U_\infty r \cos\theta,$$

$$\frac{\partial \psi}{\partial r} = U_\infty \sin\theta \qquad \text{at large } r. \tag{9.53}$$

In a way similar to the treatment of the flow around the sphere, Eq. (9.53) may be integrated to

$$\psi = U_\infty r \sin\theta,$$

and, again, one is tempted to try

$$\psi = f(r)\sin\theta \qquad \text{for any } r.$$

Substitution in Eq. (9.51) yields

$$\left[\frac{1}{r}\frac{d}{dr}\left(r\frac{d}{dr}\right) - \frac{1}{r^2}\right]^2 f = 0, \tag{9.54}$$

which is analogous to Eq. (9.35). Application of standard methods of ordinary differential equations yields

$$f(r) = Ar^3 + Br\left(\ln r - \tfrac{1}{2}\right) + Cr + Dr$$

and

$$\psi = \left[Ar^3 + Br\left(\ln r - \tfrac{1}{2}\right) + Cr + Dr\right]\sin\theta. \tag{9.55}$$

One might try to proceed here along the way which proved successful for the flow around the sphere, i.e., by applying the boundary conditions Eqs. (9.53) at $r \to \infty$ to Eq. (9.55). This would result in $A = 0$, $B = 0$, $C = U_\infty$, and only a single constant, i.e., D, is left to satisfy both boundary conditions at the surface of the cylinder. Indeed, application of Eqs. (9.52) results in the two contradictory equations:

$$D = a^2 \quad \text{and} \quad D = -a^2.$$

Thus no solution is obtained.

This result is known as the *Stokes Paradox*. Similar mathematical difficulties do occur in some other *two-dimensional* problems of mathematical physics where boundary conditions at *infinity* are imposed, e.g., in the solution for the temperature field outside a given isothermal cylinder. Once it is recognized that the paradox is associated with the combination of *two-dimensionality* and of boundary conditions at *infinity*, a way is indicated to still solve for the flow field:

There is nothing one can do about the two-dimensionality of the cylinder. If this is modified, one does not solve the given problem. But then the boundary conditions, Eqs. (9.53), may be satisfied at a large r, say at $r = b < \infty$, instead of at infinity. This approach would permit the completion of the solution. The details are presented in the following example.

Example 9.4

Complete the solution of Stokes flow around a cylinder, with the boundary conditions, Eq. (9.53), satisfied at a finite R, i.e., at $r = R$, $\infty > R > a$. Find expressions for the velocities.

Solution

It is helpful to rewrite the general solution, Eq. (9.55), and the boundary conditions, Eqs. (9.48) and (9.53), in dimensionless forms, using a as the characteristic length and U_∞ as the characteristic velocity. The dimensionless equations become

$$\psi = \left[Ar^3 + Br\left(\ln r - \tfrac{1}{2}\right) + cr + \frac{D}{r}\right]\sin\theta,$$

where r is now dimensionless and A, B, C, D are modified accordingly, and

$$q_r = \frac{1}{r}\frac{\partial \psi}{\partial \theta} = 0, \qquad q_\theta = -\frac{\partial \psi}{\partial r} = 0 \qquad \text{at} \qquad r = 1,$$

$$\frac{\partial \psi}{\partial \theta} = 1 \times r \times \cos\theta = R\cos\theta \qquad \text{and}$$

$$\frac{\partial \psi}{\partial r} = 1 \times \sin\theta = \sin\theta \qquad \text{at} \qquad r = R,$$

with R now measured in the cylinder's radii, i.e., R is a number stating at how many cylinder's radii away from the cylinder's center this boundary condition is satisfied (e.g., at a distance of 100 radii). The boundary conditions at $r = 1$, substituted in the general solution, yield

$$A - \tfrac{1}{2}B + C + D = 0, \tag{i}$$

$$3A + \tfrac{1}{2}B + C - D = 0. \tag{ii}$$

The boundary conditions at $r = R$ yield

$$AR^3 + BR\left(\ln R - \tfrac{1}{2}\right) + CR + \frac{D}{R} = R, \tag{iii}$$

$$3AR^2 + B\left(\ln R + \tfrac{1}{2}\right) + C - \frac{D}{R^2} = 1. \tag{iv}$$

This set of four linear algebraic equations is solved by elimination, and the results are:

$$A = \frac{-1}{2G(R)}, \qquad B = \frac{R^2 + 1}{G(R)}, \qquad C = \frac{1}{G(R)}, \qquad D = \frac{R^2}{2G(R)},$$

with

$$G(R) = \left(R^2 + 1\right)\ln R - \left(R^2 - 1\right).$$

Of course, R is large, and for $R^2 \gg 1$ an approximation for $G(R)$ may be obtained by

$$G(R) \approx H(R) = R^2 \ln R - R^2 = R^2 (\ln R - 1).$$

The expressions for the dimensionless velocity components are now obtained:

$$q_r = \frac{1}{r}\frac{\partial \psi}{\partial \theta} = \left[-\frac{1}{2}r^2 + \left(R^2 + 1\right)\left(\ln r - \frac{1}{2}\right) + 1 + \frac{R^2}{2r^2}\right]\frac{\cos\theta}{G(R)},$$

$$q_\theta = -\frac{\partial \psi}{\partial r} = \left[-\frac{3}{2}r^2 + \left(R^2 + 1\right)\left(\ln r + \frac{1}{2}\right) + 1 - \frac{R^2}{r^2}\right]\frac{\sin\theta}{G(R)}.$$

Example 9.5

For a cylinder of radius $a = 0.01$ m set in an undisturbed parallel flow of $U_\infty = 0.1$ m/s, find the velocities when the boundary conditions "at infinity" are satisfied:

a. At a distance of 1 m from the cylinder's center, i.e., at $r = R = 100$.
b. At a distance of 2 m, i.e., at $r = R = 200$.

Solution

The dimensionless radius, r, is measured here in units of a (see Example 9.4), which happens to be 1 cm. Accordingly, in part a, $R = 100$, and in part b, $R = 200$.

a. With reference to Example 9.4:
$R = 100$, $G(R) = 36{,}057.3$, $H(R) = 36{,}051.7$,

$$q_r = \left[-0.5r^2 + 10{,}001\left(\ln r - \tfrac{1}{2}\right) + 1 + \frac{10{,}000}{2r^2} \right]\frac{\cos\theta}{G(R)},$$

$$q_\theta = \left[-1.5r^2 + 10{,}001\left(\ln r + \tfrac{1}{2}\right) + 1 - \frac{10{,}000}{r^2} \right]\frac{\sin\theta}{G(R)},$$

b. $R = 200$, $G(R) = 171{,}939$, $H(R) = 171{,}933$,

$$q_r = \left[-0.5r^2 + 40{,}001\left(\ln r - \tfrac{1}{2}\right) + 1 + \frac{40{,}000}{2r^2} \right]\frac{\cos\theta}{G(R)},$$

$$q_\theta = \left[-1.5r^2 + 40{,}001\left(\ln r + \tfrac{1}{2}\right) + 1 - \frac{40{,}000}{r^2} \right]\frac{\sin\theta}{G(R)}.$$

The results of part a may be put as

$q_r = F_{r,100}(r)\cos\theta$,
$q_\theta = F_{\theta,100}(r)\cos\theta$.

while those of part b may be put as

$q_r = F_{r,200}(r)\cos\theta$,
$q_\theta = F_{\theta,200}(r)\sin\theta$,

with F_r and F_θ being the expressions in the square brackets divided by $G(R)$. Some values of F_r and F_θ are given in Table 9.1.

r	1	2	4	8	16	32	64	100
$F_{r,100}$	0	0.088	0.254	0.439	0.627	0.809	0.908	1.000
$F_{\theta,100}$	0	0.296	0.514	0.711	0.897	1.057	1.122	1.000
$F_{r,200}$	0	0.074	0.213	0.369	0.528	0.687	0.839	0.926
$F_{\theta,200}$	0	0.248	0.431	0.598	0.759	0.914	1.048	1.100

Table 9.1. Velocity dependence on r in Stokes cylinder flow.

It is instructive to note that although $R = 100$ and $R = 200$ both seem to be very far from the cylinder, the differences obtained between the two flows are quite significant. Thus one is advised to choose an R, i.e., to satisfy the boundary conditions, as close to the real physical situation as possible.

References

J. Happel and H. Brenner, "Low Reynolds Number Hydrodynamics with Special Application to Particulate Media," Prentice-Hall, Englewood Cliffs, NJ, 1965.

W.F. Hughes, "An Introduction to Viscous Flow," Hemisphere, Washington, DC, 1979, Chapter 3.

H. Lamb, "Hydrodynamics," Dover, New York, 1945.

D. Pincus and B. Sternlicht, "Theory of Hydrodynamic Lubrication," McGraw-Hill, New York, 1961.

Problems

9.1   A spherical balloon filled with hydrogen flies away off a child's hands. The radius of the balloon is 0.2 m, the density of the air is 1.18 kg/m³ and its viscosity is 2×10^{-5} kg/m·s. Find the terminal velocity of the balloon. Find the Reynolds number at this terminal velocity, and decide whether this is a creeping flow.

9.2   A copper ball has a radius of 5 mm and the density of 8,700 kg/m³. The ball is dropped into a tank full of a fluid whose density is 833 kg/m³. The ball descends at a constant velocity of 10 cm/s. Find the value of the Reynolds number at this terminal velocity. Find the viscosity of the fluid.

9.3   Crop dusting is done from an airplane flying at the altitude of 5 m. The solid dust particles of insecticide are approximately spherical with a diameter of $d = 10^{-2}$ cm and a density of 2,000 kg/m^3. Estimate the time required for the dust particles to fall to the ground.

The nearest population center is located 1 km from the dusted field. At what wind velocity, directed from the field toward the town, must the dusting operation be stopped?

9.4   Dust particles have approximately the density of water. Find the diameter of the dust particles that have a terminal velocity in air of

a. 1 m/s,   b. 0.1 m/s,   c. 0.01 m/s,   d. 0.001 m/s.

Calculate the approximate Reynolds numbers for each case. Does the simple Stokes formula apply in each of these cases?

9.5   Apply the boundary conditions Eq. (9.53) at $r = R = 300$ and then the boundary conditions Eq. (9.52) and complete the solution for the Stokes two-dimensional creeping flow around a cylinder. Obtain an expression for the pressure field and for the drag force on the cylinder.

9.6   Use the results of Problem 9.5 and compute the drag force on a cylinder when the boundary condition at large r, Eq. (9.53), is applied at $r = b = 4a$, at $b = 5a$, at $b = 7a$, at $b = 10\ a$. Draw schematically some streamlines for the four cases. Which do you consider a better approximation?

9.7   The linear velocity of the slide bearing in Example 9.1 is increased to 100 m/s. As a result the temperature of the nitrogen is raised to 350 K ($\mu = 19.9 \times 10^{-6}$ kg/m·s). Find h_1, h_2 and the drag force.

9.8   A wide slide bearing is constructed of a steel shaft with a diameter of 0.1 m and a bronze sleeve with a diameter of 0.1008 m. The gap between shaft and sleeve is filled with oil, $\rho = 880$ kg/m^3, $\mu = 2 \times 10^{-2}$ kg/m·s. The shaft turns at 3,000 rpm. To avoid contact between the surfaces of the shaft and the sleeve, they must be separated by at least 0.1 mm. Find the transverse load, in N/m (per meter width), that may be applied to the shaft.

9.9   A plastic cup 0.12 m long and 0.08 m in diameter is manufactured by injecting molten plastic into a flat gap inside a mold. The mold looks like an annulus between two cylinders and may be approximated as a 0.0005-m-wide gap between two parallel plates. The molten plastic has the apparent viscosity $\mu = 80$ kg/m·s.

a. Find the pressure necessary to have the liquid plastic front advance in the mold at a rate of at least 0.1 m/s.

b. What is the flow rate of the molten plastic at these conditions?

9.10 Glycerin flows at a creeping flow in a rectangular channel having the sides 0.05 m by 0.02 m. The maximum velocity in the channel is 0.2 m/s. Find the pressure gradient along the channel.

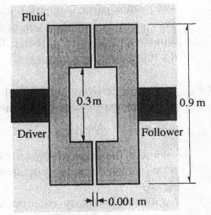

9.11 A simple hydraulic transmission is shown in Fig. P9.11. The transmission fluid has a viscosity of $\mu = 1.0\,\text{kg/m·s}$. The flow in the gap is a creeping flow and is assumed to be approximated by a shear flow between two flat plates. The drive disk turns at 3,000 rpm.

Figure P9.11

a. Find the moment transmitted to the follower disk as a function of the number of turns per minute of the follower disk.

b. Find the efficiency of the transmission at the various speeds.

9.12 A two-dimensional rectangular cavity is shown in Fig. P9.12. Three sides of the cavity are enclosed by stationary walls, and a plane slides along the fourth side. Write, in terms of the stream function, the differential equation and the boundary conditions that govern this flow.

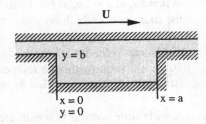

Figure P9.12

9.13 A shock absorber is shown in Fig. P9.13. It is common engineering practice to designate shock absorbers by a constant C in the form $F = C\,V$, where F is the force the absorber transmits, and V is the velocity of the piston. Find C for this absorber.

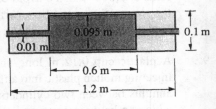

Figure P9.13

9.14 The hydraulic oscillation damper shown in Fig. P9.14 consists of a bent tube section of 0.06 m in diameter filled with oil of viscosity 0.01 kg/m·s, in which a tube of 0.05 m in diameter moves concentrically at an arc of 0.4 m radius. The flow in the annular gap is creeping. Find the braking moment as a function of the angular velocity.

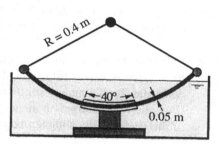

Figure P9.14

9.15 Creeping flow is associated with small Reynolds numbers. Is a fully developed flow in a pipe different for creeping flows and for flows where the Reynolds number is not very small?

9.16 A pipe ends with a nozzle shaped like a cone, Fig. P9.16 . The flow in the cone is creeping and is approximated by a fully developed pipe flow, i.e., at each cross section the flow is like in a pipe of that cross section. The cone is 0.1 m long, and the pipe feeds fluid to the cone with a maximum velocity of 1.0 m/s . The fluid viscosity is 0.1 kg/m·s. Find the pressure drop in the cone.

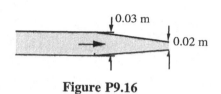

Figure P9.16

9.17 The flow in Problem 9.16 is reversed, i.e., the cone feeds fluid to the pipe. The velocity field is exactly reversed. Find the pressure drop in the cone.

9.18 Three spheres are connected by a string, as shown in Fig. P9.18. The three are dragged through a fluid at a constant velocity, and the flow is creeping. It is suggested to estimate the total drag force on the three spheres as three times the drag force on a single sphere in Stokes flow. Find the distance between any two spheres, expressed in sphere diameters, necessary to make the approximation of the order of 1% error.

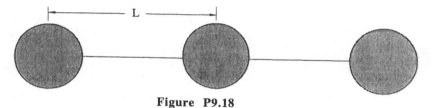

Figure P9.18

9.19 Can the Rayleigh problem be solved for a creeping flow? Are the results the same as those obtained for noncreeping flows?

9.20 It can be shown that in a creeping flow the velocity field is such that the total viscous dissipation over the whole field is minimized. Show that when the fluid contains solid particles, which are moving freely with the flow, the total dissipation increases. In other words, the apparent viscosity of the seeded fluid is higher than that of the pure fluid.

10. HIGH REYNOLDS NUMBER FLOWS – REGIONS FAR FROM SOLID BOUNDARIES

Negligible Shear – The Euler Equations

The interpretation of the physics of flows with vanishing viscosity has already been attempted in Chapter 5, where a simple formal substitution of $v \to 0$ in the Navier–Stokes equations has led to the Euler equation (5.54). The more elaborate formulation in Chapter 8 has transferred the role of a perturbation parameter from the kinematic viscosity, v, to the Reynolds number, $\mathrm{Re} = UL / v$, with the condition $v \to 0$ being expressed by $\mathrm{Re} \to \infty$. We have also been fore-warned that for $\mathrm{Re} \to \infty$ the second order Navier–Stokes equations yield the Euler equations, which are of the first order and therefore cannot satisfy both boundary conditions

$$\left. \begin{array}{l} q_n = 0 \\ q_t = 0 \end{array} \right\} \text{ at solid boundaries.} \tag{5.46}$$

Still, a legitimate question may come to mind at this point: Have the Euler equations no physical meaning at all? Are the solutions of these equations not even approximations to real flows? We know that near rigid boundaries, where the boundary conditions dominate the flow, the Euler equations yield wrong solutions; but are there no other regions, far away from the boundaries, where they correctly predict the physics?

We do not set out on this path of enquiry just as adventurers. We follow the footsteps of such giants as Euler and Lagrange, who also wondered where it led, and Prandtl, who walked it and found the answer.

Before embarking on the analysis of flows for which $v \to 0$, we consider an analogous case of such perturbations, which is simpler because the equations in it are linear. In the following example we ask under what conditions does the solution of the second order differential equation which describes the deflection of a membrane approximate the deflection of a thin plate, which is governed by a fourth order differential equation.

Example 10.1

The vertical deflection of a two-dimensional thin plate clamped at both ends and subject to a uniform continuous load, Fig. 10.1, is given by the dimensionless ordinary differential equation

$$B \frac{d^4 y}{dx^4} - M \frac{d^2 y}{dx^2} = -f,$$

with the boundary conditions

$$y = 0 \quad \text{at} \quad x = \pm 1,$$

$$\frac{dy}{dx} = 0 \quad \text{at} \quad x = \pm 1.$$

The dimensionless parameters in the equation are: B *the beam parameter* and M *the membrane parameter.*[*]
For very thick plates $B >> M$ and for membranes $B << M$:

a. Find the exact deflection of the plate.

b. Consider plates which become thinner and thinner, i.e., $B \to 0$. This causes the differential equation to reduce to the second order and then it cannot satisfy both boundary conditions. Find the approximate deflection of the plate: apply the regular perturbation $B \to 0$ and solve the perturbed equation. Consistent with the perturbation, which boundary condition is ignored? Where does the approximation hold?

---

[*] A thick beam bends according to

$$B \frac{d^4 y}{dx^4} = -f$$

with the same boundary conditions, while a membrane deflects according to

$$M \frac{d^2 y}{dx^2} = f$$

with the boundary conditions $y = 0$ at $x = \pm 1$.

Solution

a. A particular solution of the full equation is

$$y_p = \frac{f}{2M}x^2,$$

and the homogeneous solution is

$$y_h = C_1 + C_2 x + C_3 \sinh \lambda x + C_4 \cosh \lambda x, \qquad \lambda^2 = M/B.$$

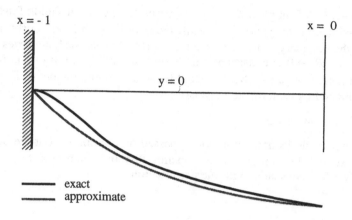

x = −1

x = 0

y = 0

———— exact

━━━━ approximate

Figure 10.1 Deflection of a two-dimensional clamped plate.

Addition of the two solutions and the use of the boundary conditions yield

$$y = \frac{f}{M}\left[\frac{1}{2}(x^2 - 1) + \frac{\cosh \lambda - \cosh \lambda x}{\lambda \sinh \lambda}\right],$$

which for large λ may be put as

$$y = \frac{f}{M}\left[\frac{1}{2}(x^2 - 1) + \frac{1}{\lambda}\left(1 - \frac{e^{\lambda |x|}}{e^\lambda}\right)\right]$$

and

$$\frac{dy}{dx} = \frac{f}{M}\left[x - \frac{e^{\lambda |x|}}{e^\lambda}\right].$$

b. With $B \to 0$ the equation undergoes a regular perturbation into the membrane equation,

$$-M\frac{d^2y}{dx^2} = -f,$$

with the solution again constructed of a particular part and a homogeneous one,

$$y = \frac{f}{2M}(x^2 - 1), \qquad \frac{dy}{dx} = \frac{f}{M}x.$$

The kept boundary condition is that for the membrane,

$$y = 0 \quad \text{at} \quad x = \pm 1.$$

Because $B \to 0$ implies $\lambda \to \infty$, the perturbed equation yields fairly good approximations for y; however, while the error in dy/dx is still fairly small far from the boundary, it is very large (about 100 %) near the boundaries.

The limit $B \to 0$ is consistent with a very thin plate and in the limit with a membrane. Now a membrane cannot transmit bending moments. Even if the membrane had been forced to satisfy

$$dy/dx = 0 \quad \text{at} \quad x = \pm 1,$$

this angular deflection could not be passed on to points not at the boundary because it takes moments to transmit this information. Thus ignoring $dy/dx = 0$ is consistent with the perturbation $B \to 0$.

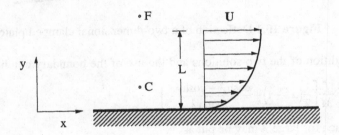

Figure 10.2 High-Reynolds-number flow.
Point F far from boundary. Point C close to boundary.

Fortified by Example 10.1 we return now to the implication of $\nu \to 0$. Consider a real fluid, which satisfies both boundary conditions, Eq. (5.46), at a rigid boundary but which has a very small viscosity. The boundary condition $q_t = 0$ at the wall states that the fluid layer touching the wall does not slip with respect to it. But if the second fluid layer, adjacent to the first, may slide freely with respect to the first, a third layer further away from the wall would have no way of learning that the first layer did not slip. In other words, the transmission of the no-slip boundary condition to points not touching the boundary is done through shear stresses, and for sufficiently small ν this transmission decays very quickly. Thus regions far from rigid boundaries in small-ν large-Reynolds-number flows are not affected by the no-slip boundary condition. In these regions, e.g., at point F in Fig. 10.2, the Euler equations with the boundary condition $q_n = 0$ only yield good approximations. At points very close to the boundaries, point C, Fig. 10.2, both boundary conditions prevail and the Euler equations are no approximations at all. The layer of fluid very close to the rigid boundaries is denoted the *Boundary Layer*. This region is considered in the next chapter.

The region outside the boundary layer, where the Euler equations serve as good approximations, is sometimes referred to as the *region of negligible shear*, in reminiscence of the $\nu \to 0$ formulation. The Euler equations are solved there with the remaining boundary condition $q_n = 0$ satisfied at the rigid boundaries, Fig. 10.3. The geographical fact that thin boundary layers do lie near the rigid boundaries is ignored and does not affect the solutions far from the boundaries. One remembers, however, that inside the boundary layers these negligible shear solutions are not valid at all.

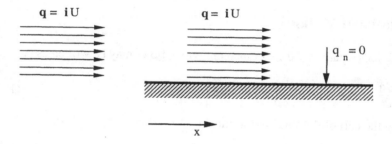

Figure 10.3 Negligible shear flow over a flat plate.

The Euler equation, (5.54), is now written in detail,

$$\Omega\frac{\partial \mathbf{q}}{\partial t} + \tfrac{1}{2}\nabla q^2 - \mathbf{q}\times\nabla\times\mathbf{q} = -\mathrm{Eu}\nabla p + \frac{1}{\mathrm{Fr}^2}\mathbf{g},$$

or dimensionally,

$$\frac{\partial \mathbf{q}}{\partial t} + \tfrac{1}{2}\nabla q^2 - \mathbf{q}\times\nabla\times\mathbf{q} = -\frac{1}{\rho}\nabla p + \mathbf{g} \qquad (10.1)$$

with the boundary condition

$\quad q_n = 0 \quad$ at rigid boundaries. $\qquad\qquad (10.2)$

The remainder of this chapter deals with solutions of the Euler equation.

Example 10.2

A two dimensional flow is given at $x \ll 0$ by

$$q = iU.$$

At $x \sim 0$ the flow encounters a flat plate parallel to it as shown in Fig. 10.3. Assume this to be a flow with negligible shear. How is the flow modified at $x > 0$?

Solution

Euler equation (10.1) is satisfied by

$$q = iU.$$

This velocity also satisfies $q_n = 0$ on the flat plate. Therefore $q = iU$ is the solution everywhere and the negligible shear flow is not modified at all. Note, however, that $q = iU$ is a good approximation to the physical flow only outside the plate boundary layer and certainly not for $x > 0$, $y = 0$.

Irrotational Motion

Euler's equation (10.1) for incompressible flows may be written as[*]

$$\frac{\partial q}{\partial t} - q \times \nabla \times q = -\nabla \left[\frac{p}{\rho} + \tfrac{1}{2} q^2 + gh \right]. \tag{10.3}$$

Taking the curl of this equation results in

$$\frac{\partial}{\partial t}(\nabla \times q) - \nabla \times (q \times \nabla \times q) = 0. \tag{10.4}$$

Define a *vorticity vector* $\xi = \nabla \times q$ and Eq. (10.4) becomes

$$\frac{\partial \xi}{\partial t} - \nabla \times (q \times \xi) = 0. \tag{10.5}$$

One way to satisfy Eq. (10.5) is by $\xi = 0$. Flows for which $\xi = 0$ are called *irrotational flows*.

The vorticity vector written in cartesian coordinates is

$$\xi = \left(\frac{\partial w}{\partial y} - \frac{\partial v}{\partial z} \right) i + \left(\frac{\partial u}{\partial z} - \frac{\partial w}{\partial x} \right) j + \left(\frac{\partial v}{\partial x} - \frac{\partial u}{\partial y} \right) k ; \tag{10.6}$$

---

[*] Note that for compressible barotropic flows this form still holds; however, the term $\qquad$ on the right hand side changes to

in cylindrical coordinates it is

$$\xi = \left[\frac{1}{r}\frac{\partial q_z}{\partial \theta} - \frac{\partial q_\theta}{\partial z}\right]\mathbf{e}_r + \left[\frac{\partial q_r}{\partial z} - \frac{\partial q_z}{\partial r}\right]\mathbf{e}_\theta + \left[\frac{1}{r}\frac{\partial(rq_\theta)}{\partial r} - \frac{1}{r}\frac{\partial q_r}{\partial \theta}\right]\mathbf{e}_z; \qquad (10.7)$$

and in spherical coordinates

$$\xi = \frac{1}{r\sin\theta}\left[\frac{\partial(q_\phi \sin\theta)}{\partial \theta} - \frac{\partial q_\theta}{\partial \phi}\right]\mathbf{e}_r + \frac{1}{r\sin\theta}\left[\frac{\partial q_r}{\partial \phi} - \frac{\partial(rq_\phi \sin\theta)}{\partial r}\right]\mathbf{e}_\theta + \frac{1}{r}\left[\frac{\partial(rq_\theta)}{\partial r} - \frac{\partial q_r}{\partial \theta}\right]\mathbf{e}_\phi.$$
$$(10.8)$$

The vorticity vector equals twice the angular velocity defined in Chapter 5. Hence, the term *irrotational flow*, which implies *flows with no angular velocity*.

Returning to Eq. (10.6) we note that for a flow to be irrotational, i.e., to satisfy $\nabla \times \mathbf{q} = 0$, it must also satisfy the following conditions simultaneously:

$$\frac{\partial w}{\partial y} = \frac{\partial v}{\partial z}, \qquad \frac{\partial u}{\partial z} = \frac{\partial w}{\partial x}, \qquad \frac{\partial v}{\partial x} = \frac{\partial u}{\partial y}, \qquad (10.9)$$

and for two dimensional flows these conditions simplify to

$$\frac{\partial v}{\partial x} = \frac{\partial u}{\partial y}. \qquad (10.10)$$

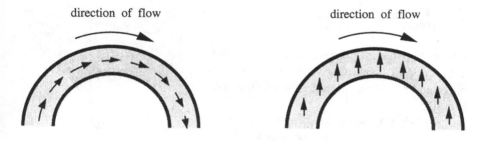

direction of flow   direction of flow

a. rotational flow   b. irrotational flow

Figure 10.4 Rotational vs. irrotational flow.

To gain some physical insight into the character of irrotational flows, we note that the motion of any solid object can always be described in terms of the motion of its center of gravity, or of some other fixed point on the object, and of its angular motion around this point. We place a small object in the flow and expect its motion to approximate that of the fluid if has displaced. In irrotational flow such a small object has no angular velocity. Figure 10.4 shows a small arrow-shaped object in rotational and in irrotational motions. In irrotational motion, though the path of the object may be curved, there is no angular velocity and the arrow retains its direction in space.

We now return to Eq. (10.3) and note that for flows that are steady and irrotational the left hand side of the equation vanishes, resulting in

$$\nabla\left[\frac{p}{\rho}+\tfrac{1}{2}q^2+gh\right]=0, \tag{10.11}$$

which implies

$$\frac{p}{\rho}+\tfrac{1}{2}q^2+gh=\text{const.} \tag{10.12}$$

Equation (10.12) holds for steady irrotational flows in the whole domain. It is the *strong Bernoulli equation* already discussed in Chapter 7.

Equation (10.5) states that all irrotational flows satisfy the Euler equation; thus solutions based on irrotationality need not be checked again on that point. These properties make it desirable to establish conditions under which real flows are irrotational.

In irrotational flows $\boldsymbol{\xi}=\nabla\times\mathbf{q}=0$. Using the Stokes theorem, we define a new concept related to fluid rotation,

$$-\int_{S_c}(\nabla\times\mathbf{q})\cdot\mathbf{n}\,dS=\oint_C\mathbf{q}\cdot d\mathbf{s}=\Gamma_c, \quad (10.13)$$

where Γ_c is the *circulation* along the closed curve c, S_c is a surface supported by this curve and $\mathbf{n}$ is the outer normal to this surface, Fig. 10.5.

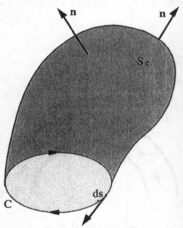

In what follows we prove that when Euler's equation describes correctly the flow, the fluid carries with it its circulation, which is preserved. This is known as the Kelvin theorem, which states:

Figure 10.5 Curve C supporting surface S_C.

In a barotropic flow which satisfies Euler's equation the circulation does not change.

A corollary of this theorem is that a flow which is irrotational at one moment is always irrotational. It starts with $\Gamma_c=0$ on any c and therefore must remain with zero circulation.

Proof:

Let AB be a curve consisting of a string of fluid elements which move as a part of the motion of the whole fluid, Fig. 10.6.

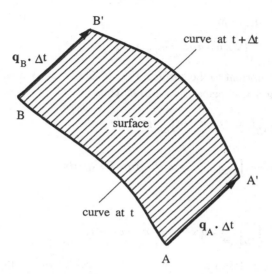

Figure 10.6 Surface $A'B'BA$ generated by the curve AB moving with the fluid.

Consider

$$\frac{D}{Dt}\int_{AB} \mathbf{q}\cdot d\mathbf{s},$$

where D/Dt denotes the rate of change *while following the same fluid elements*. This concept of the *material derivative* has already been used in deriving the *Reynolds transport theorem* in Chapter 4. Referring to Fig. 10.6,

$$\frac{D}{Dt}\int_{AB} \mathbf{q}\cdot d\mathbf{s} = \lim_{\Delta t\to 0}\frac{1}{\Delta t}\left\{\int_{A'B'} \mathbf{q}(t+\Delta t)\cdot d\mathbf{s} - \int_{AB}\mathbf{q}(t)\cdot d\mathbf{s}\right\}$$

$$= \lim_{\Delta t\to 0}\frac{1}{\Delta t}\left\{\int_{A'B'}\mathbf{q}(t+\Delta t)\cdot d\mathbf{s} - \int_{A'B'}\mathbf{q}(t)\cdot d\mathbf{s} + \int_{A'B'}\mathbf{q}(t)\cdot d\mathbf{s} - \int_{AB}\mathbf{q}(t)\cdot d\mathbf{s}\right\},$$

$$(10.14)$$

$$\lim_{\Delta t\to 0}\frac{1}{\Delta t}\int_{A'B'}[\mathbf{q}(t+\Delta t)-\mathbf{q}(t)]\cdot d\mathbf{s} = \int_{AB}\frac{\partial\mathbf{q}}{\partial t}\cdot d\mathbf{s}, \qquad (10.15)$$

$$\int_{A'B'}\mathbf{q}\cdot d\mathbf{s} - \int_{AB}\mathbf{q}\cdot d\mathbf{s} = \oint_{A'B'BAA'}\mathbf{q}\cdot d\mathbf{s} - \int_{B'B}\mathbf{q}\cdot d\mathbf{s} - \int_{AA'}\mathbf{q}\cdot d\mathbf{s}, \qquad (10.16)$$

$$\lim_{\Delta t\to 0}\frac{1}{\Delta t}\int_{AA'}\mathbf{q}\cdot d\mathbf{s} = \lim_{\Delta t\to 0}\frac{1}{\Delta t}\left[\Delta t q_A^2\right] = q_A^2. \qquad (10.17)$$

Similarly

$$\lim_{\Delta t \to 0} \frac{1}{\Delta t} \int_{BB'} \mathbf{q} \cdot d\mathbf{s} = q_B^2. \tag{10.18}$$

Now, by the Stokes theorem

$$\oint \mathbf{q} \cdot d\mathbf{s} = -\int_{S_{A'B'BA}} (\nabla \times \mathbf{q}) \cdot \mathbf{n} \, dS = -\int \boldsymbol{\xi} \cdot \mathbf{n} \, dS, \tag{10.19}$$

where $\mathbf{n}$ is the unit normal to the surface, Fig. 10.6. For $\Delta t \to 0$ $\mathbf{n} dS = -\Delta t \, \mathbf{q} \times d\mathbf{s}$ with $d\mathbf{s}$ taken along AB, directed from A to B. Thus

$$\oint \mathbf{q} \cdot d\mathbf{s} = -\Delta t \int_{AB} \boldsymbol{\xi} \cdot \mathbf{q} \times d\mathbf{s} = -\Delta t \int_{AB} \mathbf{q} \times \boldsymbol{\xi} \cdot d\mathbf{s} \tag{10.20}$$

and

$$\lim_{\Delta t \to 0} \frac{1}{\Delta t} \oint \mathbf{q} \cdot d\mathbf{s} = -\int_{AB} \mathbf{q} \times \boldsymbol{\xi} \cdot d\mathbf{s} = -\int_{AB} (\mathbf{q} \times \nabla \times \mathbf{q}) \cdot d\mathbf{s}. \tag{10.21}$$

Collecting all terms,

$$\frac{D}{Dt} \int_{AB} \mathbf{q} \cdot d\mathbf{s} = \int_{AB} \left[\frac{\partial \mathbf{q}}{\partial t} - \mathbf{q} \times \nabla \times \mathbf{q} \right] \cdot d\mathbf{s} + q_B^2 - q_A^2. \tag{10.22}$$

Now let AB be a closed curve, i.e., point B coincides with point A, and let this curve be denoted c. For this case $\oint_c \mathbf{q} \cdot d\mathbf{s} = \Gamma_c$ is the circulation of $\mathbf{q}$ along c, and Eq. (10.22) becomes

$$\frac{D\Gamma_c}{Dt} = \oint \left[\frac{\partial \mathbf{q}}{\partial t} - \mathbf{q} \times \nabla \times \mathbf{q} \right] \cdot d\mathbf{s} = \int_{S_c} \nabla \times \left[\frac{\partial \mathbf{q}}{\partial t} - \mathbf{q} \times \nabla \times \mathbf{q} \right] \cdot \mathbf{n} \, dS, \tag{10.23}$$

where the Stokes theorem has been used again. However, inspection of Eq. (10.3) reveals that the integrand in the surface integral in Eq. (10.23) has the structure of a curl of a gradient and therefore vanishes. Thus

$$\frac{D\Gamma_c}{Dt} = 0, \tag{10.24}$$

which completes the proof.

To describe what happens in *rotational flows* which satisfy Euler's equation, we need a new concept, that of the vortex lines. Let *vortex lines* be defined as the field lines of the vectorial field $\nabla \times \mathbf{q}$, i.e., as curves that are everywhere tangent to $\boldsymbol{\xi} = \nabla \times \mathbf{q}$. We now consider a vortex tube, Fig. 10.7, i.e., a surface that consists of vortex lines and encloses a tube within. Let three curves be chosen on this surface, one which is not linked with the tube, C, and two which are linked with it, A and B. Now the Helmholtz theorem states:

The circulation along a curve lying on the surface of a vortex tube and not linked with the tube is zero and remains so. The circulation along all curves lying on the surface of a vortex tube and linked with it is the same and remains constant in time.

The proof of the Helmholtz theorem is as follows:

For curve C:

$$\Gamma_c = \oint_C \mathbf{q} \cdot d\mathbf{s} = -\int_{S_c} \boldsymbol{\xi} \cdot \mathbf{n} \, dS = 0, \tag{10.25}$$

and this is so because $\boldsymbol{\xi}$ lies on the tube surface while $\mathbf{n}$ is normal to it. Because of Kelvin's theorem, however, $\Gamma_c = 0$ always.

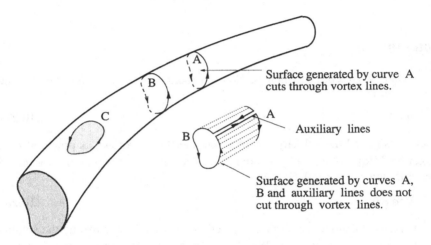

Surface generated by curve A cuts through vortex lines.

Auxiliary lines

Surface generated by curves A, B and auxiliary lines does not cut through vortex lines.

Figure 10.7 A vortex tube.

Curve A is considered together with curve B. For clarity these curves are drawn again in Fig. 10.7 without the tube. Now let one point be cut out of curve A and one point be cut out of curve B. Also let two auxiliary lines be drawn as shown, and compute $\oint \mathbf{q} \cdot d\mathbf{s}$ over the whole figure, as indicated by the arrows. This new figure comprises a curve not linked with the vortex tube, and therefore $\oint \mathbf{q} \cdot d\mathbf{s} = 0$. The integration on the auxiliary lines, which coincide in the limit, cancels out, and thus

$$\oint_B \mathbf{q} \cdot d\mathbf{s} - \oint_A \mathbf{q} \cdot d\mathbf{s} = 0,$$

or for all times

$$\Gamma_B = \Gamma_A, \tag{10.26}$$

which completes the proof.

Thus while the fluid particles move according to the velocity field, those lying on the surfaces of vortex tubes remain on those surfaces, and as they move, they take with them the tubes. Denoting the integral over the tube cross section, $\int \boldsymbol{\xi} \cdot \mathbf{n} \, dA$, the *intensity* of the tube, this intensity persists in time. Furthermore, a vortex tube cannot start or end in the fluid. It must either be linked into itself in a toroid shape or start and end at surfaces of discontinuity, such as rigid walls or free surfaces.

Armed with Kelvin and Helmholtz theorems we now proceed to consider irrotational solutions to Euler's equation, and for simplicity we assume these barotropic flows to be also incompressible.

Potential Flow

When $\nabla \times \mathbf{q} \equiv 0$, i.e., when the flow is irrotational, there always exists a function ϕ such that

$$\mathbf{q} = \nabla \phi, \tag{10.27}$$

where ϕ is called the velocity potential and the flow is known as potential flow. Irrotational flow is thus also potential flow. The equation of continuity for incompressible flows is $\nabla \cdot \mathbf{q} = 0$. Hence

$$\nabla^2 \phi = 0. \tag{10.28}$$

The velocity potential is a harmonic function and can be obtained by the solution of the Laplace equation (10.28). A proper set of boundary conditions must guarantee no normal flow relative to rigid surfaces. It is noted that the solution of Eq. (10.28) is really a direct integration of the equation of continuity, while the Euler equation is implicitly satisfied by irrotationality. Euler's equation is still used through the Bernoulli equation to evaluate the pressure. The momentum and continuity equations thus become uncoupled. Since the continuity equation is linear, as is Eq. (10.28), superposition of solutions is permissible, and elaborate flows may be constructed by superpositions of simpler ones. It is noted, however, that Euler's equation is not linear, and neither is Bernoulli's equation. Therefore the pressure in a flow obtained by superposition of two flows is not the sum of the pressures of the two partial flows.

We now consider two-dimensional flows. Equation (10.27) states

$$u = \frac{\partial \phi}{\partial x}, \qquad v = \frac{\partial \phi}{\partial y}.$$

Another differentiation yields

$$\frac{\partial^2 \phi}{\partial x \partial y} = \frac{\partial u}{\partial y} = \frac{\partial v}{\partial x},$$ (10.29)

which can be identified as a condition for irrotationality, Eq. (10.10). Equation (10.29) can be combined with Eq. (5.13) for the stream function

$$u = \frac{\partial \psi}{\partial y}, \qquad v = -\frac{\partial \psi}{\partial x}$$

to become

$$\frac{\partial^2 \psi}{\partial x^2} + \frac{\partial^2 \psi}{\partial y^2} = 0.$$ (10.30)

Hence, in two-dimensional irrotational flow the stream function is also harmonic. Furthermore,

$$u = \frac{\partial \phi}{\partial x} = \frac{\partial \psi}{\partial y}, \qquad v = \frac{\partial \phi}{\partial y} = -\frac{\partial \psi}{\partial x},$$ (10.31)

or in polar coordinates

$$q_r = \frac{\partial \phi}{\partial r} = \frac{1}{r}\frac{\partial \psi}{\partial \theta}, \qquad q_\theta = \frac{1}{r}\frac{\partial \phi}{\partial \theta} = -\frac{\partial \psi}{\partial r}.$$ (10.32)

Equations (10.31) and (10.32) are known as the Cauchy–Riemann conditions.

The family of $\phi = \text{const}$ lines and that of $\psi = \text{const}$ lines intersect orthogonally. This is easily shown by noting that

$$\nabla\phi = \mathbf{i}u + \mathbf{j}v, \qquad \nabla\psi = -\mathbf{i}v + \mathbf{j}u, \qquad \nabla\phi \cdot \nabla\psi = 0.$$ (10.33)

Example 10.3 – Source Flow

Consider a steady two-dimensional axisymmetrical flow from a line source. Such a flow can be approximated by a radial flow emitted through the wall of a long slender tube made of porous material. The tube lies along the z-axis and the fluid flows radially in the x-y plane. Let Q be the fluid flow rate per unit length of tube. This flowrate is called the *intensity* of the line source. Continuity yields the fluid velocity,

$$q_r = \frac{Q}{2\pi r}, \qquad q_\theta = 0.$$ (10.34)

The stream function for source flow can be calculated from Eq. (10.32),

$$\frac{Q}{2\pi r} = \frac{1}{r}\frac{\partial \psi}{\partial \theta}, \qquad \frac{\partial \psi}{\partial r} = 0.$$

Integration of $\partial\psi/\partial\theta$ yields

$$\psi = \left(\frac{Q}{2\pi}\right)\theta + f(r),$$

and $\partial\psi/\partial\theta = 0$ yields $f = $ const, which is chosen to be zero. Thus

$$\psi = \left(\frac{Q}{2\pi}\right)\theta. \tag{10.35}$$

Constant ψ lines are therefore radii, as shown in Fig. 5.6. Similarly, the flow potential is evaluated from

$$\frac{\partial\phi}{\partial r} = \frac{Q}{2\pi r}, \qquad \frac{1}{r}\frac{\partial\phi}{\partial\theta} = 0,$$

which results in

$$\phi = \left(\frac{Q}{2\pi}\right)\ln r. \tag{10.36}$$

The origin itself is a singular point, but everywhere else the flow is well defined.

A source with a reversed flow is a *sink*. Accordingly, for the flow into a sink

$$\psi = -C\theta, \qquad \phi = -C\ln r, \tag{10.37}$$

where $C = Q/2\pi$.

Example 10.4 – Potential Vortex

In potential vortex flow the radial velocity component q_r is zero and only q_θ exists. The condition of irrotationality results in $\xi = 0$. Hence, Eq. (10.7) requires

$$\frac{1}{r}\frac{\partial(rq_\theta)}{\partial r} = 0 \tag{10.38}$$

which upon integration yields $q_\theta = C/r + f(\theta)$, and because nothing depends on θ, $f = $ const $= 0$ is chosen. Thus

$$q_\theta = \frac{C}{r}. \tag{10.39}$$

Substitution into Eq. (10.32) results in expressions for ψ and ϕ,

$$\psi = -C\ln r, \qquad \phi = C\theta. \tag{10.40}$$

Comparison with the source flow, Fig. 10.8, shows that ϕ and ψ lines have simply changed their roles. Indeed, changing the roles of ϕ and ψ lines in any potential flow yields a new flow pattern.

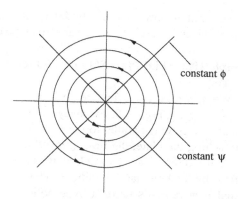

constant ϕ

constant ψ

The constant C in Eq. (10.40) is related to the circulation around the vortex, as can be shown using Eq. (10.25):

Figure 10.8 Vortex flow.

$$\Gamma = \oint_c \mathbf{q} \cdot ds = \int \left(\frac{C}{r} \right) \mathbf{e}_\theta \cdot \left(\mathbf{e}_r dr + \mathbf{e}_\theta r d\theta \right) = \int_0^{2\pi} C d\theta = 2\pi C.$$

Hence

$$C = \frac{\Gamma}{2\pi} \qquad\qquad (10.41)$$

and

$$\psi = -\frac{\Gamma}{2\pi} \ln r, \qquad \phi = \frac{\Gamma}{2\pi} \theta. \qquad\qquad (10.42)$$

There is no contradiction between $\Gamma \neq 0$ and the irrotationality of this flow. The circulation along any closed path not linked with the origin is still zero. The singular point at the origin contributes the Γ value in Eq. (10.42), and the same Γ is obtained along any contour linked with the origin.

Example 10.5 – Parallel Flow

Consider uniform flow with a constant velocity, U, parallel to the x-axis. Thus,

$$q_x = u = U , \qquad q_y = v = 0. \qquad\qquad (10.43)$$

Equation (10.31) yields

$$\psi = Uy, \qquad \phi = Ux. \qquad\qquad (10.44)$$

Since in potential flows any streamline may be interpreted as a rigid boundary, the parallel flow may also be interpreted as a flow over a flat plate (say the x-axis between $x = -a$ and $x = a$) at zero angle of incidence; see Example 10.2.

Example 10.6 – Flat Plate Stagnation Flow

We now describe a flow perpendicular to a flat plate, Fig. 10.9. Far away from the plate the flow is parallel to the x-axis. Approaching the plate, the flow gradually turns until it becomes parallel to it, as required by the boundary conditions. This gradual turn reminds us of a hyperbolic arc, which we may try as a streamline. We are also guided toward this choice by the fact that an expression quadratic in x and y may be amenable to satisfy the Laplace equation. Trying

$$\psi = Cxy , \qquad (10.45)$$

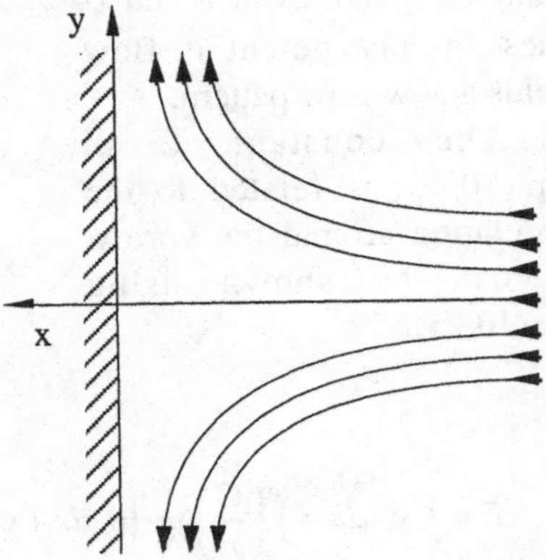

Figure 10.9 Flat plate stagnation flow.

we find that indeed $\nabla^2 \psi = 0$ and the resulting flow field is the two-dimensional stagnation flow on a flat plate. The velocity components are obtained as

$$u = \frac{\partial \psi}{\partial y} = Cx = \frac{\partial \phi}{\partial x} ,$$
$$v = -\frac{\partial \psi}{\partial x} = -Cy = \frac{\partial \phi}{\partial y} . \qquad (10.46)$$

The flow potential may now be computed as

$$\phi = \int^x \frac{\partial \phi}{\partial x} dx + f(y) = \int^x Cx \, dx + f(y) = \tfrac{1}{2}Cx^2 + f(y). \qquad (10.47)$$

Differentiating Eq. (10.47), we obtain

$$\frac{\partial \phi}{\partial y} = \frac{df}{dy} = -Cy.$$

Hence

$$f = -\tfrac{1}{2}Cy^2 \qquad (10.48)$$

and

$$\phi = \tfrac{1}{2}C(x^2 - y^2), \qquad (10.49)$$

which is also a hyperbola.

Example 10.7 – Doublet Flow

Consider the superposition of two flows: $\psi_a = C\theta_a$, which is a source located at point a, Fig. 10.10, and $\psi_b = -C\theta_b$, which is a sink located at point b. The combined flow has the stream function

$$\psi = \psi_a + \psi_b = C(\theta_a - \theta_b).$$
(10.50)

From Fig. 10.10, $\theta = \theta_b - \theta_a$, and thus lines of constant ψ are also lines of constant θ, i.e., circles. We may, of course, use the expressions (10.36) and also obtain

$$\phi = \left(\frac{Q}{2\pi}\right)(\ln r_a - \ln r_b).$$
(10.51)

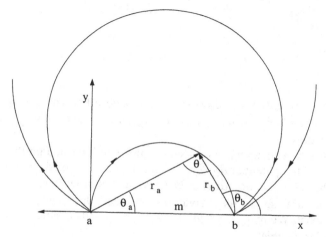

Figure 10.10 Doublet flow.

Let point a coincide with the origin of the coordinates, and let point b be located at $x = m$, $y = 0$. Thus

$$\ln r_a = \tfrac{1}{2}\ln(x^2 + y^2)$$

and

$$\ln r_b = \tfrac{1}{2}\ln\left[(x-m)^2 + y^2\right].$$

Hence

$$\phi = \frac{Q}{4\pi}\ln\frac{x^2 + y^2}{(x-m)^2 + y^2}.$$
(10.52)

We note that the argument showing the streamlines to be circles does not depend on m. It is interesting therefore to check what happens as we let $m \to 0$. Substitution of $m = 0$ in Eq. (10.52) leads of course to $\phi = 0$, as the source and the sink counteract and cancel each other. However, we may imagine a process by which as $m \to 0$, $Q \to \infty$. We want Q to increase at such a rate that the limit obtained for ϕ is well defined and finite. Thus,

$$\phi = \frac{1}{4\pi} \lim_{m \to 0} Q\left\{ \ln(x^2 + y^2) - \ln\left[(x - m)^2 + y^2\right] \right\}, \tag{10.53}$$

and we note that if $Q \to \infty$ as M/m does when $m \to 0$, the expression becomes

$$\phi = \frac{1}{4\pi} \lim_{m \to 0}\left[\frac{M}{m}\left(\ln(x^2 + y^2) - \ln[(x - m)^2 + y^2]\right) \right] = \frac{M}{4\pi} \frac{\partial}{\partial x}\left[\ln\left(x^2 + y^2 \right) \right] \tag{10.54}$$

or

$$\phi = \frac{M}{4\pi} \frac{\partial}{\partial x} \ln(x^2 + y^2) = \frac{M}{2\pi} \frac{x}{x^2 + y^2} = \frac{M}{2\pi} \frac{r\cos\theta}{r^2}.$$

Hence

$$\phi = \frac{M}{2\pi} \frac{\cos\theta}{r}. \tag{10.55}$$

This new flow potential, ϕ, describes a flow pattern known as *doublet flow*. Its streamlines are still circles, and it has one singular point at the doublet itself, i.e., in this example the origin.

Similarly, the coalescence of a source and a sink located on the y-axis leads to a doublet with a potential $\partial\phi_a/\partial y$. The results thus obtained could be expected, at least in the sense that $\partial\phi_a/\partial x$ and $\partial\phi_a/\partial y$ yield new flow fields. If ϕ satisfies the Laplace equation, so do its derivatives provided the required higher partial derivatives exist. We may also find that $\partial^2\phi_a/\partial x\,\partial y$ leads to a new field describing two coalescing doublets.

Example 10.8 – Flow over Streamlined Bodies

First we consider the superposition of a parallel flow and a two-dimensional source flow, Fig. 10.11. The source is located at the origin and has the stream function

$$\psi_S = \frac{Q}{2\pi}\theta. \tag{10.56}$$

The parallel flow is taken to proceed from left to right and thus has the stream function

$$\psi_p = Uy = Ur\sin\theta. \tag{10.57}$$

The combined flow has the stream function

$$\psi = \frac{Q}{2\pi}\theta + Ur\sin\theta \qquad (10.58)$$

and therefore the velocity components

$$q_r = \frac{1}{r}\frac{\partial\psi}{\partial\theta} = \frac{Q}{2\pi r} + U\cos\theta,$$

$$\qquad (10.59)$$

$$q_\theta = -\frac{\partial\psi}{\partial r} = -U\sin\theta.$$

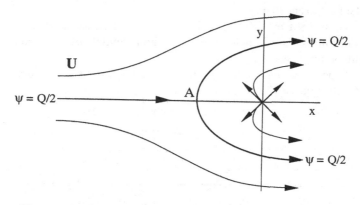

Figure 10.11 Superposition of source flow and parallel flow.

Equation (10.58) may require pointwise numerical calculation to actually give the shape of the streamlines, i.e., lines of constant ψ. Still some qualitative information can be obtained by general considerations. We recognize the need for the stagnation point A. Inspecting Eq. (10.59), we see that for $\theta = \pi$ and $r = Q/2\pi U$ both q_r and q_θ vanish, and a stagnation point is obtained. We could have reached the same conclusion by the argument that the velocity along the splitting streamline, $\psi = Q/2$, as it comes from the left towards the source could not proceed into the source, and therefore the streamline must split. At the splitting point the velocity vector has more than one direction, and its magnitude must therefore be zero.

We also note that the streamline $\psi = Q/2$ encircling the source cannot close again. Far to the right the flow becomes parallel again, but the splitting streamline does not close because if it did, the output of the source would have nowhere to go.

Can we superimpose another flow on what we already have, so as to close the splitting streamline? We already know that in flows with negligible shear any streamline may represent a solid boundary, simply because it satisfies the boundary condition of no normal velocity, i.e., the only boundary condition required in such flows. If we could close the splitting streamline, it would represent a finite solid two-dimensional body immersed in the flow. This can be done in a number of ways. We could place a sink of the same intensity Q somewhere to the right of the source. This sink would swallow up all the output of the source and thus allow the splitting streamline to close. Another possibility is to place several sinks to the right of the source, such that the sum of their intensities equals Q. We could also replace the source by several sources and have a series of sources and sinks, which, as long as the sum total of their intensities is zero, would yield a closed splitting streamline. By varying the distributions of these sources and sinks, we could obtain different shapes.

The whole idea may be carried into three dimensions, where the stream function does not satisfy Laplace equation. The potential still does, and superposition is still permissible.

The shapes thus obtained, in both two and three dimensions, are known as *Rankine Ovals*, Fig. 10.12.

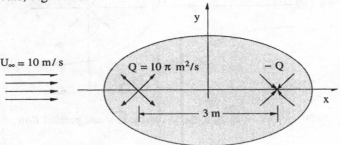

Figure 10.12 Rankine oval.

Example 10.9 – Flow over a Circular Cylinder

A particular case of a two-dimensional Rankine oval is that of parallel flow, one source and one sink. We let the sink coalesce with the source in a limiting process to become a doublet. The superposition of the parallel flow and the doublet, Eqs. (10.43) and (10.55), yields the potential

$$\phi = Ur\cos\theta + \frac{M}{2\pi}\frac{\cos\theta}{r}. \tag{10.60}$$

Thus,

$$q_r = \frac{\partial\phi}{\partial r} = \left[1 - \frac{M}{2\pi r^2 U}\right]U\cos\theta = \frac{1}{r}\frac{\partial\psi}{\partial\theta},$$

$$q_\theta = \frac{1}{r}\frac{\partial\phi}{\partial\theta} = -\left[1 + \frac{M}{2\pi r^2 U}\right]U\sin\theta = -\frac{\partial\psi}{\partial r}, \tag{10.61}$$

from which

$$\psi = \left[1 - \frac{M}{2\pi r^2 U}\right] Ur \sin \theta.$$ (10.62)

Setting $R^2 = M/2\pi U$ gives

$$\psi = \left[1 - \left(\frac{R}{r}\right)^2\right] Ur \sin \theta$$ (10.63)

and

$$q_r = \left[1 - \left(\frac{R}{r}\right)^2\right] U \cos \theta,$$

(10.64)

$$q_\theta = -\left[1 + \left(\frac{R}{r}\right)^2\right] U \sin \theta.$$

This is a flow around a circular cylinder. Indeed we see that for $\theta = 0$ and $\theta = \pi$, $\psi = 0$. Also, for $r = R$, $\psi = 0$, and the $\psi = 0$ line is as shown in Fig. 10.13a.

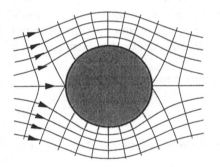

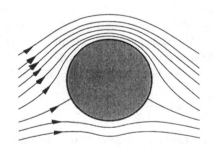

a. Without circulation   b. With circulation

Figure 10.13 Flow around a cylinder.

Far from the cylinder, i.e., for large r/R, $r/R^2 \gg 1/r$, and

$$\psi = \lim_{\frac{r}{R} \to \infty} \left[1 - \left(\frac{R}{r}\right)^2\right] Ur \sin \theta = Ur \sin \theta = Uy,$$ (10.65)

which is parallel flow.

Inspection of Eq. (10.64) reveals that on the cylinder, i.e., at $r = R$, $q_r = 0$ as required by the boundary condition, and

$$q_\theta = -2U \sin \theta. \tag{10.66}$$

We may use this q_θ in Bernoulli's equation (10.12) to obtain the pressure distribution on the cylinder. Neglecting gravitational effects and performing a Bernoulli balance between $r \to \infty$ and $r = R$, we obtain

$$\frac{p_\infty}{\rho} + \frac{U^2}{2} = \left(\frac{p}{\rho} + \frac{q_\theta^2}{2} \right)_{r=R} = \text{const.} \tag{10.67}$$

Hence

$$p_R = p_\infty + \frac{\rho U^2}{2}\left(1 - 4\sin^2\theta\right), \tag{10.68}$$

and we find that because $\sin^2\theta$ is symmetric with respect to both the x-axis and the y-axis, the flow exerts no drag force or lift force on the cylinder.

We may try to destroy this symmetry by the superposition of yet another flow, the potential vortex flow. In this choice we are guided by the following observation:

The streamlines in potential vortex flow are circles. One of these circles will coincide with the $r = R$ circle, which is considered as the rigid cylinder. Thus, while the flow pattern in the field is modified by the addition of the potential vortex, that particular circular streamline $r = R$ remains circular. Hence we still have a flow around a cylinder.

We note, however, that for the potential vortex rotating clockwise,

$$\psi_V = \frac{\Gamma}{2\pi}\ln r, \tag{10.69}$$

and the field in general is affected nonsymmetrically as required. Thus the combined stream function becomes

$$\psi = +\frac{\Gamma}{2\pi}\ln r + \left[1 - \left(\frac{R}{r}\right)^2\right]Ur\sin\theta \tag{10.70}$$

and

$$q_\theta = -\frac{\Gamma}{2\pi r} - \left[1 + \left(\frac{R}{r}\right)^2\right]U\sin\theta, \tag{10.71}$$

which on the cylinder reduces to

$$q_\theta\big|_{r=R} = -\frac{\Gamma}{2\pi R} - 2U\sin\theta. \tag{10.72}$$

Substitution of this q_θ in Eq. (10.67) now yields

$$p_R = p_\infty + \frac{\rho U^2}{2}\left[1 - 4\sin^2\theta\right] - \frac{\rho}{2}\left(\frac{\Gamma}{2\pi R}\right)^2 - \rho\frac{\Gamma U}{\pi R}\sin\theta \ . \tag{10.73}$$

This pressure distribution is symmetrical with respect to the y-axis but not so with respect to the x-axis. The nonsymmetrical term $\rho(\Gamma U/\pi R)\sin\theta$ now yields a lift force. From Fig. 10.12b

$$dL = -pds\sin\theta = -pR\sin\theta\,d\theta = \rho\frac{\Gamma U}{\pi}\sin^2\theta\,d\theta.$$

Hence

$$L = \frac{\rho\Gamma U}{\pi}\int_0^{2\pi}\sin^2\theta\,d\theta = \rho\Gamma U \tag{10.74}$$

and a lift force has been obtained.

One is tempted to look for another potential flow such that when superimposed on the one already considered will destroy the symmetry with respect to the y-axis and thus produce drag. No such flow exists. The only mechanism which can produce entropy in incompressible flows is viscous dissipation. Viscosity effects are barred from potential flows. Drag force, i.e., a force in the direction of the undisturbed flow far from the cylinder, is always associated with entropy production; we therefore must not expect drag in any potential flow. To produce drag we must admit viscosity back into the field, at least in some thin layers near the boundaries of solid bodies, as will be done in the next chapter.

Example 10.10 – Magnus Effect

A cylinder of radius R is dragged sidewise under water and is made to rotate at the same time. It is believed that the flow around it is potential and that the stream function and the circumferential velocity are given by Eqs. (10.70) and (10.71), i.e.,

$$\psi = -\frac{\Gamma}{2\pi}\ln r + \left[1 - \left(\frac{R}{r}\right)^2\right]Ur\sin\theta,$$

$$q_\theta = -\frac{\Gamma}{2\pi r} - \left[1 + \left(\frac{R}{r}\right)^2\right]U\sin\theta.$$

Find the lift force acting on the cylinder.

Solution

This example seems the same as Example 10.8. However, the density of water is high and we, therefore, include the corresponding term in the Bernoulli polynomial, Eq. (10.67), which becomes for this case

$$\frac{p_\infty}{\rho} + \frac{U^2}{2} + 0 = \left(\frac{p}{\rho} + \frac{q_\theta^2}{2} + gy\right)_{r=R} = \text{const.}$$

Because $y|_{r=R} = K \sin\theta$, the pressure on the cylinder is [see Eq. (10.73)]

$$p_R = p_\infty + \rho\frac{U^2}{2}\left[1 - 4\sin^2\theta\right] - \frac{\rho}{2}\left(\frac{\Gamma}{2\pi R}\right)^2 - \rho\frac{\Gamma U}{\pi R}\sin\theta - \rho g R \sin\theta.$$

The integral of the first three terms in this expression is still zero, as in Example 10.8. However, the last two terms yield

$$L = \left(\rho\frac{\Gamma U}{\pi R} + \rho g R\right)\int_0^{2\pi} R\sin^2\theta\,d\theta = \rho\Gamma U + g\rho\pi R^2.$$

Note that the addition to the lift force in this case, i.e., $g\rho\pi R^2$, equals the buoyancy force one would expect to find in a hydrostatic situation.

References

L.M. Milne–Thompson, "Theoretical Aerodynamics," 4th ed., Macmillan, London, 1960.

A.H. Shapiro, "Shape and Flow," Doubleday, New York, 1961.

Problems

10.1 Suppose that in Fig. 10.12 we have two line sources perpendicular to the plane of the paper.
   a. Find the stream function.
   b. Find the potential function.
   c. Find the velocity components.
   d. Are the streamlines and equipotential lines orthogonal?
   e. Sketch the streamlines and equipotential lines (surfaces).

10.2   A two-dimensional source of strength $Q = 100\pi$ (m³/s)/m is located at $A(-2, 3)$. A sink of the same strength is located at $B(2, 3)$. All lengths are measured in meters.
Add a flow parallel to the x-axis and find a streamline that may represent a body of maximum width of 1.5 m.

10.3   A flow is given in Fig. 10.12 consisting of a superposition of a source–sink and uniform flow.
a.   Find the stagnation point
b.   Sketch to scale the streamline that represents the rigid body.
c.   Determine the pressure distribution along the surface of this body.
d.   Is there a drag force?

10.4   Instead of the sink in Fig. 10.12 there are two sinks at $x = 0$ and $x = 1.5$. The intensity of the sinks is $Q = -5\pi$ m² / s each. Repeat Problem 10.3 for this case.

10.5   In two-dimensional incompressible flow the velocity vector is given as

$$q = iu + jv,$$

where

$$u = 3y \text{ m/s}, \qquad v = -3x \text{ m/s}.$$

a.   What is the equation of the stream function?
b.   Sketch a few streamlines and calculate their values.

10.6   In two-dimensional incompressible flow the velocity vector is

$$q = e_r q_r + e_\theta q_\theta,$$

where

$$q_r = \frac{Q}{2\pi r}, \qquad q_\theta = 0.$$

Derive an expression for the streamlines.

10.7   A solid wall is added to the field in Problem 10.2 along the x-axis, Fig. P10.7.
a.   Find the velocity at point C (0,1) with and without the parallel flow.
b.   Find the volumetric flux between the origin $(0,0)$ and point $C(0,1)$ with and without the parallel flow.
Hint: Consider the x-axis to be an axis of symmetry, or a mirror.

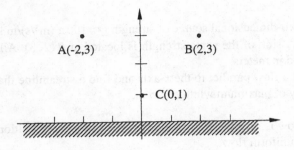

Figure P10.7 Source, sink and wall.

10.8 Consider a two-dimensional flow field with the potential ϕ and the stream function ψ formulated in the complex form $\phi(z) = \phi + i\psi$. For example $\phi(z) = z = x + iy = \phi + i\psi$. Draw some streamlines and interpret the flow (e.g., flow around a circle?)

Repeat for $\phi(z) = z^2 = x^2 y^2 + i2xy = \phi + i\psi$. Do ϕ and ψ satisfy the Laplace equation?

10.9 Repeat Problem 10.8 for $\phi(z) = z^3$, $\phi(z) = z^4$.

10.10 Repeat Problem 10.2 for a three-dimensional source and sink of intensity 100π m³/s.

10.11 Repeat Problem 10.7 for a three-dimensional source and sink for the following cases:
   a. With parallel flow at $C(0, 1, 0)$.
   b. Without the parallel flow.

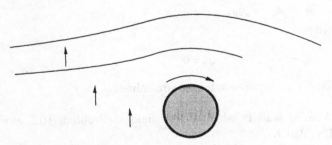

Figure P10.12 Flow around a rotating cylinder.

10.12 Figure P10.12 shows the cylinder of Example 10.10, with the flow around it. The other curves and the arrows were drawn by someone who wanted to show the rotationality (or irrotationality) of the field, the same as in

Fig. 10.4. The curves represent streamlines. Draw the arrows as they look in other locations in the field.

10.13 Given a potential function in spherical coordinates

$\phi = A / R,$

show that it is harmonic. Find its velocity components. Sketch this flow. Does this flow have a stream function? If yes, find it.

10.14 Figure P10.14 shows a two-dimensional flow inside a sharp bend. A stream function is suggested:

$$\psi = A r^{\frac{\pi}{\alpha}} \sin\left(\frac{\theta \pi}{\alpha}\right).$$

a. Show that indeed this is a stream function and that it satisfies the Laplace equation.
b. Show that it is the stream function of this flow, that its velocity satisfies the boundary conditions.
c. Find the potential of the flow.

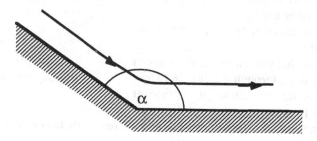

Figure P10.14 Two-dimensional flow in a bend.

10.15 a. A three-dimensional point source of intensity $Q = 5$ m³/s is located at $z = 1$. Find the stream function of the resulting flow. Sketch the flow. Are there stagnation points?
b. Superimpose a parallel flow of the speed $w = 2$ m/s in the z-direction. Repeat part a.

10.16 a. A three-dimensional point source of intensity $Q = 5$ m³/s is located at the origin. A sink of the same strength is located at $z = 1$. Find the stream function of the resulting flow. Sketch the flow. Are there stagnation points?
b. Superimpose a parallel flow of a speed $w = 2$ m/s in the z-direction. Repeat part a.

10.17 A circular rod with rounded tips has a diameter of 0.05 m and a length of 1 m from tip to tip. The rod is fixed in a parallel flow of the speed of 3 m/s, with the rod axis in the direction of the flow. Using one source and one sink find the "best fit" to the rod and the flow resulting from it.

A "best fit" is defined here as a closed stream surface with stagnation points at the tips of the rod and with a girdle that does not bulge out of the cylinder prescribed by the rod yet touches this cylinder.

10.18 A two-dimensional source of intensity $Q_1 = 100$ m³/m·s is located at $x = -1$, $y = 1$. A similar source is located at $x = 1$, $y = 1$. A two-dimensional sink of intensity $Q_3 = 200$ m³/m·s is located at $x = 0$, $y = -1$. The field is two dimensional and contains the two points A (−0.5; 0.5) and B(−0.5; −0.5).

a. Sketch the flow.

b. Find the velocity vector at point A.

c. Find the volume flux between point A and point B.

10.19 The angle between two diverging walls is 36°, Fig. P10.19. A two-dimensional source of strength $Q = 2$ m³/s·m is located at $x = 1$ m.

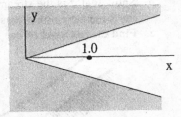

a. Find the velocity field between the walls.

b. Find the volumetric flux between the walls through a plane $x = 15$ m.

Figure P10.19

c. Find the flux through a plane at $x = 16$ m.

d. Find the force acting on the part of one wall between $x = 15$ m and $x = 16$ m.

10.20 Put a two-dimensional source at the origin such that the volume flux through a sector of 36° is 2 m³/s·m.

a. Is this a possible solution of the Euler equation for the flow between two diverging walls?

b. Find for this flow all quantities sought in Problem 10.19 .

10.21 The walls in Fig. P10.21 are two dimensional. A three-dimensional source of strength $Q = 4$ m³/s is located at $x = 4$ m, $y = 1$, $z = 0$.

a. Find the velocity at $x = 8$ m, $y = 1$, $z = 0$.

b. Find the velocity at $x = y = z = 1$.

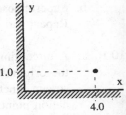

Figure P10.21

10.22 Find the total force on the wall $x = 0$ and on the wall $y = 0$, in Fig. P10.21, under the flow conditions of Problem 10.21. Use analytic or numerical integration.

10.23 A two-dimensional wing is schematically shown in Fig. P10.23. The length of the underside arc of the wing is 1 m, and the length of the convex arc is 1.2 m. The velocity of the air streaming toward the wing is 50 m/s. Assuming the air velocities on the wing to be proportional to the arcs lengths, estimate the lift force generated by the wing.

Figure P10.23

10.24 A potential flow is obtained by the superposition of a two-dimensional sink of strength $-Q = -2$ m^3 / s·m and of a potential vortex of circulation $\Gamma = 3$ m^2/s, both located at the origin.
   a. Find the flow velocity at $r = 1$ m, at $r = 0.5$ m.
   b. Find the pressure difference between these points and a point at $r = 20$ m .

10.25 The flow field around a sharp rod sticking in front of an airplane may be obtained by the superposition of three flows: a parallel flow at 45 m/s in the x-direction, a three-dimensional source $(Q_a = 10 \, m^3/s)$ located at the origin, and a source $(Q_b = 1 \, m^3/s)$ located on the x-axis at $x = -0.2$ m. Find the stagnation point and draw the streamline representing the rod.

10.26 A parallel flow is superimposed on a two-dimensional source–sink pair flow, resulting with a Rankine oval flow. The Rankine oval is denoted an obstacle. Find how far form the obstacle the modification of the parallel flow by the obstacle is less than 1%.

10.27 Use the results of Problem 10.26 to formulate the potential flow field in a two-dimensional channel with an obstacle in it.

10.28 Repeat Problem 10.26 with a three-dimensional source–sink pair.

10.29 A 10-m-high circular cylinder has the diameter of 1 m and rotates at 800 rpm. A wind blows across the cylinder with the velocity of 10 m/s. Find the sideways lift force on the cylinder.

10.30 A three-dimensional source of strength 20 m^3/s is located on the x-axis at $x = 5$ m. Another source of strength 10 m^3/s is also located on the x-axis, at $x = 10$ m. Draw the flow field at the x-y plane.

11. HIGH REYNOLDS NUMBER FLOWS – THE BOUNDARY LAYER

In this chapter we still consider flows in which Re → ∞. In the previous chapter we sought the flow field in regions which were far away from rigid boundaries. Now we want to complete the description and treat regions quite close to rigid boundaries. The geometry of these boundaries is generally flat, e.g., a flat plate, and the large Re = UL/ν associated with the flow is based on the large velocity U, which is approximately parallel to the rigid surface. The Navier–Stokes equations for these flows may be approximated by the boundary layer equations, Eqs. (8.38) - (8.42). We already know that in large Re flows the flow not near the boundaries is approximated rather well by solutions of the Euler equation. The deviations from this approximation become pronounced in a very thin layer close to the boundary, in the boundary layer.

Given a boundary layer flow we start with the solution of the potential flow equation (10.28), which is linear. Once the flow potential, ϕ, is found, we derive u and v and hence U by differentiation. We then use u and v in the Bernoulli equation (7.5) to obtain p. We now have p everywhere, including at $y = 0$, because $\partial p/\partial y = 0$ there, as given by Eq. (8.41).

Inside the potential flow Bernoulli's equation holds:

$$\frac{p}{\rho} + gh + \tfrac{1}{2}U^2 = \text{const.}$$

Differentiation for cases where h is not important yields

$$-\frac{1}{\rho}\frac{dp}{dx} = U\frac{dU}{dx}, \tag{11.1}$$

which makes Eq. (8.38) read

$$u\frac{\partial u}{\partial x}+v\frac{\partial u}{\partial y}=U\frac{dU}{dx}+v\frac{\partial^2 u}{\partial y^2}. \qquad (11.2)$$

Here U is the potential flow velocity just outside the boundary layer. Note that when h is important, Eqs. (8.37) and (8.38) should both include gdh/dx; Eq. (11.2), however, remains unchanged.

Assuming the potential flow solved, we proceed to solve the boundary layer equations. The solution is easier than that of the original Navier–Stokes equations because there are two equations only, the pressure being known, and because these equations are parabolic rather than elliptic. The boundary condition at $y\mathrm{Re}^{1/2} \to \infty$ is sometimes approximated at finite $y = \delta$, where δ is defined as a *boundary layer thickness*. We now illustrate this procedure by some examples.

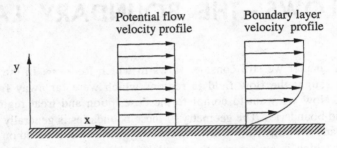

Figure 11.1 Flow over a flat plate.

Flow over a Flat Plate – The Blasius Solution

Consider a flow in the x-direction parallel to a flat plate, as shown in Fig. 11.1. The potential flow for this example has already been solved, Eq. (10.43):

$$\phi = Ux, \qquad u = U = \text{const.} \qquad (11.3)$$

Equation (11.3) yields $UdU/dx = 0$, and the remaining set of equations to be solved is

$$u\frac{\partial u}{\partial x}+v\frac{\partial u}{\partial y}=v\frac{\partial^2 u}{\partial y^2}, \qquad (11.4)$$

$$\frac{\partial u}{\partial x}+\frac{\partial v}{\partial y}=0, \qquad (11.5)$$

with the boundary conditions

$$u = v = 0 \quad \text{at} \quad y = 0, \qquad (11.6)$$

$$u = U \qquad \text{at} \qquad y\,\mathrm{Re}^{\frac{1}{2}} \to \infty. \tag{11.7}$$

Physically $y\,\mathrm{Re}^{1/2} \to \infty$ also requires $y \to \infty$. At the plate $u = 0$ because of the no-slip boundary condition of Eq. (11.6). Adjacent layers are also slowed down because of shear stress. Thus the velocity profile approaches that of potential flow only asymptotically, Fig. 11.1, and an exact definition of the boundary layer thickness as that y for which there is no more modification of the potential flow, i.e., as y for $u = U$, requires $y \to \infty$.

The continuity equation (11.5) can be eliminated from the set of Eqs. (11.4) - (11.7) by the use of the stream function ψ,

$$u = \frac{\partial \psi}{\partial y}, \qquad v = -\frac{\partial \psi}{\partial x},$$

resulting in

$$\frac{\partial \psi}{\partial y} \cdot \frac{\partial^2 \psi}{\partial x \partial y} - \frac{\partial \psi}{\partial x}\frac{\partial^2 \psi}{\partial y^2} = v\frac{\partial^3 \psi}{\partial y^3}. \tag{11.8}$$

$$\frac{\partial \psi}{\partial y} = U \quad \text{at} \quad y \to \infty, \tag{11.9}$$

$$\frac{\partial \psi}{\partial x} = 0, \qquad \frac{\partial \psi}{\partial y} = 0 \quad \text{at} \quad y = 0. \tag{11.10}$$

To solve these equations, some additional insight is necessary. We review Chapter 8 and find the dimensionless form of the x-component of the full Navier–Stokes equation, Eq. (8.12), with the similarity parameter Re. We have therefore the dimensionless form of Eq. (11.4) as

$$u\frac{\partial u}{\partial x} + v\frac{\partial u}{\partial y} = \frac{1}{\mathrm{Re}}\frac{\partial^2 u}{\partial y^2}.$$

The dimensionless continuity equation (8.11) becomes $\partial u / \partial x + \partial v / \partial y = 0$ (see Eq. (11.5)) and contributes no similarity parameter, and we find that the only similarity parameter, as far as similitude is concerned, is the Reynolds number, $\mathrm{Re} = LU/v$.

Now we read in Chapter 8 that the characteristic physical quantities used to construct the similarity parameters can be chosen arbitrarily, provided they are *well defined* and correspond to the *same geometrical locations* in both cases. Therefore, to construct $\mathrm{Re}_1 = \mathrm{Re}_2$ for two boundary layer flows, we should choose U_1 and U_2 and v_1 and v_2; but what are the characteristic lengths?

Inspection of the geometry reveals that the only well-defined length in this flow is x itself, i.e., the distance along the plate measured from its forward tip. Thus for two such flows to be similar,

$$\text{Re}_1 = \text{Re}_2, \qquad \frac{U_1 x_1}{\nu_1} = \frac{U_2 x_2}{\nu_2}.$$

To emphasize the fact that this Reynolds number has x as its characteristic length, we denote it Re_x.

This interesting result indicates that the flow may be viewed as similar to itself: We can cover a whole range of Re_x by considering one x and several U values, but also by considering one U and several x values. There seems to be nothing special about the x value we have chosen. Could it be that the velocity profiles are *always* similar, with U and v just scales of magnitude? If so, then there must be a similarity solution. Armed with this hint, we now define a new single independent variable which we call a similarity variable,

$$\eta = Ayx^b,$$

and try a solution of the form $\psi = \psi(\eta)$. The boundary condition at $y \to \infty$ must of course be satisfied by

$$u(x,\infty) = \left.\frac{\partial \psi}{\partial y}\right|_{y\to\infty} = \left.\frac{d\psi}{d\eta}\right|_{\eta\to\infty} \cdot \frac{\partial \eta}{\partial y} = U,$$

or

$$Ax^b \psi'(\infty) = U. \tag{11.11}$$

Because U is constant while x^b is not, the chosen form $\psi(\eta)$ must be modified such that the left-hand side of Eq. (11.11) also becomes a constant. We therefore try the form

$$\psi = Bx^{-b} f(\eta), \tag{11.12}$$

which now satisfies the boundary condition at infinity in the form

$$u(x,\infty) = ABf'(\eta) = U, \tag{11.13}$$

and by letting $AB = U$, we obtain

$$f'(\infty) = 1 \tag{11.14}$$

and f' is identified as the dimensionless velocity u/U. The boundary conditions at $y=0$, Eq. (11.10) become

$$u(x,0) = \left.\frac{\partial \psi}{\partial y}\right|_{y=0} = ABf'(\eta)\big|_{\eta=0} = 0,$$

or

$$f'(0) = 0,$$

and

$$v(x,0) = -\left.\frac{\partial \psi}{\partial x}\right|_{y=0} = -Bbx^{-b-1}(\eta f' - f)\big|_{\eta=0} = 0.$$

Hence

$$f(0) = 0.$$

Substitution of ψ from (11.12) into the differential equation (11.8) with the necessary algebra and the selection of the free constants A, B and b such that x and y do not appear in the equation result in

$$ff'' + 2f''' = 0 \tag{11.15}$$

with the boundary conditions

$$f(0) = f'(0) = 0, \tag{11.16}$$

$$f'(\infty) = 1 \tag{11.14}$$

and with

$$\psi = \sqrt{v x U} f(\eta), \tag{11.17}$$

$$\eta = y\sqrt{\frac{U}{vx}} = \frac{y}{x} \operatorname{Re}_x^{\frac{1}{2}}, \qquad \operatorname{Re}_x = \frac{Ux}{v}, \tag{11.18}$$

$$u = \frac{\partial \psi}{\partial y} = U f'(\eta), \qquad f'(\eta) = \frac{u}{U}, \tag{11.19}$$

$$v = -\frac{\partial \psi}{\partial x} = \frac{1}{2}\sqrt{\frac{vU}{x}}(\eta f' - f). \tag{11.20}$$

For a particular x, η may be interpreted as a dimensionless distance y from the plate. The function $f'(\eta)$, Eq. (11.19), is the normalized velocity u along the plate, while $f(\eta)$, which is

$$f(\eta) = \int_0^\eta f'(\eta)d\eta,$$

counts the flow rate between the plate and the considered η value, i.e., it is the dimensionless stream function.

η	f''	f'	f
0	0.332	0	0
1	0.323	0.330	0.166
2	0.267	0.630	0.650
3	0.161	0.846	1.397
4	0.0642	0.956	2.306
5	0.0159	0.992	3.283
6	0.0024	0.999	4.280
∞	0	1.000	∞

Table 11.1 $f(\eta)$ in the boundary layer.

Equation (11.15), known as the Blasius equation, must now be solved with the boundary conditions (11.14) and (11.16). Blasius solved this equation in 1908 by matching a power series expansion for small η with an asymptotic form for large η. A digital computer, however, would solve this equation in seconds. The results of the numerical solution are summarized in Table 11.1 and in Fig. 11.2.

One important result of the numerical computation is $f''(0) = 0.332$, which is related to the shear rate on the plate. From Eq. (11.19)

$$\frac{\partial u}{\partial y} = U f''(\eta) \frac{\partial \eta}{\partial y} = U \sqrt{\frac{U}{vx}} f''(\eta).$$

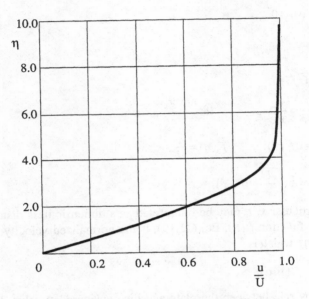

Figure 11.2 Velocity in boundary layer.

On the plate where $y = 0$,

$$\left.\frac{\partial u}{\partial y}\right|_{y=0} = U \sqrt{\frac{U}{vx}} f''(0) = 0.332 \frac{\rho U^2}{\mu} \mathrm{Re}_x^{-\frac{1}{2}}.$$

The shear stress that the fluid exerts on the plate at a given x is

$$\tau_o = \mu \left.\frac{\partial u}{\partial y}\right|_{y=0} = \frac{0.332\rho U^2}{\sqrt{\mathrm{Re}_x}}, \tag{11.21}$$

or in its dimensionless form

$$C_f = \frac{\tau_o}{\frac{1}{2}\rho U^2} = \frac{0.664}{\sqrt{\mathrm{Re}_x}}, \tag{11.22}$$

where C_f is defined as the skin friction coefficient.

The average shear stress over the region $0 \leq x \leq L$ is also found,

$$\bar{\tau}_o = \frac{1}{L} \int_0^L \tau_o \, dx = 0.664 \mu \, U \sqrt{\frac{U}{\nu L}} = \frac{0.664 \rho \, U^2}{\sqrt{\mathrm{Re}_L}}, \qquad \mathrm{Re}_L = \frac{UL}{\nu}, \qquad (11.23)$$

and the average friction coefficient, $\bar{C}_f$, becomes

$$\bar{C}_f = \frac{1.328}{\sqrt{\mathrm{Re}_L}}. \qquad (11.24)$$

We now consider more closely the concept of the boundary layer thickness. If the boundary layer is defined to extend to where $u = U$, we already know that this would result in $\delta \to \infty$. We therefore seek other definitions. Suppose we define δ as that y at which

$$u = 0.99 \, U,$$

or

$$f'(\eta) = 0.99.$$

Table 11.1 reveals that this corresponds to $\eta \approx 5$, and a workable definition has been obtained, which by Eq. (11.18), for $y = \delta$, becomes

$$\frac{\delta}{x} = \frac{5.0}{\sqrt{\mathrm{Re}_x}}. \qquad (11.25)$$

Of course, values different from 0.99 may be chosen to define δ, resulting in different coefficients in Eq. (11.25).

Other useful definitions for the thickness of the boundary layer are the *displacement thickness*

$$\delta_d = \frac{1}{U} \int_0^\infty (U - u) dy = \int_0^\infty \left(1 - \frac{u}{U}\right) dy \qquad (11.26)$$

and the *momentum thickness*

$$\delta_m = \frac{1}{U^2} \int_0^\infty u(U - u) dy = \int_0^\infty \frac{u}{U}\left(1 - \frac{u}{U}\right) dy \qquad (11.27)$$

Figure 11.3 shows the velocity distribution in the boundary layer. Region A is the difference between the flows as assumed by the potential model and as presented by the boundary layer model. The difference in the velocities at some y is $U - u$, and the total volumetric flux deficiency is $\int_0^\infty (U - u) dy$, i.e., because of the presence of the plate, the flux between $y = 0$ and $y = \infty$ is smaller by the amount $\int_0^\infty (U - u) dy$ than it would have been had the flow really been potential.

One could cause this same difference in a purely potential flow by moving the plate parallel to itself upward by the amount δ_d; hence

$$U\delta_d = \int_0^\infty (U-u)dy\,.$$

Thus δ_d, Eq. (11.26), may be interpreted as the distance by which the flat plate must be displaced into a potential flow to cause the same mass flux deficiency as that caused by the boundary layer. Going through an analogous argument for the momentum, δ_m may be forced into an analogous interpretation with respect to momentum flux deficiency. The real importance of δ_d and δ_m, however, is their reappearance in later expressions.

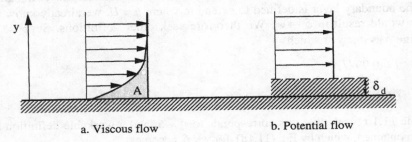

a. Viscous flow   b. Potential flow

Figure 11.3 Displacement thickness.

Having the Blasius solution for the flat plate at our disposal, it is interesting to compute δ_d and δ_m just defined. Equation (11.26) can be rewritten in a dimensionless form as

$$\frac{\delta_d}{x} = \mathrm{Re}_x^{-\frac{1}{2}} \int_0^\infty (1-f')d\eta. \tag{11.28}$$

Similarly, the dimensionless momentum thickness, Eq. (11.28), becomes

$$\frac{\delta_m}{x} = \mathrm{Re}_x^{-\frac{1}{2}} \int_0^\infty f'(1-f')d\eta. \tag{11.29}$$

Numerical integrations, using values of f' vs. η from Table 11.1, yield

$$\frac{\delta_d}{x} = \frac{1.73}{\sqrt{\mathrm{Re}_x}} \tag{11.30}$$

and

$$\frac{\delta_m}{x} = \frac{1.328}{\sqrt{\mathrm{Re}_x}}\,. \tag{11.31}$$

Comparison of Eq. (11.31) and Eq. (11.24) suggests the interpretation that the plate which causes this loss of momentum must indeed withstand a corresponding shear stress.

Example 11.1

A thin metal plate of dimensions $a = 0.2$ m, $b = 0.5$ m is held in water ($\nu = 10^{-6}$ m²/s, $\rho = 1000$ kg/m³) flowing parallel to it. The water velocity is $U = 1$ m/s. What is the drag force acting on the plate? Which edge of the plate should be in the direction of the flow to obtain a lower drag force?

Solution

The drag force may be directly calculated using Eq. (11.23):

$$F = \bar{\tau}_o \cdot A = 2ab\bar{\tau}_o = \frac{2 \times 0.664 \, ab\rho U^2}{\sqrt{\mathrm{Re}_L}} .$$

The area is taken twice, because the force is exerted on both sides of the plate. Hence,

$$F = \frac{2 \times 0.664 \times 0.2 \times 0.5 \times 1000 \times 1^2}{\sqrt{\mathrm{Re}_L}} = \frac{132.8}{\sqrt{\mathrm{Re}_L}} .$$

For the case $L = a = 0.2$ m

$$\mathrm{Re}_L = \frac{LU}{\nu} = \frac{0.2 \times 1}{10^{-6}} = 200,000$$

and

$$F = \frac{132.8}{\sqrt{200,000}} = 0.297 \text{ N}.$$

For $L = b = 0.5$ m

$$F = \frac{132.8}{\sqrt{500,000}} = 0.188 \text{ N}.$$

Hence, holding the plate with the longer edge parallel to the flow results in a lower drag.

Von Karman–Pohlhausen Integral Method

Up to this point we have seen just one exact solution to the boundary layer equations, that corresponding to the flat plate in parallel flow. Foreseeing situations where exact solutions to the boundary layer equations cannot be found, an approximate method due to von Karman and Pohlhausen is now introduced. The approximation is that the boundary layer equation is not satisfied pointwise but rather on the average over the region.

The method consists of first transforming the differential boundary layer equations into an integral equation. This stage is exact. We then proceed to satisfy this integral equation by the selection of an appropriate velocity profile inside the boundary layer. This stage introduces an approximation into the method, because satisfying the integral relation is a necessary condition for the solution, but not a sufficient condition.

Let h denote some constant value of y well outside the boundary layer,

$h > \delta$.

Consider again the boundary layer equation

$$u\frac{\partial u}{\partial x} + v\frac{\partial u}{\partial y} = U\frac{dU}{dx} + v\frac{\partial^2 u}{\partial y^2}. \tag{11.18}$$

Let the continuity equation be multiplied by u,

$$u\left(\frac{\partial u}{\partial x} + \frac{\partial v}{\partial y}\right) = 0,$$

and added to Eq. (11.18), which results in

$$\frac{\partial u^2}{\partial x} + \frac{\partial (uv)}{\partial y} = U\frac{dU}{dx} + v\frac{\partial^2 u}{\partial y^2}. \tag{11.32}$$

Equation (11.32) is called the *conservative* form of the boundary layer equation. Integrating Eq. (11.32) between $y = 0$ and $y = h$, we obtain

$$\int_0^h \frac{\partial u^2}{\partial x}dy + \int_0^h \frac{\partial(uv)}{\partial y}dy = \int_0^h U\frac{dU}{dx}dy + v\int_0^h \frac{\partial^2 u}{\partial y^2}dy.$$

Using Leibnitz's rule on the first term and performing the integration on the second and fourth terms result in

$$\frac{d}{dx}\int_0^h u^2 dy + uv\Big|_h = \int_0^h U\frac{dU}{dx}dy - v\frac{\partial u}{\partial y}\Big|_{y=0}. \tag{11.33}$$

In

$$v\frac{\partial u}{\partial y}\Big|_{y=0} = \frac{\tau_o}{\rho}$$

we identify τ_o, the shear stress on the plate. We also use the equation of continuity, Eq. (11.5), to express v as

$$v = -\int_0^y \frac{\partial u}{\partial x}dy,$$

leading to

$$(uv)\Big|_h = -U\int_o^h \frac{\partial u}{\partial x}dy = -U\frac{d}{dx}\int_o^h u\,dy.$$

Rearranging Eq. (11.33) accordingly, we obtain

$$\frac{\tau_o}{\rho} = \frac{d}{dx}\int_0^h (U-u)u\,dy + \frac{dU}{dx}\int_0^h (U-u)\,dy \tag{11.34}$$

or

$$\frac{\tau_o}{\rho} = \frac{d}{dx}(U^2\delta_m) + (U\delta_d)\frac{dU}{dx} \tag{11.35}$$

with δ_m and δ_d defined by Eqs. (11.27) and (11.26), respectively.

Example 11.2

Apply the von Karman–Pohlhausen integral method to the flow over a flat plate and compare your results with the exact ones. For the velocity profile in the boundary layer assume:
a. A second order polynomial.
b. A third order polynomial.

Solution

The boundary conditions for both cases are

$$u = 0 \qquad\qquad \text{at} \qquad y = 0, \tag{11.36}$$

$$u = U, \quad \frac{\partial u}{\partial y} = 0 \qquad \text{at} \qquad y = \delta, \text{ the edge of the boundary layer.}$$

a. The assumed profile is

$$\frac{u}{U} = a + b\eta + c\eta^2,$$

where $\eta = y/\delta$. The boundary conditions of Eq. (11.36) are satisfied for $a = 0$, $b = 2$, $c = -1$. Hence

$$u = U\left(2\eta - \eta^2\right)$$

Equations (11.26) and (11.27), with $dy = \delta d\eta$, yield

$$\delta_d = \frac{1}{3}\delta, \qquad \delta_m = \frac{2}{15}\delta,$$

while

$$\frac{\tau_o}{\rho} = v\frac{\partial u}{\partial y}\bigg|_{y=0} = v\frac{U}{\delta}\times 2.$$

Equation (11.35) now becomes

$$2v\frac{U}{\delta}=\frac{2}{15}U^2\frac{d\delta}{dx}$$

or

$$30\frac{v}{U}=\frac{d\delta^2}{dx}.$$

Now, since $\delta = 0$ at $x = 0$, integration yields

$$\delta=\sqrt{30\frac{v}{U}x}=\frac{5.48}{\sqrt{\text{Re}_x}}x,\tag{11.37}$$

This compares favorably with Eq. (11.25). The local shear stress on the plate is

$$\tau_o=\mu\frac{\partial u}{\partial y}\Big|_{y=0}=\mu\frac{U}{\delta}\times2=0.365\mu U\sqrt{\frac{U}{vx}}=0.365\rho U^2\text{Re}_x^{-1/2}$$

and the average one is

$$\bar{\tau}_o=\frac{1}{L}\int_0^L\tau_o dx=0.73\rho U^2\text{Re}_L^{-1/2}.\tag{11.38}$$

b. The assumed profile is

$$\frac{u}{U}=a+b\eta+c\eta^2+d\eta^3.$$

Equation (11.36) yields only three boundary conditions. However, the boundary layer equation (11.4) itself, with $u = v = 0$ at $y = 0$, yields

$$\frac{\partial^2 u}{\partial y^2}\Big|_{y=0}=0.$$

The boundary conditions of Eq. (11.36) and this additional condition are satisfied by $a=c=0$, $b+d=1$, $b+3d=0$. Hence

$$u=U\left(\tfrac{3}{2}\eta-\tfrac{1}{2}\eta^3\right).\tag{11.39}$$

Again, Eqs. (11.26) and (11.27), with $dy = \delta\, d\eta$, yield

$$\delta_d=\frac{3}{8}\delta,\qquad \delta_m=\frac{39}{280}\delta$$

while

$$\frac{\tau_o}{\rho}=v\frac{\partial u}{\partial y}\Big|_{y=0}=v\frac{U}{\delta}\times\frac{3}{2},$$

and Eq. (11.35) now becomes

$$\frac{3}{2}v\frac{U}{\delta} = \frac{39}{280}U^2\frac{d\delta}{dx}$$

or

$$\frac{280}{13}\frac{v}{U} = \frac{d\delta^2}{dx}.$$

Since for $x = 0$, $\delta = 0$, integration yields

$$\delta = \sqrt{\frac{280}{13}\frac{v}{U}x} = \frac{4.64}{\sqrt{\text{Re}_x}}x.$$

Compare this with Eq. (11.25). The shear stress on the plate is now computed as

$$\tau_o = \mu\frac{\partial u}{\partial y}\bigg|_{y=0} = \frac{3}{2}\mu\frac{U}{\delta} = 0.323\mu U\sqrt{\frac{U}{vx}}$$

or

$$\tau_o = 0.323\rho U^2\text{Re}_x^{-\frac{1}{2}}. \tag{11.40}$$

Similarly

$$\bar{\tau}_o = \frac{1}{L}\int_0^L \tau_o\,dx = 0.646\rho U^2\text{Re}_L^{-\frac{1}{2}} \tag{11.41}$$

which compare rather favorably with the exact results, Eqs. (11.21) and (11.23).

It is noted that applications of higher order polynomials, with additional conditions of the form

$$\frac{\partial^2 u}{\partial y^2}\bigg|_{y=\delta} = 0, \qquad \frac{\partial^3 u}{\partial y^3}\bigg|_{y=\delta} = 0,$$

i.e., "smoother" transition to the potential flow, yield worse and worse results. We do not know that these higher derivatives even exist and the assumption of a very large number of derivatives vanishing at $y = \delta$ completely destroys the profile of the boundary layer.

The Two-dimensional Laminar Jet

Another example of an exact solution of the boundary layer equations is the two-dimensional jet, Fig. 11.4. It has a mass flux $Q = \rho b u_o$ and a momentum flux $M = \rho b u_o^2$. We assume the width of the slit through which the jet emerges, b, to approach zero, while the momentum M is kept constant. This would make $u_o \rightarrow \infty$ there, but because M is constant, $Q = M/u_o \rightarrow 0$. Thus for small b it is reasonable to consider a two-dimensional momentum source, which is a very weak mass source.

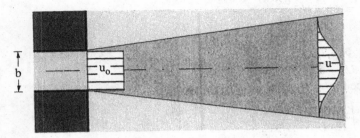

Figure 11.4 Two-dimensional jet.

For large y, $u = U = 0$, and since the fluid extends to $y = \pm\infty$, the original momentum flux must be conserved:

$$M_x = M_o = \int_{-\infty}^{\infty} \rho u^2 dy. \tag{11.42}$$

The system to be solved is

$$u\frac{\partial u}{\partial x} + v\frac{\partial u}{\partial y} = v\frac{\partial^2 u}{\partial y^2}, \tag{11.43}$$

$$v = 0, \quad \frac{\partial u}{\partial y} = 0 \quad \text{at} \quad y = 0, \tag{11.44}$$

$$u = 0 \quad\quad \text{at} \quad y \to \infty. \tag{11.45}$$

To completely specify the problem, the momentum of the jet, Eq. (11.42), has to be given.

We note that this flow also has no characteristic length except for x, and hence we follow Bickley (1939) and Schlichting (1933) in seeking a similarity transformation:

$$\psi = \left(\frac{Mvx}{\rho}\right)^{1/3} f(\eta), \qquad \eta = \left(\frac{M}{\rho v^2 x^2}\right)^{1/3} y, \tag{11.46}$$

whence

$$u = \frac{\partial \psi}{\partial y} = \left(\frac{M^2}{\rho^2 vx}\right)^{1/3} f'(\eta), \qquad v = -\frac{\partial \psi}{\partial x} = \frac{1}{3}\left(\frac{Mv}{\rho x^2}\right)^{1/3} (2\eta f' - f), \tag{11.47}$$

which transforms Eq. (11.43) to

$$3f'' + f'^2 + ff' = 0 \tag{11.48}$$

and the boundary conditions to

$$f(0) = f''(0) = 0, \tag{11.49}$$

$$f'(\pm\infty) = 0. \tag{11.50}$$

Integration of Eq. (11.48) yields

$$3f'' + ff' = 0$$

with the constant of integration $A = 0$ because of Eq. (11.49). A second integration yields

$$6f' + f^2 = B^2$$

or

$$d\eta = 6\frac{df}{B^2 - f^2} = \frac{3}{B}\left[\frac{df}{B + f} + \frac{df}{B - f}\right],$$

which is integrated to

$$\frac{B}{3}\eta = \ln\frac{B + f}{B - f} = \ln\frac{1 + f/B}{1 - f/B}$$

or

$$f = \sqrt{2}B\tanh\frac{B\eta}{3\sqrt{2}}. \tag{11.51}$$

Substitution in u, Eq. (11.47), and then in (11.42) yields

$$B^3 = \frac{9}{4\sqrt{2}},$$

resulting in

$$f = \left(\frac{9}{2}\right)^{1/3}\tanh\left[\frac{\eta}{48^{1/3}}\right].$$

Finally

$$\psi = \left(\frac{9Mx\nu}{2\rho}\right)^{1/3}\tanh\left(\frac{\eta}{48^{1/3}}\right)$$

and

$$u = \left(\frac{3M^2}{32\rho^2\nu x}\right)^{1/3}\frac{1}{\cosh^2\left(\eta/48^{1/3}\right)},$$

$$\upsilon = \left(\frac{M\nu}{36\rho x^2}\right)^{1/3}\left[\frac{\eta}{\cosh^2\left(\eta/48^{1/3}\right)}\right] - \tanh\left(\frac{\eta}{48^{1/3}}\right).$$

References

W. Bickley, "The Plane Jet," Phil. Mag., Ser. 7, **23**, 727 (1939).

N. Curle, "The Laminar Boundary Layer Equations," Oxford University Press, London, 1938.

S.I. Pai, "Viscous Flow Theory," Van Nostrand, Princeton, 1957.

H. Schlichting, "Laminare Strahlenausbreitung," ZAMM, **13**, 260 (1933).

H. Schlichting, "Boundary-Layer Theory," 7th ed., McGraw-Hill, New York, 1979.

Problems

11.1   Assume a velocity distribution in the boundary layer in the form $u = U\sin(\pi y/2\delta)$ and apply the von Karman–Pohlhausen method to find the shear stress on a flat plate.

11.2   A flat plate 1 m by 0.5 m is set in a stream of air flowing at 1 m/s parallel to the plate. What is the minimal drag force on the plate?

11.3   A thin tube, as shown in Fig. P11.3, is put in a parallel flow. The fluid velocity is 0.1 m/s, the kinematic viscosity is 1.5×10^{-5} m²/s and its density is 1.3 kg/m³. Find the force required to hold the tube in place if its length is
a. 1 m,   b. 3 m.

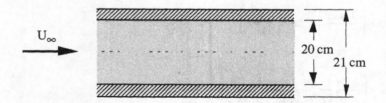

Figure P11.3

11.4   Is the information that the boundary layer flow admits a similarity solution used in the von Karman–Pohlhausen integral method? If yes, where?

11.5   A flat plate of dimensions 0.1 m by 0.05 m is inserted into a flow such that a boundary layer is believed to develop along its 0.1-m side. The shear force on the plate is 0.1 N. Find the force expected to act on a similar plate of dimensions 0.2 m by 0.1 m.

11.6 The stream function in a potential flow into a 90° corner shown in Fig. P11.6 is

$$\psi = xyU.$$

Estimate the shear stress on the L long part of the structure along the x-axis.

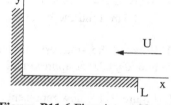

Figure P11.6 Flow into a 90° corner.

11.7 A two-dimensional flow field is given approximately by

$$\psi = x^2 - y^2.$$

A small metal flag with the dimensions a and b is inserted into the flow field, where it arranges itself such that the flat plane lies (approximately) tangent to a stream sheet, as shown in Fig. P11.7. The force with which the flag must be held in place is then measured. The

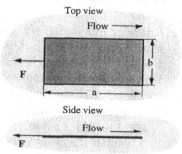

Figure P11.7 Small flag in a flow field.

viscosity of the fluid is $\mu = 10$ kg/m·s, and its density is $\rho = 1000$ kg/m³. Neglect the change of the flow field velocity along the flag, assume laminar flow and compute the force acting on the flag when its leading edge is located at

a. $(1;2)$,   b. $(2;3)$,   c. $(10;1)$.

Hint: The flow along the flag is a boundary layer flow, with ψ representing the potential flow outside the boundary layer.

11.8 Consider the flow along the flag in Problem 11.7. Assume the velocity to change linearly along the length of the flag. Obtain the differential equation for the boundary layer thickness resulting from the integral approach.

11.9 Two wide, parallel flat plates are shown in Fig. P11.9. The flow streaming toward the plates is at $U = 0.25$ m/s, and the flow is assumed two dimensional. The den-sity of the fluid is $\rho = 1,000$ kg/m³, and its vis-

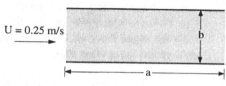

Figure P11.9

cosity is $\mu = 10^3$ kg/m·s. The two plates are part of a structure, and $a = 1$ m, $b = 0.5$ m. Find the force transferred from the plates to the structure.

11.10 What is the smallest dimension b in Problem 11.9 which still permits a boundary layer approximation for this configuration.

11.11 Figure P11.9 now represents a cross section of a thin-walled cylinder. Is the smallest diameter which still admits a boundary layer approximation smaller or greater than the value of b obtained in Problem 11.10. Explain.

11.12 Consider the boundary layer flow shown in Fig. P11.12. Use the von Karman–Pohlhausen integral method, with a third order polynomial.

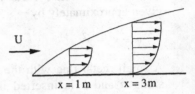

Figure P11.2

a. Obtain the velocity profile and the boundary layer thickness at $x = 1$ m and $x = 3$ m.
b. Calculate the mean velocity inside the boundary layer, at $x = 1$ m. Now use this mean velocity as the free stream velocity and obtain the velocity profile and the boundary layer thickness for a flat plate at $x = 2$ m. Compare these results with those obtained in part a for $x = 3$ m.

11.13 Oil with the kinematic viscosity of 0.05 m²/s enters a 0.1-m-diameter pipe. Using boundary layer approximation, find how far downstream will the boundary layers meet.

11.14 Repeat Problem 11.13 for a flow in an annulus with the diameters 0.2 m and 0.1 m.

11.15 Calculate the total shear force on the entrance section of the pipe in Problem 11.13 between the entrance and up to the point where the boundary layers meet, Fig. P11.15. Compare the result

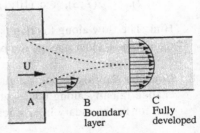

Figure P11.15

with the shear force on the same pipe section for a fully developed laminar flow of the same mass flux.

11.16 Let the oil in Problem 11.15 enter the pipe, as shown in Fig. P11.15. The entrance section, from point A to point B, approximates the boundary layer flow, with B the location where the boundary layers meet. The section from

B to C, which is of the same length as that from A to B, represents the fully developed flow. Three manometers read the pressures at points A, B and C. The pipe is horizontal. Express the readings of the manometers in terms of the shear forces calculated in Problem 11.15 and of the pipe geometry.

11.17 Express the friction head losses between points A and B and B and C in Problem 11.16.

11.18 Consider the boundary layer flow over a porous flat plate, with constant suction applied along the plate. The boundary conditions in this case are $u = U$ at $y \to \infty$ and $u = 0$, $v = -S = \text{const}$ at $y = 0$. Try $\partial u/\partial x = 0$, and obtain an exact solution of the boundary layer equations.

11.19 An airplane flies at 1,000 km/h. An air scoop is designed in the belly of the airplane to take in air to cool electronic equipment. Air is to be taken from the potential flow, avoiding the boundary layer region. Compute b, Fig. P11.19.

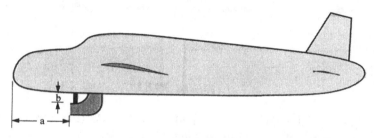

Figure P11.19

11.20 An old way to determine the speed of a ship was to drop into the water a log of wood tied with a rope. The length of the rope that the log dragged off the ship per some predetermined time was then measured, e.g., by counting knots made in the rope. The force needed to drag a rope off a ship is 20 N. The log is a wooden cylinder 1.2 m long and 0.12 m in diameter. Neglect the contribution of the pressure field, and compute the speed with which the log must be dragged to begin measuring, i.e., the log drift correction.

11.21 An estimate of the forces of wind on a tall rectangular building is obtained as follows: The front of the building is assumed to be a 20 m wide by 50 m high flat plate perpendicular to the flow, subjected to the stagnation pressure of the wind. The back side is at the static pressure. The sides of the building are flat plates 50 m high by 60 m long in the direction of the wind, which suffer shear forces as in a boundary layer flow. The top is a flat plate 20 m wide by 60 m long, also subjected to shear forces. Compute the total force on the building for winds of 10, 20 and 40 km/h.

11.22   A glass sinker has the tear-drop shape shown in Fig. P11.22, which is drawn to scale. The density of glass is 2,000 kg/m³. The drag force on the sinker is caused by viscous friction only. Measure the figure and compute the terminal speed of the sinker in water.

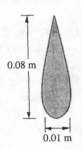

0.08 m

0.01 m

Figure   P11.22

11.23   An arrow is drawn in Fig. P11.23, to scale. Find the distribution of the drag force along the arrow. Find the correcting moment for small deviation angles, i.e., that moment which keeps the arrow tip pointing forward, for arrow speeds of 10 and 30 m/s.

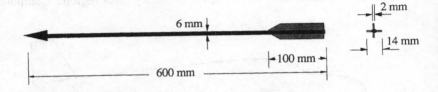

6 mm

2 mm

14 mm

100 mm

600 mm

Figure   P11.23

11.24   A sleeve of cloth is towed behind a ship. The sleeve is 2 m long and 0.5 m in diameter. The ship sails at 3 m/s. Find the drag force on the sleeve.

11.25   A bullet fired into a pond enters the water at the speed of 300 m/s. The bullet has a diameter of 0.009 m and is 0.015 m long. Its mass is 0.008 kg. Assume just viscous friction as in a boundary layer flow and estimate the bullet's deceleration in the water.

11.26   An amphibian car floats on the water with its rear wheels spinning. The outer radius of the tires is 0.8 m, and they are 0.3 m wide; the wheels are just half submerged. The speedometer shows that on land the speed would be 80 km/h. Estimate the force driving the car forward in the water.

11.27   A javelin has an average diameter of 0.03 m and is 2 m long, with an apparent density of 800 kg/m³. A man throws that javelin at an elevation angle of 45° and reaches a distance of 50 m. Find the distance he could have reached in a vacuum.

11.28 A streamlined boat has most of its drag derived from viscous friction on its wet skin. It is suggested to use results known for an old well-designed boat to construct a new double-hulled boat that has the same total carrying capacity as the old one. Using boundary layer considerations, estimate by how much must the power driving the new boat be increased relative to that driving the old boat.

11.29 A rope which has a diameter of 0.05 m and is 20 m long dangles in the water behind a boat sailing at 5 m/s. Find the drag caused by the rope.

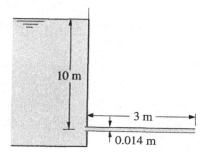

11.30 A 0.014-m-diameter pipe 3 m long is set in the side of a 10-m-high water tank, Fig. 11.30. Find the mass flowrate of the water, including boundary layer effects in the pipe.

Figure 11.30

Figure 11.39

12. TURBULENT FLOW

Reynolds Experiment

Consider the experimental apparatus shown in Fig. 12.1. It consists of a small tube inserted inside a large transparent pipe. The inner tube has a small diameter and its opening is located far from the bend, i.e., L_B is sufficiently long to guarantee fully developed flow. The fluid in the inner tube is the same as that in the outer pipe, and they have the same ρ and μ. However, the inner fluid is dyed while the outer one is clear. In all experiments the velocities of the two fluids are adjusted so as to have as little shear strain between them as possible.

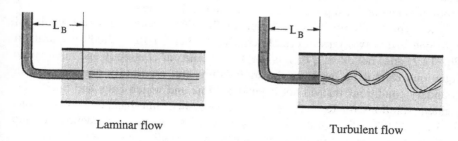

Laminar flow   Turbulent flow

Figure 12.1 Reynolds experiment.

As the experiment proceeds, a dyed stream cylinder is seen to form downstream from the opening of the inner tube. The stream tube which makes the envelope of this stream cylinder is well defined, and the dye is restricted to the inside of this stream tube. The dye begins to spread out by molecular diffusion only far downstream.

The experiment is repeated for various fluids and for various velocities, and as discussed in Chapter 8, all these can be summarized using the Reynolds number as the single parameter involved. It is found that the description given above fits flows whose Reynolds number, calculated using the pipe diameter D as the characteristic length, is below a certain numerical value, namely,

$$\mathrm{Re}_D = \frac{VD}{\nu} < 2,000, \tag{12.1}$$

where V is the average velocity in the pipe, i.e. the flowrate per unit cross-sectional area. Above this Reynolds number the envelope of the dyed stream tube is not clearly defined. The dye is seen to mix with the outer fluid right from the opening of the inner tube. The laminar character of the flow is lost and the flow becomes turbulent.

At this point it is necessary to reconsider the definition of the velocity vector. Consider a body of fluid, with a system S within, and with the point A inside the system, Fig. 1.1. Let the momentum of the system be $\mathbf{M}$ and let its mass be m. An average velocity for the system may be defined as

$$\mathbf{q} = \frac{\mathbf{M}}{m}$$

and the velocity at point A may be defined as

$$\mathbf{q} = \lim_{m \to m_\varepsilon} \frac{\mathbf{M}}{m}. \tag{12.2}$$

The definition of m_ε (or rather of the volume V_ε which contains the mass m_ε) as well as the limitations on this definition are presented in the sections on the continuum and on local properties in a continuum, in Chapter 1.

Returning now to the description of Reynolds' experiment, we are faced with one of the two uncomfortable choices: Either for $\mathrm{Re} > 2,000$ the limit in Eq. (12.2) does not exist; $\mathbf{q}$ then is not defined and our continuity and momentum equations cannot be used; or the momentum and continuity equations, with boundary conditions that do not depend on time and which have had time-independent solutions for $\mathrm{Re} < 2,000$, do not have such time-independent solutions for $\mathrm{Re} > 2,000$. Although the boundary conditions remain time independent, the resulting flows persist to be time dependent, nonperiodic and non-decaying.

While one should always bear in mind the first possibility, the second choice is the correct one. To explain the appearance of the turbulent flow, use must be made of the theory of hydrodynamic stability, which is outside the scope of this book. Still it is important to note that the explanation relies on the amplification of random perturbations, and therefore a certain degree of randomness in the ensuing turbulent flow is to be expected.

Consider the turbulent flow in a pipe. If one was to plot the velocity of the fluid at a given point in the pipe, a pattern similar to that of Fig. 12.2 would result. As seen, there are small rapid velocity fluctuations about some average value. This average value may or may not depend on time. The velocity then is a superposition of an average velocity $\bar{u}$ and a velocity fluctuation u',

$$u = \bar{u} + u'. \tag{12.3}$$

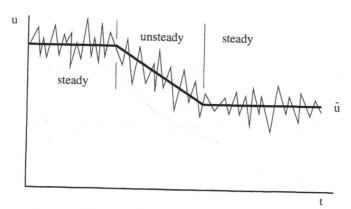

Figure 12.2 Velocity in turbulent flow.

Similar relations may be written for the other velocity components as well as for other properties, such as the pressure, the temperature, etc.:

$$v = \bar{v} + v', \qquad p = \bar{p} + p', \qquad T = \bar{T} + T'. \tag{12.4}$$

The average quantities are defined according to the form

$$\bar{\mathbf{q}} = \frac{1}{\Delta t} \int_t^{t+\Delta t} \mathbf{q} \, dt = \frac{1}{\Delta t} \int_t^{t+\Delta t} (\bar{\mathbf{q}} + \mathbf{q}') dt \tag{12.5}$$

and

$$\int_t^{t+\Delta t} \mathbf{q}' dt = 0, \tag{12.6}$$

i.e., the time average of the fluctuations vanishes.

The Nature of Turbulence

The concept of the coefficient of viscosity has already been used to obtain the Navier–Stokes equations. It was treated as a coefficient, i.e., as a multiplier which serves to adjust units and to formally yield good approximations to physical phenomena. Some insight into the mechanism of viscosity is helpful to explain its modification by turbulence.

Consider the differential layer shown in Fig. 12.3 and note that any point in a three-dimensional flow can be locally described by that scheme. In addition to the continuum velocity u there exists the microscopic random thermal motion of the molecules. This thermal motion has an average zero mass flux, and a molecule A traveling a distance $+dy$ is compensated for by a molecule B traveling $-dy$. The A molecules have an average x-wise velocity of u, while the B ones have $u + du$ velocities.

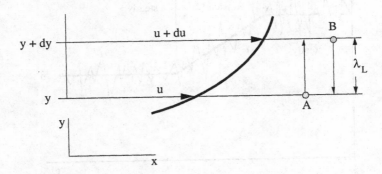

Figure 12.3 Exchange of momentum on the molecular scale.

As seen directly from Fig. 12.3, the shear stress between the layers is proportional to both du/dy and the speed of the random motion of the molecules. In turbulent flow, in addition to the random motion of the individual molecules, whole chunks of fluid move at random. For a given du/dy the transfer of momentum is thus enhanced by turbulence.

The increase in shear stresses in turbulent flows is sometimes said to result from *eddy viscosity*, which is just another statement describing the action of the random turbulent fluctuations.

The Equations of Motion in Turbulent Flow

Consider the x-component of the momentum equation for incompressible flow. Let the equation of continuity be multiplied by u,

$$u\left(\frac{\partial u}{\partial x} + \frac{\partial v}{\partial y} + \frac{\partial w}{\partial z}\right) = 0,$$

and added to the x-component of the momentum equation, to yield:

$$\rho\left(\frac{\partial u}{\partial t} + \frac{\partial (u^2)}{\partial x} + \frac{\partial (uv)}{\partial y} + \frac{\partial (uw)}{\partial z}\right) = -\frac{\partial p}{\partial x} + \rho g_x + \mu\left[\frac{\partial^2 u}{\partial x^2} + \frac{\partial^2 u}{\partial y^2} + \frac{\partial^2 u}{\partial z^2}\right]. \quad (12.7)$$

This form of the equation of motion is sometimes referred to as the conservative form.

In turbulent flow, for each term in Eq. (12.7) we substitute the sum of its mean value and the corresponding fluctuations. Thus

$$q = \bar{q} + q', \qquad p = \bar{p} + p', \tag{12.8}$$

$$\frac{\partial(u^2)}{\partial x} = \frac{\partial(\bar{u}^2)}{\partial x} + 2\frac{\partial(\bar{u}u')}{\partial x} + \frac{\partial(u'^2)}{\partial x}, \tag{12.9}$$

$$\frac{\partial(uv)}{\partial y} = \frac{\partial(\bar{u}\bar{v})}{\partial y} + \frac{\partial(u'\bar{v})}{\partial y} + \frac{\partial(\bar{u}v')}{\partial y} + \frac{\partial(u'v')}{\partial y}. \tag{12.10}$$

We now time-average the resulting equation by integration over the time interval Δt. Thus each term in the equation is replaced by its mean value. We note that

$$\bar{u}' = \frac{1}{\Delta t}\int_t^{t+\Delta t} u'\,dt = 0$$

and

$$\bar{v}' = \bar{w}' = \bar{p}' = 0.$$

Also

$$\frac{\partial \overline{\bar{u}u'}}{\partial x} = \frac{\partial}{\partial x}\int_t^{t+\Delta t}\overline{\bar{u}u'}\,dt = 0$$

and

$$\frac{\partial \overline{u'\bar{v}}}{\partial y} = \frac{\partial \overline{\bar{u}v'}}{\partial y} = 0.$$

However

$$\frac{\partial \overline{(u'^2)}}{\partial x} = \frac{\partial}{\partial x}\left[\frac{1}{\Delta t}\int_t^{t+\Delta t}(u'^2)\,dt\right] \qquad \text{does not vanish.}$$

Using these expressions, Eq. (12.9) yields

$$\frac{\partial \overline{(u^2)}}{\partial x} = \frac{\partial(\bar{u}^2)}{\partial x} + \frac{\partial \overline{(u'^2)}}{\partial x}. \tag{12.11}$$

Time averaging eliminates all the terms in which the fluctuations are linear. However, the terms nonlinear in the fluctuations remain in the time-averaged equations. Time smoothing of Eq. (12.7) results in

$$\rho\left(\frac{\partial u}{\partial t} + \frac{\partial(u^2)}{\partial x} + \frac{\partial(uv)}{\partial y} + \frac{\partial(uw)}{\partial z}\right) = -\frac{\partial p}{\partial x} + \rho g_x + \mu\left[\frac{\partial^2 u}{\partial x^2} + \frac{\partial^2 u}{\partial y^2} + \frac{\partial^2 u}{\partial z^2}\right]$$
$$-\rho\left(\frac{\partial \overline{(u'^2)}}{\partial x} + \frac{\partial \overline{(u'v')}}{\partial y} + \frac{\partial \overline{(u'w')}}{\partial z}\right). \tag{12.12}$$

Comparison of Eqs. (12.7) and (12.12) reveals a new group of terms in Eq. (12.12) which do not have corresponding ones in Eq. (12.7). These terms account for the turbulent momentum transfer due to the random velocity fluctuations. Their role is similar to that of momentum transfer by molecular viscosity resulting in the laminar shear stresses. These terms are known as the turbulent Reynolds stresses and may be written as

$$\tau_{xx}^t = -\rho \overline{u'^2},\tag{12.13}$$

$$\tau_{xy}^t = -\rho \overline{u'v'},\tag{12.14}$$

$$\tau_{xz}^t = -\rho \overline{u'w'}.\tag{12.15}$$

Repeating the procedure for the y- and z-components of the time-averaged Navier–Stokes equation, three additional terms of the turbulent stress tensor $\overline{\tau}^t$ are obtained:

$$\tau_{yy}^t = -\rho \overline{v'^2},\tag{12.16}$$

$$\tau_{yz}^t = -\rho \overline{v'w'},\tag{12.17}$$

$$\tau_{zz}^t = -\rho \overline{w'^2}.\tag{12.18}$$

Let Eq. (12.7) be divided by ρ and rewritten with a slight modification, i.e., instead of the kinematic viscosity, $\nu = \mu/\rho$, we substitute $\nu^t + \nu$,

$$\frac{\partial u}{\partial t} + \frac{\partial(u^2)}{\partial x} + \frac{\partial(uv)}{\partial y} + \frac{\partial(uw)}{\partial z} = -\frac{1}{\rho}\frac{\partial p}{\partial x} + g_x + \left(\nu^t + \nu\right)\left[\frac{\partial^2 u}{\partial x^2} + \frac{\partial^2 u}{\partial y^2} + \frac{\partial^2 u}{\partial z^2}\right].\tag{12.19}$$

Equation (12.19) is now compared with Eq. (12.12), and we see that Eq. (12.19) has the extra terms resulting from ν^t, while Eq. (12.12) has the extra terms of the Reynolds stresses. May Eq. (12.19) be used instead of Eq. (12.12)? In many practical problems Eq. (12.19) yields good approximations to measured results, which justifies its use. The term ν^t is called *turbulent viscosity* or *eddy viscosity*.

Prandtl Mixing Length, Generalized Profiles

In early work on turbulent flows close to rigid boundaries Prandtl (1925) attempted to obtain the Reynolds stresses from an a priori estimate of the $\overline{u'v'}$ quantities. He defined the macroscopic analog of the microscopic mean free path, i.e., that distance traveled by a lump of fluid laterally to the mean velocity direction, before it randomly changed direction. This distance is called the *Prandtl Mixing Length*. This mixing length is not a property of the fluid, as is the mean free path, but rather a property of the flow. Certainly the distance traveled by a chunk of fluid cannot exceed the distance from some nearby wall. Prandtl chose his mixing length as proportional to that distance, i.e.,

$$L = k_L y,$$ (12.20)

with k_L the proportionality factor. We look again at Fig. 12.3 and consider it now as corresponding to turbulent flows. Instead of λ_L, the molecular mean free path, we substitute L, the Prandtl mixing length, i.e., that length traveled by a chunk of fluid laterally to the direction of the average flow. We note from the figure that

$$u' \sim v' \sim L \left(\frac{d\bar{u}}{dy} \right).$$ (12.21)

Hence the turbulent stresses are

$$\tau^t = \rho \left(L \frac{d\bar{u}}{dy} \right)^2.$$

Close to a solid surface we may take

$$\tau^t \approx \tau_w = \text{const}$$

so that

$$\rho \left(L \frac{d\bar{u}}{dy} \right)^2 = \tau_w$$

or

$$\frac{d\bar{u}}{dy} = \frac{1}{L} \sqrt{\frac{\tau_w}{\rho}},$$

which together with (12.20) gives

$$\frac{d\bar{u}}{dy} = \frac{1}{k_L y} \sqrt{\frac{\tau_w}{\rho}}$$

or

$$d\bar{u} = \frac{dy}{y} \frac{1}{k_L} \sqrt{\frac{\tau_w}{\rho}}.$$

Integration yields

$$\frac{\bar{u}}{\sqrt{\tau_w / \rho}} = \frac{A}{k_L} \ln y + B.$$ (12.22)

A general velocity profile near a rigid wall has been obtained. Though this treatment seems rather qualitative, the resulting profile is confirmed experimentally for many classes of turbulent flows, at least in the vicinity of the wall.

The constants of Eq. (12.22) must be found experimentally for each flow. The equation recast in dimensionless form for pipe flow becomes

$$u^+ = 2.5 \ln y^+ + 5.5, \tag{12.23}$$

where $u^+ = u/u*$ is the dimensionless velocity with $u^* = \sqrt{\tau_w/\rho}$ known as the shear velocity. The dimensionless distance from the wall is defined in the form of a Reynolds number as

$$y^+ = \frac{yu^*}{\nu} = \frac{y}{\nu}\sqrt{\frac{\tau_w}{\rho}}. \tag{12.24}$$

At the wall itself, because of the viscous boundary conditions, all velocities must damp out and the flow cannot remain turbulent. Equation (12.23) does not hold at the wall, as can be seen from the $\ln y^+$ term.

A more elaborate treatment divides the flow in a pipe into a laminar sublayer near the wall, a turbulent core and a buffer layer in between. The velocity profile obtained is given in the form of three equations, one for each region. It is known as the universal profile and it holds for the whole cross section of the pipe.

For the laminar sublayer, which extends from the wall to $y^+ = 5$, the profile is linear:

$$u^+ = y^+, \qquad 0 \le y^+ < 5. \tag{12.25}$$

For the buffer layer the profile is given by

$$u^+ = 5.0 \ln y^+ - 3.05, \qquad 5 \le y^+ < 30, \tag{12.26}$$

while for the turbulent core Prandtl's simple expression still applies,

$$u^+ = 2.5 \ln y^+ + 5.5, \qquad y^+ \ge 30. \tag{12.23}$$

The universal profile, given by Eqs. (12.23), (12.25) and (12.26), applies to turbulent flows in smooth pipes. It is particularly useful in the analysis of heat and mass transfer to walls in turbulent conduit flows, where the processes depend strongly on the flow field details near the wall. The use of this profile, ignoring the laminar sublayer and the buffer zone, yields very good results in many cases.

The Moody Diagram

Inspection of the time-averaged momentum equation for turbulent flow, Eq. (12.12), reveals that for the same number of equations as in laminar flow we now have three additional dependent variables, u', v', w'. Attempts to construct a sufficient number of additional equations were not successful, and solutions of turbulent flows rely largely on experiments.

While Eq. (12.12) cannot be solved, it still implies that the only similarity parameter in the flow is the Reynolds number, in addition, of course, to geometrical similarity.

All round pipes are geometrically similar. The fact that the pipe walls may not be smooth leads to the use of the concept of *wall roughness*. Wall roughness is defined as the average height, ε, of the asperities on the wall surface, made dimensionless with respect to the pipe diameter. We thus expect the dimensionless pressure drop along the pipe to depend on the *relative roughness, ε/D*, and on the Reynolds number.

A convenient way to formulate the problem is to define a friction factor f as

$$f = \frac{\Delta P}{L/D}\frac{1}{\frac{1}{2}\rho V^2}, \qquad f = f(\mathrm{Re}, \varepsilon/D), \tag{12.27}$$

where $V = Q/A$ is the average velocity in the pipe. Using Eq. (6.49) one may write for a round pipe

$$\frac{\Delta P}{L} = \frac{1}{r}\frac{\partial}{\partial r}\left(r\,\tau_{rz}\right)$$

with the boundary conditions at the centerline ($r = 0$) and at the pipe wall ($r = R$), respectively,

$$\tau_{rz} = 0 \quad \text{at} \quad r = 0,$$
$$\tau_{rz} = \tau_w \quad \text{at} \quad r = R.$$

Integration yields

$$\tau_w = \frac{R}{2}\frac{\Delta P}{L}$$

or

$$\tau_w = \frac{D}{4}\frac{\Delta P}{L}. \tag{12.28}$$

This result, combined with Eq. (12.27), defines the friction factor in terms of the wall shear stress, τ_w,

$$\tau_w = \frac{f}{8}\rho V^2, \qquad f = \frac{8\tau_w}{\rho V^2} \tag{12.29}$$

Comparison of Eq. (12.29) with Eq. (11.22) shows that the definition of the friction factor is similar to that of the skin friction coefficient, C_f, for a flat plate. Equation (12.27) is now rewritten in terms of the friction head loss, h_f,

$$h_f = f\frac{L}{D}\left(\frac{V^2}{2g}\right). \tag{12.30}$$

This result may be used in the modified Bernoulli Equation, Eq. (7.23).

Generally, the friction factor has to be evaluated experimentally, except in rare cases where an analytical solution exists.

For the special case of laminar flow Eq. (12.27) may be combined with the Poiseuille Equation, Eq. (6.58),

$$V = \frac{\Delta P D^2}{32 \mu L},$$

to yield

$$f = 64\left(\frac{\mu}{DV\rho}\right)$$

or

$$f = \frac{64}{\text{Re}}. \tag{12.31}$$

The friction factor, f, is presented for the general case as a function of the Reynolds number and the relative roughness, ε/D. This was done graphically by Moody in 1944 in what is called the Moody diagram, Fig. 12.4. The diagram contains three distinct zones:

a. Laminar flow, for which an analytical solution was given by Eq. (12.31).

b. Transient zone, where head losses depend both on the molecular viscosity (linear with the velocity) and on Reynolds stresses (proportional to the square of the velocity).

c. Turbulent zone, where Reynolds stresses prevail and f hardly changes with Re. It is really the existence of this region which justifies the choice of V^2 in the definition of f in Eq. (12.27).

Example 12.1

The universal velocity profile for turbulent flow in a smooth pipe is given by Eqs. (12.25), (12.26) and (12.23). The profile is written in terms of

$$u^+ = u / \sqrt{\tau_w / \rho} \quad \text{and} \quad y^+ = \left(y\sqrt{\tau_w / \rho}\right)/v.$$

To perform calculations it is convenient to have the profile put in terms of the parameters used in the Moody diagram. Rewrite the universal profile in terms of the friction factor, f, and the Reynolds number, Re.

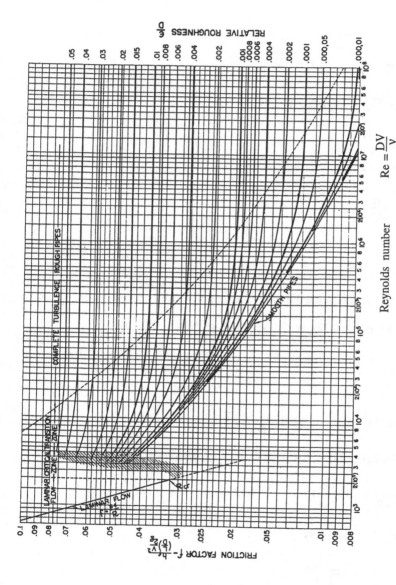

Figure 12.4 Moody diagram of friction loss characteristics for laminar and turbulent flow in pipes. (L. F. Moody, Trans. ASME **66** (8), p. 671, 1944.)

$$Re = \frac{DV}{\nu}$$

Reynolds number

Solution

The velocity profile is given by

$$u^+ = f(y^+),$$

where

$$u^+ = \frac{u}{\sqrt{\tau_w/\rho}}, \qquad y^+ = \frac{y\sqrt{\tau_w/\rho}}{\nu}.$$

We substitute $\sqrt{\tau_w/\rho}$ from Eq. (12.29),

$$\sqrt{\frac{\tau_w}{\rho}} = V\sqrt{\frac{f}{8}},$$

and obtain

$$u^+ = \frac{u}{V\sqrt{f/8}}.$$

Also,

$$y^+ = \frac{yV}{\nu}\sqrt{\frac{f}{8}} = \frac{y}{D}\text{Re}\sqrt{\frac{f}{8}},$$

and for the various regions the universal velocity profile becomes:

a. For the laminar sublayer, $0 \le y^+ < 5$, from Eq. (12.25),

$$\frac{u}{V\sqrt{f/8}} = \frac{y}{D}\text{Re}\sqrt{f/8}$$

or

$$\frac{u}{V} = \frac{y}{D}\text{Re}\left(\frac{f}{8}\right) \qquad \text{at} \qquad 0 \le \frac{y}{D}\text{Re}\sqrt{\frac{f}{8}} < 5.$$

b. For the buffer layer, $5 \le y^+ < 30$, Eq. (12.26) yields

$$\frac{u}{V\sqrt{f/8}} = 5.0\ln\left[\frac{y}{D}\text{Re}\sqrt{\frac{f}{8}}\right] - 3.05 \qquad \text{at} \qquad 5 \le \frac{y}{D}\text{Re}\sqrt{\frac{f}{8}} < 30.$$

c. For the turbulent core, $y^+ \ge 30$, from Eq. (12.23)

$$\frac{u}{V\sqrt{f/8}} = 2.5\ln\left[\frac{y}{D}\text{Re}\sqrt{\frac{f}{8}}\right] + 5.5 \qquad \text{at} \qquad \frac{y}{D}\text{Re}\sqrt{\frac{f}{8}} \ge 30.$$

Example 12.2

Water (density $\rho = 1,000$ kg/m³, kinematic viscosity $v = 10^{-6}$ m²/s) flows with the average velocity of $V = 10$ m/s in a smooth pipe of diameter $D = 10$ cm.
a. Find the thickness of the laminar sublayer.
b. Where does the turbulent core begin?
c. Write the equations of the velocity distribution in the pipe.
d. What are the velocities at the edge of the laminar sublayer, at the beginning of the turbulent core, and at the pipe centerline?

Solution

We first establish the flow regime by computing the Reynolds number

$$\text{Re} = \frac{VD}{v} = \frac{10 \times 0.1}{10^{-6}} = 10^6.$$

Hence, the flow is turbulent. From the Moody diagram, Fig. 12.4, we obtain $f = 0.0117$.
The variable y^+ is

$$y^+ = \frac{y}{D}\text{Re}\sqrt{\frac{f}{8}} = y\frac{10^6\sqrt{0.0117/8}}{0.1} = 382,400y$$

or $y = 2.615 \times 10^{-6} y^+$.

a. The thickness of the laminar sublayer, $y^+ = 5$, is

$$y = 2.615 \times 10^{-6} \times 5.$$

b. The turbulent core begins at $y^+ = 30$, i.e.,

$$y = 0.0000784 \text{ m} = 0.078 \text{ mm}.$$

Hence in this pipe the buffer zone extends only up to 0.16% of the radius; and the turbulent core covers 99.8% of the radius.

c. The velocity profile is obtained from the results of Example 12.1. Hence, for the laminar sublayer,

$$u = \frac{V}{D}\text{Re}\left(\frac{f}{8}\right)y = \frac{10 \times 10^6 \times 0.0117}{0.1 \times 8}y,$$

$$u = 146,000y.$$

For the buffer zone

$$u = 5V\sqrt{\frac{f}{8}} \times \ln\left[\frac{y}{D}\operatorname{Re}\sqrt{\frac{f}{8}}\right] - 3.05V\sqrt{\frac{f}{8}},$$

where $\sqrt{f/8} = \sqrt{0.0117/8} = 0.03824.$

Thus,

$$u = 5 \times 10 \times 0.03824 \times \ln\left[\frac{10^6 \times 0.03824\,y}{0.1}\right] - 3.05 \times 10 \times 0.03824,$$

$$u = 1.912 \ln(382{,}400\,y) - 1.166.$$

For the turbulent core

$$u = 2.5V\sqrt{\frac{f}{8}}\ln\left[\frac{y}{D}\operatorname{Re}\sqrt{\frac{f}{8}}\right] + 5.5V\sqrt{\frac{f}{8}},$$

$$u = 2.5 \times 10 \times 0.03824 \ln(382{,}400\,y) + 5.5 \times 10 \times 0.03824,$$

$$u = 0.956 \ln(382{,}400y) + 2.103.$$

d.   The velocity at the edge of the laminar sublayer, where $y = 1.307 \times 10^{-5}$ m, is

$$u = 146000 \times 1.307 \times 10^{-5} = 1.91 \text{ m/s}.$$

The velocity at the beginning of the turbulent core, where $y = 0.0000784$ m, is

$$u = 0.956 \ln(382{,}400 \times 7.84 \times 10^{-5}) + 2.103 = 5.35 \text{ m / s}.$$

At the centerline $y = 0.05$ m, and the velocity is

$$u = 0.956 \ln(328{,}400 \times 0.05) + 2.103 = 11.38 \text{ m/s}.$$

It is worthwhile to note that although the velocity at the centerline seems reasonable, the velocity gradient is not. Differentiation shows that it does not vanish at the centerline, as it should.

Example 12.3

It has been found experimentally that turbulent flow in rough pipes is not affected by the roughness elements as long as the latter do not protrude outside the laminar sublayer; i.e., the pipe walls are said to be hydraulically smooth for

$$\varepsilon^+ = \frac{\varepsilon u^*}{\nu} \le 5,$$

where ε is the height of the roughness elements.

For turbulent flow in a pipe of 5 cm diameter and Re = 50,000, find the maximum height of roughness elements allowing flow under hydraulically smooth conditions.

Solution

We are essentially looking for the thickness of the laminar sublayer for a flow with Re = 50,000. We use the expression for y derived in Example 12.2

$$y^+ = \frac{y}{D}\mathrm{Re}\sqrt{\frac{f}{8}}\,, \qquad y = \frac{y^+D}{\mathrm{Re}\sqrt{f/8}}\,,$$

and let $y^+ = 5$. For Re = 50,000 we read, from Fig. 12.4, $f = 0.021$. Thus,

$$\varepsilon_{max} = y = \frac{5 \times 5}{50,000 \times \sqrt{0.021/8}} = 9.75 \times 10^{-3} \text{ cm} \approx 0.1 \text{mm}.$$

Hence, for roughness elements smaller than ε_{max} the pipe walls can be considered hydraulically smooth.

One should also note how thin the laminar sublayer is. It is not surprising that Prandtl's formula, Eq. (12.23), which neglects it altogether, is nevertheless quite accurate for hydrodynamic calculations.

Flow in Pipes

The problems encountered in the use of the Moody diagram to compute pipe flow are of two kinds:

1. The direct problem: given the mass flow in a pipe, what is the head loss?
2. The indirect problem: given the pressure drop, what is the flowrate?

The indirect problem is nonlinear. Most problems in water distribution belong to the indirect class, where solutions are obtained by iterations and the use of a computer.

Example 12.4

A cast iron pipe 15 cm in diameter was installed in a water system 15 years ago. Experience indicated that this type of pipe will increase its wall roughness with time according to the relation

$$\varepsilon = \varepsilon_o(1 + 0.25t),$$

where $\varepsilon_o = 25 \times 10^{-5}$ m is the wall roughness of a new pipe and ε is the roughness after t years. A new pump is to be installed in the system which is capable of delivering a maximum discharge of 0.03 m³/s of water at 27°C. Estimate the pressure drop per meter of horizontal pipe at the maximum discharge.

Solution

After 15 years of service,

$$\varepsilon = 25 \times 10^{-5}(1 + 0.25 \times 15) = 0.00119 \, \text{m}.$$

The relative roughness is then given by

$$\frac{\varepsilon}{D} = \frac{0.00119}{0.15} = 0.0079.$$

The average velocity of the water flow in the pipe is

$$V = \frac{Q}{A} = \frac{0.03}{(0.15)^2 (\pi/4)} = 1.698 \, \text{m/s}.$$

Now, at 27°C, $v = 0.0086 \times 10^{-4}$ m²/s and $\rho = 1,000$ kg/m³. Thus, the Reynolds number is

$$\text{Re} = \frac{VD}{v} = \frac{1.698 \times 0.15}{0.0086 \times 10^{-4}} = 2.96 \times 10^5.$$

For the values of ε/D and Re calculated above the Moody diagram, Fig. 12.4, yields $f = 0.036$. The friction loss is then given by

$$h_f = f \frac{L}{D} \frac{V^2}{2g} = 0.036 \left(\frac{1}{0.15} \right) \frac{1.698^2}{2 \times 9.8} = 0.035 \, \text{m}.$$

Example 12.5

A pipe carries water from a reservoir and discharges it as a free jet, Fig. 12.5. Find the flowrate to be expected through a 20-cm-inner-diameter commercial steel pipe for which $\varepsilon = 4.6 \times 10^{-5}$ m.

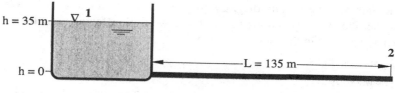

Figure 12.5

Solution

Points 1 and 2 are both at atmospheric pressure. Thus, Bernoulli's equation, Eq. (7.23), which applies between these two points is simplified to

$$h_1 = \frac{V_2^2}{2g} + h_f,$$

where the velocity at which the reservoir surface recedes, V_1, has been neglected. Now,

$$h_f = f \frac{L}{D} \frac{V_2^2}{2g}.$$

Combining this with the Bernoulli equation leads to

$$h_1 = \frac{V_2^2}{2g} \left(1 + f \frac{L}{D}\right).$$

Substituting the data results in

$$35 = \frac{V_2^2}{2 \times 9.8} \left(1 + f \frac{135}{0.2}\right)$$

or

$$V_2 = \sqrt{\frac{686}{1 + 675f}}.$$

The problem cannot be solved directly in the case of turbulent flow, since f as a function of V is not known, and so V cannot be isolated algebraically. Instead, an iterative process involving the Moody diagram must be used.

We use the Moody diagram and as a first guess estimate the friction factor for the roughness of $\varepsilon/D = 4.6 \times 10^{-5}/0.2 = 0.00023$ to be $f = 0.014$. We then solve for V_2:

$$V_2 = \sqrt{\frac{686}{1 + 675 \times 0.014}} = 8.1 \, \text{m/s}.$$

The Reynolds number is then

$$Re = \frac{V_2 D}{\nu} = \frac{8.1 \times 0.2}{0.0086 \times 10^{-4}} = 1.88 \times 10^6.$$

For this number, the Moody diagram gives a friction factor of 0.0155. We now repeat the above computations using this new friction factor until there is no appreciable change in the factor. We find

$$V_2 = 7.736 \text{ m / s},$$
$$f = 0.0155,$$

the friction factor remaining essentially unchanged. The volume flow rate is then

$$Q = V_2 A = 7.736 \left(\frac{\pi}{4}\right) \times 0.2^2 = 0.243 \text{ m}^3 \text{ / s},$$

and the pressure drop is

$$\Delta p = f \frac{L}{D} \frac{V_2^2}{2} \rho = 0.0155 \times \frac{135}{0.2} \times \tfrac{1}{2} \times 7.736^2 \times 1,000 = 3.13 \times 10^5 \text{N / m}^2.$$

Example 12.6

Figure 12.6 shows a line diagram of a piping system: Pipe AB is 500 m long and has an inner diameter of 0.1 m. Pipe BC is 200 m long and has an inner diameter of 0.05 m. Pipe BD is 300 m long and has an inner diameter of 0.025 m. The pressures were measured at points A, D, C and found to be $p_A = 600$ kPa, $p_C = p_D = 100$ kPa. Find the flowrates in the pipes for a fluid of $v = 10^{-6}$ m²/s, $\rho = 1000$ kg/m³ and $\varepsilon/D = 0.002$ for all pipes.

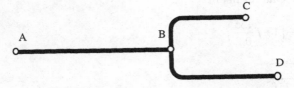

Figure 12.6 Piping system.

Solution

All flows are assumed turbulent. From Eq. (12.27)

$$\Delta p = f \frac{L}{D} \times \tfrac{1}{2} \rho V^2.$$

Therefore

$$\Delta p_{AB} = 2.5 \times 10^6 f \, V_{AB}^2,$$
$$\Delta p_{BC} = 2 \times 10^6 f \, V_{BC}^2,$$
$$\Delta p_{BD} = 6 \times 10^6 f \, V_{BD}^2.$$

Also $\Delta p_{BC} = \Delta p_{BD}$, and because of conservation of mass

$$V_{AB} = 0.5^2 V_{BC} + 0.25^2 V_{BD}.$$

The solution is iterative:

a.   Assume: $V_{BD} = 5$ m/s $\Rightarrow$ $Re_{BD} = 125{,}000$, $f = 0.0185$,
     $\Delta p_{BD} = 2{,}775$ kPa > 500 kPa.

b.   Assume: $V_{BD} = 1$ m/s $\Rightarrow$ $Re_{BD} = 25{,}000$, $f = 0.03$,
     $\Delta p_{BD} = 180$ kPa.
     Assume $V_{BC} = 2$ m/s $\Rightarrow$ $Re_{BC} = 100{,}000$, $f = 0.025$,
     $\Delta p_{BC} = 200$ kPa > 180 kPa.
     Assume $V_{BC} = 1.9$ m/s $\Rightarrow$ $Re_{BC} = 95{,}000$, $f = 0.025$,
     $\Delta p_{BC} = 180.5$ kPa ~ 180 kPa

   $V_{AB} = 0.5^2 \times 1.9 + 0.25^2 \times 1 = 0.5375$ m/s,
   $Re_{AB} = 53{,}750$, $f = 0.025$, $\Delta p_{AB} = 1.95$ kPa,
   $\Delta p_{AB} + \Delta p_{BC} = 182$ kPa < 500 kPa.

c.   Assume: $V_{BD} = 1.67$ m/s $\Rightarrow$ $Re_{BD} = 41{,}665$,   f = 0.027,
     $\Delta p_{BD} = 450$ kPa.
     Assume $V_{BC} = 3.17$ m/s $\Rightarrow$ $Re_{BC} = 158500$,   f = 0.0245,
     $\Delta p_{BC} = 492$ kPa > 450 kPa.
     Assume $V_{BC} = 3.03$ m/s $\Rightarrow$ $Re_{BC} = 151500$,   f = 0.0245
     $\Delta p_{BC} = 450$ kPa.

   $V_{AB} = 0.5^2 \times 3.03 + 0.25^2 \times 1.67 = 0.86$ m/s
   $Re_{AB} = 86{,}000$, $f = 0.025$; $\Delta p_{AB} = 46$ kPa.
   $\Delta p_{AB} + \Delta p_{BC} = 496$ kPa ~ 500 kPa.

Results: $V_{AB} = 0.86$ m/s ; $V_{BC} = 3.03$ m/s ; $V_{BD} = 1.67$ m/s.

Flow through Pipe Fittings

Pipe fittings are such parts as connections, flow dividers, bends and elbows, reducers, valves and many more, all of which have fluids flowing through them. Pipes used to transfer fluids invariably involve pipe fittings, and the combination of pipes and fittings is sometimes referred to as a piping network. The municipal water supply in a city is an example of a piping network. In the previous section we have considered friction losses in pipes. We now extend our treatment to the contribution of the various fittings to the friction losses in the piping network.

Suppose we had only one type of fitting, say a 90° short bend, which we use when we want to connect two pipes perpendicular to one another. We could look for similarity conditions and find that in addition to similarity in geometry there is one similarity parameter, the Reynolds number. Closer scrutiny shows that surface roughness is also important and that a dimensionless roughness parameter may be defined in exactly the same manner as in pipe flow. Armed with this information we could perform experiments and sum up our results in a chart somewhat like the Moody diagram. We could denote this diagram as the friction loss chart for a 90° short bend.

We could now repeat the procedure for a 90° long bend, for a tee-shaped flow divider, etc. Indeed under certain circumstances this elaborate treatment is adopted. In general, however, one tries to avoid such a detailed investigation for each and every fitting; there are very many fittings, and the amount of the experimental work involved can become considerable.

Suppose we had obtained the friction loss diagram for a 90° short bend. The friction coefficient is denoted by F, and by analogy to Eqs. (12.30) and (12.27), the friction head loss may be expressed as

$$h_f = F \frac{V^2}{2g} \tag{12.32}$$

with

$$F = F(\mathrm{Re}, \varepsilon / D). \tag{12.33}$$

Simplification of treatment now depends on the behavior of F.

Some experimental data published in the literature show this F to be a constant for each particular fitting, i.e., $F = k$,

$$h_f = k \frac{V^2}{2g} \tag{12.34}$$

with k being referred to as a *loss coefficient*. A partial explanation for this somewhat surprising simplification is that many fittings are designed for particular forms of use, and the range of the Reynolds numbers at which they are used is not large. In other words, the function F does not vary much over the range in which its argument, Re, varies. Another possible explanation is that the losses result from turbulent eddies which are activated by deceleration and draw their energy at a rate proportional to the kinetic energy of the flow. In such cases the loss coefficient k in Eq. (12.34) is simply the proportionality coefficient, and the losses may be interpreted as a certain percentage of the kinetic energy. Some k values are presented in Table 12.1.

Fitting	Loss coefficient, Eq. (12.34) $k = h_f \big/ \left(V^2/2g\right)$	Equivalent length, Eq. (12.36) $L_e = \frac{L}{D} = h_f \big/ f\left(V^2/2g\right)$
Elbow, 90°	0.9	30 - 40
Elbow, 45°	0.4	15
Tee, flow turns 90°	1.8	60
Tee, flow straight through	0.9	30
Return bend	2.2	
Gate valve, fully open	0.2	10
Globe valve, fully open	10	100 - 300
Sharp-edged pipe entrance	1.0	
Rounded pipe entrance	0.02 - 0.1	
Inward projecting pipe entrance	0.8	
Sharp-edged pipe exit	0.5	
Square expansion from d to D		
$\quad d/D = 0.2$	0.92	
$\quad d/D = 0.4$	0.72	
$\quad d/D = 0.8$	0.16	
Square contraction from D to d		
$\quad d/D = 0.2$	0.49	
$\quad d/D = 0.4$	0.42	
$\quad d/D = 0.8$	0.18	

Table 12.1 Friction losses in pipe fittings.

The total friction losses of a fluid flowing through a pipe with fittings is found by combining Eq. (12.30) with Eq. (12.34),

$$h_f = \left(f\frac{L}{D} + \sum_i k_i \right)\frac{V^2}{2g}. \tag{12.35}$$

Another approach to an approximation relies on the speculation that perhaps the F function, Eq. (12.33), may behave like the Moody friction factor f, Eq. (12.27), in the form

$$F(\mathrm{Re}, e/D) = L_e \cdot f(\mathrm{Re}, e/D),$$

with $L_e = L/D$, in which L_e is a dimensionless equivalent length, expressed as an equivalent number of pipe diameters, which is a constant. Experimental results confirm this speculation for the 90° short bend and for many other pipe fittings, with a constant L_e value for each fitting. To compute friction losses under these circumstances, the F function is obtained for a particular fitting by reading its L_e value from an appropriate table and multiplying it by the friction factor f read from the Moody diagram for a pipe of the corresponding diameter. Thus the friction head loss due to a fitting may be computed as

$$h_f = F\frac{V^2}{2g} = L_e \cdot f\frac{V^2}{2g} , \qquad (12.36)$$

and comparison with Eq. (12.30) suggests that L_e may be indeed expressed as $L_e = L/D$, in which L is a dimensionless equivalent length, i.e., the losses associated with the fitting are the same as for a pipe of the same diameter and an equivalent length expressed in a number of pipe diameters. Table 12.1 also lists the equivalent lengths for some fittings. It should be noted that for some flow patterns the use of the loss coefficient k yields better approximations, while for other patterns the L_e approximation is better. When one approach is definitely better, the table does not list the other parameter at all.

Example 12.7

A piping network is shown in Fig. 12.7. The pipe from A to E and from F to G has a diameter of 0.05 m and a total length of 60 m. The pipe section EF is 0.025 m in diameter and 20 m long. All the pipes have a relative roughness $\varepsilon/D = 0.001$. The mean velocity needed at G, when the globe valve D is fully open, is 2 m/s. Find the water level h needed to sustain such a flow of water.

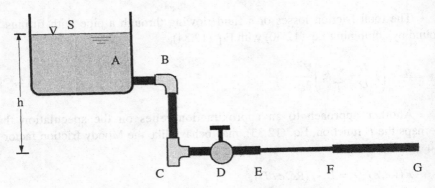

Figure 12.7 Piping system.

Solution

The Bernoulli equation between points S and G yields

$$h = h_f + \frac{V_G^2}{2g}.$$

To find the friction factor, we first calculate the Reynolds number for the pipes with $D = 0.05\,\text{m}$,

$$\text{Re}_G = \frac{VD}{\nu} = \frac{2 \times 0.05}{10^{-6}} = 100,000.$$

The Moody diagram yields

$$f = 0.0220.$$

Similarly, for the pipe with the 0.025 m diameter,

$$f = 0.0215.$$

The friction losses due to piping alone are

$$h_{f_p} = \sum f \frac{L}{D}\left(\frac{V^2}{2g}\right) = 0.022 \times \frac{60}{0.05} \times \frac{2^2}{2g} + 0.0215 \times \frac{20}{0.025} \times \frac{8^2}{2g} = 61.5.$$

From Table 12.1, the losses due to the fittings are given in terms of k-values as:
A (sharp pipe entrance) 1.0; B (90° elbow) 0.9; C (tee, 90°) 1.8; D (globe valve) 10;
E (contraction, $d/D = 0.5$) 0.36; G (sharp pipe exit) 0.5; F (expansion, $d/D = 0.5$) 0.58.
Hence,

$$h_{f_F} = \left(\sum_i k_i\right)\frac{V^2}{2g} = (1 + 0.9 + 1.8 + 10 + 0.36 + 0.5) \times \frac{2^2}{2g} + 0.58 \times \frac{8^2}{2g} = 4.86\,\text{m}.$$

The total friction losses are

$$h_f = h_{f_p} + h_{f_F} = 61.5 + 4.86 = 66.4 \text{ m},$$

and the water level h required is

$$h = h_f + \frac{V_G^2}{2g} = 66.4 + \frac{2^2}{2g} = 66.6 \text{ m}.$$

Noncircular Pipes

A concept which emerges from the application of the idea of the universal velocity distribution to the similarity of turbulent flows is that of the *hydraulic diameter*.

Suppose we seek the head loss in a turbulent flow in an elliptical pipe. In the main part of the pipe cross section the velocity profile is practically flat, Fig.12.8.

Near the walls, in a narrow layer, the velocity is given by the universal profile, which must support the forces corresponding to the head loss. This situation is common to the elliptical pipe and to a circular pipe, and the mechanisms acting in both seem to be the same, provided both have the same ratio of cross-sectional area to circumference, i.e., the same shear loading of the universal velocity profile.

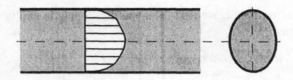

Figure 12.8 Turbulent velocity profile in elliptical pipe.

For a circular pipe the ratio of cross section to the "wetted" circumference is

$$A / C = \frac{\frac{\pi}{4}D^2}{\pi D} = \frac{D}{4}$$

and

$$D = 4\frac{A}{C}.$$

We therefore define the hydraulic diameter as

$$D_H = 4\frac{A}{C}. \tag{12.37}$$

Thus for an ellipse $A = \pi ab$, $C \approx \pi (a + b)$ and

$$D_H = 4\frac{A}{C} = 4\frac{ab}{a+b}.$$

The Moody diagram may now be used for the elliptical pipe. The same treatment may be used for rectangular pipes and for pipes of other cross sections.

Coefficient of Drag

In turbulent flows around bluff bodies the use of the friction coefficient is not convenient, and a better concept is that of the *coefficient of drag*, C_D, defined by

$$F_D = C_D A \rho \frac{U^2}{2}, \tag{12.38}$$

where F_D is the drag force acting on the body, or the force per unit span for infinite span bodies and A is the cross-sectional area normal to the flow. The drag coefficient, C_D, is usually found experimentally. Some representative values of C_D are given in Fig. 12.9 and in Table 12.2.

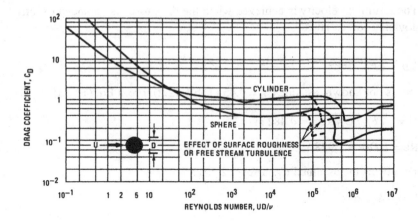

Figure 12.9 Drag coefficient of a cylinder and a sphere.

Shape and orientation to flow	Reynolds number range	C_D
Flat strip, normal to flow	Greater than 10^3	1.95
Disk, normal to flow	Greater than 10^3	1.10
Hemispherical shell:		
Hollow upstream	Greater than 10^3	1.33
Hollow downstream	Greater than 10^3	0.34
Cylinder, normal to flow	10^3 - 2×10^5	1.10
	Greater than 5×10^5	~ 0.35
Sphere	10^3 - 2×10^5	0.45
	Greater than 3×10^5	~ 0.20
Model of airship	Greater than 2×10^5	0.04
Airfoil at zero angle of attack	Boundary layer transition	0.006

Table 12.2 Coefficients of drag of various shapes.
[Compiled from data in Hunsaker and Rightmire (1947) and Schlichting (1968).]

Example 12.8

A spherical particle has the density of 3,000 kg/m^3 and the diameter of 0.004 m. The kinematic viscosity of water is $v = 10^{-6}$ m^2/s. Find the terminal sinking velocity of this sphere in water.

Solution

The terminal velocity is achieved when the drag force is balanced exactly by the buoyancy force. Thus

$$F_D = \frac{4}{3}\pi \times 0.002^3 \times (\rho - \rho_w)g = 0.657 \times 10^{-3}.$$

Since Re is not known, assume $C_D = 0.45$ (Table 12.2). From Eq. (12.33),

$$U = \sqrt{2F_D / C_D A \rho_w} = 0.482 \text{ m} / \text{s},$$

$$\text{Re} = UD/\nu = 1,928,$$

which confirms the assumed C_D, Table 12.2. Thus

$$U = 0.482 \text{ m/s}.$$

Flow-induced Lift Forces

The previous section, which dealt with drag forces, treated flows around closed configurations such that far from the configuration the velocity field could be assumed undisturbed, i.e., it was the same as if the closed configuration did not exist. The undisturbed velocity vector was essentially constant. The total force which such a flow applied to the configuration had a component in the undisturbed flow direction; this component was denoted the drag force. The total force acting on the configuration may also have a component in the direction perpendicular to that of the undisturbed flow. This perpendicular component is now denoted the *lift force*. The lift force is expected to appear in all flows not symmetrical with respect to the configuration, even when the configuration itself is symmetrical. In some cases the lift force may not be as large as the drag force, but this does not necessarily reflect on its importance. The experienced engineer is always aware of the lift force, which sometimes he or she may choose to ignore.

Some attempts to calculate lift forces for nonsymmetrical flows around cylinders have been done in the chapter on potential flows. There are many examples in which lift forces may be derived analytically or numerically. The lift forces may also be obtained experimentally for cases that cannot be otherwise resolved and also as experimental verifications of theoretical results. We note that in our examples of computations of lift forces we have used potential flow considerations. Indeed this is the more frequent case in the design of lifting configurations. For such cases we conclude that the viscosity, and with it the Reynolds number, have only a second order effect on the lift forces. The exception to this conclusion is that the lift forces do depend on whether the flow is laminar or turbulent and for configurations not designed for lift, the influence of the Reynolds number need be checked.

Considerations of similarity still lead us to the formulation

$$F_L = C_L \cdot A \cdot \tfrac{1}{2}\rho V^2$$

(12.39)

in which F_L is the lift force, A is a characteristic area, ρ is the fluid density and V is the undisturbed flow velocity. However, unlike the drag coefficient, the lift coefficient C_L does not depend on the Reynolds number, at least for well-designed lifting configurations. In many cases airplane wings may be treated as two dimensional. Then A, the characteristic area, may be replaced by L, the characteristic length, which is usually the cord of the wing's profile. The lift force thus obtained is, of course, per unit width of the wing.

As long as the flow may be considered incompressible, there is only one similarity parameter, the Reynolds number. In the previous section C_D, the drag force coefficient, has been presented as a function of the Reynolds number. Since here C_L seems not to depend on the Reynolds number, it comes out to be constant for a particular configuration. The angle between the cord of a wing and the direction of the undisturbed flow is called the *angle of attack*, and during flight the same wing may operate with various angles of attack. Each angle of attack corresponds to a different flow field and therefore has a different lift coefficient. It is common practice to present the lift coefficient as a function of the angle of attack, as shown in Fig. 12.10. The drag coefficient of the same wing is also shown in Fig. 12.10 as a function of the angle of attack, and with its dependence on the Reynolds number deleted.

Example 12.9

An airplane is designed for a nominal cruising speed $V_c = 800\,\text{km/h} = 222$ m/s and take-off speed $V_o = 100\,\text{km/h} = 28$ m/s. The weight of the airplane is $G = 30,000$ N. The airplane wing is shown in Fig. 12.10. Compute the dimensions of the wing and the power expanded to propel just the wing forward.

Solution

We assume that the lift forces arise mainly from the wings. The highest lift per wing area is needed at takeoff because the air speed is then the lowest. At the same time we want to reduce drag at this stage of operation. Inspecting Fig. 12.10, we find that a lift coefficient of 1.0 is still below the point where the drag begins to become large. We therefore choose for takeoff $C_L = 1.0$. Correspondingly, from Fig. 12.10, $C_D = 0.08$.

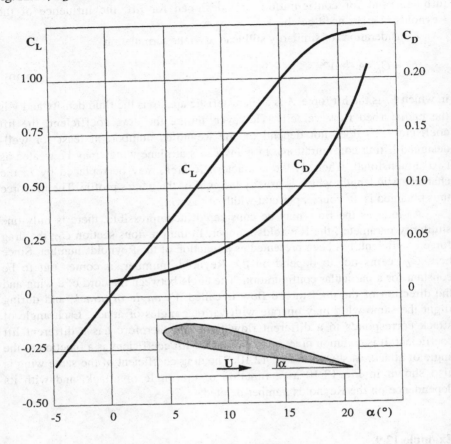

Figure 12.10 Lift and drag coefficients for a wing.

The density is assumed as $\rho = 1.1 \ \text{kg/m}^3$. Then

$$F_L = AC_L\left(\tfrac{1}{2}\rho V^2\right), \qquad A = \frac{F_L}{\tfrac{1}{2}\rho V^2 C_L}.$$

At takeoff

$$A_o = \frac{F_L}{\tfrac{1}{2}\rho V_o^2 C_L} = \frac{30,000}{\tfrac{1}{2} \times 1.1 \times 28^2 \times 1.0} = 69.6 \ \text{m}^2.$$

Choosing a wing length of 10 m on each side, with L being the wing cord,

$$L = \frac{69.6}{2 \times 10} = 3.48 \ \text{m}.$$

At cruising speed,

$$C_L = \frac{F_L}{\frac{1}{2}\rho V^2 A} = \frac{30,000}{\frac{1}{2} \times 1.1 \times 222^2 \times 69.6} = 0.016.$$

Accordingly, from Fig. 12.10, $C_D = 0.02$. The force required to overcome the drag at takeoff is

$$F_D = AC_D\left(\tfrac{1}{2}\rho V^2\right) = 69.6 \times 0.08\left(\tfrac{1}{2} \times 1.1 \times 28^2\right) = 2,400\,\mathrm{N}$$

and the power at takeoff is

$$P = F_D V_o = 2400 \times 28 = 67,200\,\mathrm{W} = 67.2\,\mathrm{kW}.$$

At cruising

$$F_D = AC_D\left(\tfrac{1}{2}\rho V^2\right) = 69.6 \times 0.02\left(\tfrac{1}{2} \times 1.1 \times 222^2\right) = 37,700\,\mathrm{N}$$

and the power is

$$P = F_D V_c = 37,700 \times 222 \times 10^{-3} = 8,380\,\mathrm{kW}.$$

Flowrate Measurement

Two devices used to measure flow velocities, the Venturi meter and the Pitot tube, have already been considered in Chapter 7. Some more devices are shown in Fig. 12.11.

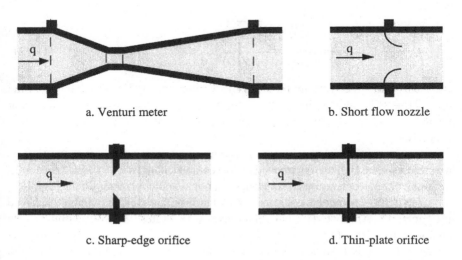

a. Venturi meter   b. Short flow nozzle

c. Sharp-edge orifice   d. Thin-plate orifice

Figure 12.11 Flow measuring devices.

The measurements in all these devices rely on relations between velocities and differences in pressure, which are derived from the Bernoulli equation. We note that with the exception of the Pitot tube, the other devices presented may be arranged in a descending order of elaboration, the Venturi meter being the most elaborate, and then the measuring flow nozzle, the sharp edge orifice and the thin plate orifice.

The expression for the velocity of an incompressible fluid, measured by a Pitot tube, was derived in Example 7.4 as

$$v = \sqrt{\frac{2}{\rho} \cdot (p_1 - p_2)}, \qquad (12.40)$$

where $p_1 - p_2$ is the difference between the total and the static pressure. For a manometer attached to both legs of the Pitot tube and reading Δh_m the expression for the velocity is

$$v = \sqrt{2g\,\Delta h_m \cdot \frac{\rho_m - \rho}{\rho}}. \qquad (12.41)$$

The relations for the flow of an incompressible fluid through a Venturi meter was derived in Example 7.2 as

$$v = \sqrt{\frac{\frac{2}{\rho}(p_1 - p_2)}{1 - (d/D)^4}} \qquad (12.42)$$

and

$$Q = \frac{\pi D^2}{4} V = \frac{\pi d^2}{4} v = \frac{\pi d^2}{4} \sqrt{\frac{\frac{2}{\rho}(p_1 - p_2)}{1 - (d/D)^4}}. \qquad (12.43)$$

where v is the velocity at the throat of the Venturi meter, V is the velocity in the pipe, $Q\,[\mathrm{m^3/s}]$ is the volumetric flowrate, $A = \pi d^2/4$ is the cross-sectional area at the throat and d/D is the ratio of the diameters of the throat and the pipe, respectively.

An assumption inherent in the derivation of Eq. (12.43) is that the flow is incompressible and that friction head losses and changes of stream tube cross sections due to accelerations, sometimes denoted *vena contracta*, are negligible.

Relations for compressible flows may be formulated using equations derived in Chapter 13. Alternatively, the velocities and flowrates can still be obtained by equations like Eqs. (12.40) and (12.43), respectively, provided a *correction for compressibility* is included. For example, the velocity of a compressible fluid, measured by a Pitot tube, can be expressed as

$$v = C_c \sqrt{\frac{2}{\rho} \cdot (p_1 - p_2)} \, . \hspace{4cm} (12.44)$$

Corrections in the form of a compressibility factor C_c, are presented in Table 12.3.

Ratio of pressures		Ratio of throat diameter to pipe diameter d/D		
P_2/P_1	$(P_1 - P_2)/P_1$	0.25	0.50	0.75
For Venturi meters and flow nozzles				
0.98	0.02	0.989	0.988	0.981
0.96	0.04	0.978	0.976	0.964
0.92	0.08	0.956	0.953	0.932
0.88	0.12	0.933	0.928	0.898
0.84	0.16	0.909	0.903	0.865
0.80	0.20	0.885	0.878	0.834
For square-edged orifices				
0.98	0.02	0.994	0.994	0.993
0.96	0.04	0.986	0.986	0.985
0.92	0.08	0.974	0.974	0.970
0.88	0.12	0.964	0.963	0.956
0.84	0.16	0.952	0.950	0.940
0.80	0.20	0.941	0.938	0.925

Table 12.3 Approximate values of the compressibility factor C_C for flow of air or gases with $k = 1.4$ through a Venturi, flow nozzle or orifice. (For gases with $k = 1.3$ the error is less than 1.0%.)

Another correction coefficient, C_d, the *discharge coefficient*, may be used to include friction head losses and effects of strong curvature of the streamline near the device, i.e., of vena contracta. Some values for discharge coefficients are shown in Table 12.4 and in Fig. 12.13. Equation (12.43), thus corrected, provides the flowrate through the device as

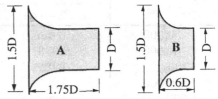

Figure 12.12 Types of flow nozzles, A – long, B – short.

$$Q = C_c C_d \frac{\pi d^2}{4} \sqrt{\frac{\frac{2}{\rho}(p_1 - p_2)}{1 - (d/D)^4}} . \tag{12.45}$$

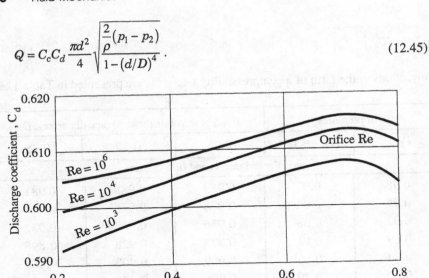

Figure 12.13 Discharge coefficients for thin plate orifice.

The compressibility correction, C_c, depends on the ratio of the diameters of the device and also, slightly, on the kind of device used. The discharge coefficient, C_d, depends also on the ratio of the diameters of the device and the kind of device used. The more elaborate devices present smaller resistance to the flow and have higher discharge coefficients. For a well-designed Venturi meter the discharge coefficient approaches 1. In principle, C_d also depends on the Reynolds number.

Reynolds number at throat	Diameter ratio, d/D		
	0.20 - 0.40	0.60	0.80
50,000	0.966	0.965	0.958
100,000	0.978	0.976	0.970
200,000	0.985	0.984	0.978
500,000	0.991	0.989	0.985
1,000,000	0.993	0.991	0.986
5,000,000	0.994	0.993	0.988

Table 12.4 Discharge coefficients for long nozzles.
(2 – 10-in. pipe; pipe-wall taps one diameter upstream and half a diameter downstream from inlet face.)

In many practical cases it is helpful to consider the discharge coefficient C_d as a multiplier obtained from similarity considerations, not unlike the drag coefficient. In some cases the measuring device, e.g, an orifice, is manufactured to fit some particular configuration and is then calibrated. It is common then to express the calibration information in the form of a calibrated discharge coefficient C_d.

Example 12.10

A Pitot tube is used to measure air flow velocity. The air pressure is 10^5 Pa, its temperature is $300\,K$ and the measured pressure difference of the air is $5,000\,Pa$. Find the velocity.

Note that this example is quite similar to Example 7.4, except for the need to correct for compressibility.

Solution

The density of the air is

$$\rho = \frac{p}{RT} = \frac{10^5}{(8,314 / 28.9) \times 300} = 1.15\,kg / m^3.$$

Using Eq. (12.44), $C_c = 0.98$, we obtain

$$v = 0.98 \times \sqrt{\frac{2}{1.15}(5,000)} = 91.4\ m / s.$$

Example 12.11

Air flows in a 6-in. pipe in which a flowrate measuring device with a diameter ratio of 2 has been installed. The air pressure is 10^5 Pa and its temperature is $300\,K$. The measured pressure drop of the air across the device is $6,000$ Pa. Find the flowrate in each of the following measuring devices:

a.   A Venturi meter.
b.   A flow nozzle.
c.   A thin plate orifice.

Solution

The flowrate is calculated by means of Eq. (12.45). The density of the air is

$$\rho = \frac{p}{RT} = \frac{10^5}{(8{,}314 / 28.9) \times 300} = 1.15\,\text{kg}/\text{m}^3.$$

The cross-sectional area at the throat of the device is

$$A = \frac{\pi}{4}d^2 = \frac{\pi}{4} \times (3 \times 0.0254)^2 = 4.56 \times 10^{-3}\,\text{m}^2.$$

Hence, Eq. (12.45) yields

$$Q = C_c C_d \frac{\pi d^2}{4} \sqrt{\frac{\frac{2}{\rho}(p_1 - p_2)}{1 - (d/D)^4}} = C_c C_d \times 4.56 \times 10^{-3} \times \sqrt{\frac{2}{1.15} \times \frac{6{,}000}{1 - (1/2)^4}}$$

$$= 0.466 C_c C_d.$$

a. For a Venturi meter:
    From Table 12.3, the compressibility correction is $C_c = 0.965$. The discharge coefficient for a well-designed Venturi meter is 1 and Eq. (12.45) yields

    $$Q = 0.466 C_c C_d = 0.466 \times 0.965 \times 1 = 0.450\,\text{m}^3/\text{s}.$$

b. For a flow nozzle:
    From Table 12.3, the compressibility correction is $C_c = 0.965$. The discharge coefficient for a nozzle, Table 12.4, depends on the Reynolds number at the throat. We find Re approximately,

    $$\text{Re} = \frac{dv}{v} = \frac{4Q}{\pi d v} = \frac{4 \times 0.450}{\pi \times (3 \times 0.0254) \times 1.57 \times 10^{-5}} = 479{,}000$$

    and the discharge coefficient is $C_d = 0.990$. Equation (12.45) now yields

    $$Q = 0.466 C_c C_d = 0.466 \times 0.965 \times 0.990 = 0.445\,\text{m}^3/\text{s}.$$

c. For a thin plate orifice:
    From Table 12.3, the compressibility correction is $C_c = 0.980$. For the Reynolds number of part b, Fig. 12.13 leads to a discharge coefficient, $C_d = 0.611$. Equation (12.45) yields

    $$Q = 0.466 C_c C_d = 0.466 \times 0.980 \times 0.611 = 0.279\,\text{m}^3/\text{s}.$$

    This leads to a lower Re. Still the discharge coefficient is roughly the same.

Turbulent Boundary Layers

In the introduction to turbulent flows in this chapter no violation of the Navier–Stokes equations has been implied. The equations are still valid; though for turbulent flows the task of finding solutions becomes even more formidable, because these are time dependent. The boundary layer equations were derived from the Navier–Stokes equations by singular perturbation and are therefore also valid for turbulent flows.

When turbulence exists inside the boundary layer, we speak of turbulent boundary layers. Turbulence in the boundary layer is determined by the local Reynolds number, $Re_x = ux/v$, where x is the distance from the leading edge. Thus for sufficiently small x the flow in the boundary layer is laminar. For larger distances from the leading edge, i.e., for Reynolds numbers exceeding 500,000, the flow in the boundary layer becomes turbulent, Fig. 12.14. It is also noted that on rigid boundaries $q = 0$ and therefore q' must also subside; hence the flow must always become laminar at the wall itself, at least within some thin sublayer. This means that velocity profiles believed to exist inside a turbulent boundary layer cannot be simply extended to the wall, and some additional information may be necessary to affect a solution.

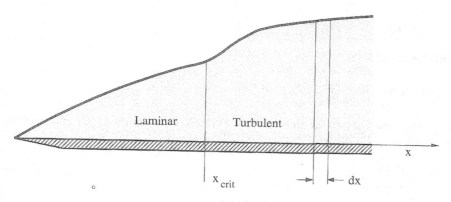

Figure 12.14 Laminar and turbulent boundary layer on a flat plate. The critical Reynolds number for transition from laminar to turbulent flow is $Re_{crit} = 500{,}000$.

Another property typical to turbulent flow is the possibility of the appearance of boundary layers in fully developed or almost fully developed flows. This is a direct result of the ability of the turbulent profile to sustain high shear stress by Reynolds stresses. The velocity profile is thus rather flat in the main flow, Fig.12.8. However, approaching the walls, the intensity of the turbulence decreases. Here the size of the eddies cannot exceed the distance from the wall, and the Prandtl mixing length must decrease. Since the shear forces must eventually be transferred to the wall and the available mechanism for shear stress is now mostly viscous stress, the velocity profile must exhibit a sharp gradient in the vicinity of the wall, i.e., a boundary layer.

One practical result of this behavior is the possibility to use measurements obtained for pipe flow in boundary layer calculations, pipe flows being considerably more amenable to experiments than flat plate flows.

Integral Solutions for Boundary Layers

We now seek an integral solution method, similar to that obtained for the laminar boundary layer in Chapter 11. We want to apply it to flows where our only justification for the use of the term boundary layer is that the velocity profile changes rapidly within a narrow region. We therefore do not integrate the boundary layer equations, as in Chapter 11, but rather start from basic concepts.

Consider the layer shown in Fig. 12.14. The mass flowrate into the dx strip in the boundary layer is just

$$\int_o^\delta \rho u\, dy,$$

and that out of it is

$$\int_o^\delta \rho u\, dy + dx\frac{d}{dx}\int_o^\delta \rho u\, dy.$$

The mass flowrate into the strip from the main flow, i.e., into the strip from above, is therefore

$$dx\frac{d}{dx}\int_o^\delta \rho u\, dy$$

with the corresponding flux of momentum

$$dxU\frac{d}{dx}\int_o^\delta \rho u\, dy.$$

The excess of momentum entering the strip at x over that leaving at $x + dx$ is

$$dx\frac{d}{dx}\int_o^\delta \rho u^2\, dy$$

and the total change in momentum is compensated by the pressure gradient δ (dP/dx) and the shear at the wall $\tau_o\, dx$. Thus

$$U\frac{d}{dx}\int_o^\delta \rho u\,dy - \frac{d}{dx}\int_o^\delta \rho u^2 dy = \delta\frac{dP}{dx} + \tau_o$$

or

$$\tau_o = \rho\frac{d}{dx}\left[\int_o^\delta (U-u)u\,dy\right] - \rho\frac{dU}{dx}\int_p^\delta u\,dy - \delta\frac{dP}{dx}, \tag{12.46}$$

and because in the main flow we still assume

$$\frac{dP}{dx} = -\rho U\frac{dU}{dx}, \tag{12.47}$$

Eq. (12.34) becomes

$$\frac{\tau_o}{\rho} = \frac{d}{dx}\left[\int_o^\delta (U-u)u\,dy\right] + \frac{dU}{dx}\int_o^\delta (U-u)\,dy, \tag{12.48}$$

which is exactly the same as Eq. (11.34). Equation (12.48) is now used to obtain an approximate solution for the flat plate turbulent boundary layer. The velocity profile is assumed, after von Karman, as the seventh-root law

$$\frac{u}{U} = \left(\frac{y}{\delta}\right)^{1/7} = \eta^{1/7} \tag{12.49}$$

while the shear stress is taken from a correlation determined experimentally by Blasius,

$$\tau_o = 0.046\left(\frac{v}{U\delta}\right)^{1/4}\tfrac{1}{2}\rho U^2. \tag{12.50}$$

Substitution in Eq. (12.48) and integration yield

$$\frac{\delta}{x} = 0.37\,\mathrm{Re}_x^{-0.2}. \tag{12.51}$$

Eliminating δ between Eqs. (12.51) and (12.50) leads to

$$\tau_o = 0.030\rho U^2\left(\frac{v}{Ux}\right)^{1/5}. \tag{12.52}$$

It should be noted that τ_o was given here, rather than computed from $\mu(\partial u/\partial y)|_{y=0}$, as for the laminar boundary layer; the turbulent profile, Eq. (12.49), does not hold down to $y=0$, where a laminar sublayer exists. Where no information on τ_o, such as Eq. (12.50), is given, this laminar sublayer must be considered with its velocity distribution, which makes the solution more involved.

Example 12.12

A thin metal plate of dimensions $a = 0.2$ m, $b = 0.5$ m is held in water ($\rho = 1000$ kg/m^3, $\nu = 10^{-6}$ m^2/s), which flows parallel to it, with a velocity $U = 4$ m /s. Find the drag force exerted on the plate for the following cases:

a. The flow being parallel to a.
b. The flow being parallel to b.

Compare your finding to those of Example 11.1.

Solution

The force exerted by the fluid on a plate of length L and width w is

$$F = w \int_o^L \tau_o \, dx.$$

For the case of Re_L being higher than the critical Reynolds number for transition from laminar to turbulent flow, namely, higher than $\mathrm{Re}_c = 500,000$, the integration has to be split into two parts. In the laminar region the wall shear stress τ_o integrated from $x = 0$ to $x = x_c$ is given by Eq. (11.23). In the turbulent region the integration is carried out from $x = x_c$ to $x = L$ using Eq. (12.49). Thus

$$F = w \int_o^{x_c} (\tau_o)_{\mathrm{lam}} \, dx + w \int_{x_c}^L (\tau_o)_{\mathrm{turb}} \, dx,$$

which upon substitution and integration yields

$$F = w \rho U^2 \left[0.664 x_c \, \mathrm{Re}_c^{-1/2} + 0.037 \left(L \, \mathrm{Re}_L^{-1/5} - x_c \, \mathrm{Re}_c^{-1/5} \right) \right]$$

In the present case x_c is found from

$$\mathrm{Re}_c = \frac{x_c U}{\nu}, \qquad x_c = \frac{\nu \mathrm{Re}_c}{U} = \frac{10^{-6} \times 500,000}{4} = 0.125 \text{ m}.$$

Hence, the force on each side of the plate is given by

$$F = 1,000 \times 4^2 \times w \left[0.664 \times 0.125 \times 500,000^{-1/2} + 0.037 \left(L \, \mathrm{Re}_L^{-1/5} - 0.125 \times 500,000^{-1/5} \right) \right]$$

or

$$F = w \left[1.878 + 592 \left(L \, \mathrm{Re}_L^{-1/5} - 0.00906 \right) \right]$$

Case a.

$L = a = 0.2$ m,
$w = b = 0.5$ m,

$$\mathrm{Re}_L = \frac{UL}{\nu} = \frac{4 \times 0.2}{10^{-6}} = 800,000.$$

The force on each side is

$$F = 0.5\,[1.878 + 592\,(\,0.2 \times 800{,}000^{-1/5} - 0.00906\,)] = 2.163\ \mathrm{N}$$

and on both sides

$$F = 4.326\ \mathrm{N}.$$

Case b.

$$L = b = 0.5\ \mathrm{m},$$
$$w = a = 0.2\ \mathrm{m},$$

$$\mathrm{Re}_L = \frac{UL}{\nu} = \frac{4 \times 0.5}{10^{-6}} = 2,000,000.$$

The force on each side is

$$F = 0.2\,[1.878 + 592\,(\,0.5 \times 2{,}000{,}000^{-1/5} - 0.00906\,)] = 2.555\ \mathrm{N}$$

and on both sides

$$F = 5.110\ \mathrm{N}.$$

As seen, the plate with the longer edge parallel to the flow yields here a higher drag force. This is contrary to what was found in Example 11.1 for the laminar case. In the case of the longer edge parallel to the flow a larger part of the plate is in turbulent flow, which leads to a higher overall drag force.

Example 12.13

Repeat Example 12.12 for a plate with the sides 2 m by 5 m, i.e., ten times as large as in Example 12.12.

Solution

The formulation of Example 12.12 is repeated, resulting in

$$F = w\left(1.878 + 592 \times L \times \mathrm{Re}_L^{-1/5} - 592 \times 0.00906\right)$$
$$= w\left(592 \times L^{0.8} \times 4{,}000{,}000^{-0.2} - 3.4855\right) = \left(28.308 \times L^{0.8} - 3.486\right) \times w.$$

Case a. $L = a = 2$ m,
$$w = b = 5\ \mathrm{m},$$
$$F = 229.0\ \mathrm{N, \ on \ one \ side}.$$

Case b. $L = a = 5$ m,
$\qquad w = b = 2$ m,
$\qquad F = 198.2$ N, on one side.

The plate with the longer side parallel to the flow now yields the lower drag force, i.e., the same behavior as in laminar flow.

References

S.K. Friedlander and L. Topper, "Turbulence, Classic Papers on Statistical Turbulence," Interscience, New York, 1961.

J.O. Hinze, "Turbulence," McGraw-Hill, New York, 1959.

J.C. Hunsaker and B.G. Rightmire, "Engineering Applications of Fluid Mechanics," McGraw-Hill, New York, 1947.

L.F. Moody, "Friction Factors for Pipe Flow," Trans. ASME, p. 671, November 1944.

H. Schlichting, "Boundary Layer Theory," 7th ed., McGraw-Hill, New York, 1979.

H. Tennekes and J.L. Lumley, "A First Course in Turbulence," MIT Press, Cambridge, MA, 1972.

Problems

12.1 Use the universal velocity profile to construct a plot of the velocity as a function of radial position when air with a density of 1.2 kg/m^3 flows at an average velocity of 20 m/s through a smooth 0.6-m-diameter pipe. The viscosity of air is 0.02 cp. What is the thickness of the laminar sublayer for this case.

12.2 Use the universal velocity distribution to calculate the local velocity at a point situated midway between the pipe axis and the pipe wall in a 0.5-m-diameter smooth pipe carrying air at an average velocity of 30 m/s and a pressure of 200 kPa gauge. Viscosity of air is 0.02 cp.

12.3 A major pipeline carrying crude oil (density 860 kg/m^3 and viscosity 5 cp) drops 600 m in elevation over the last 5 miles, as shown schematically in Fig. P12.3. The main part of the line is 0.75 m in diameter in which the average velocity of the oil is 2 m/s. The oil arrives at point A at a pressure of 300 kPa gauge and flows downhill to point B where the pressure is desired to be no more than 100 kPa gauge. What diameter of line would you recommend for the section AB? Assume the pipe to be hydrodynamically smooth.

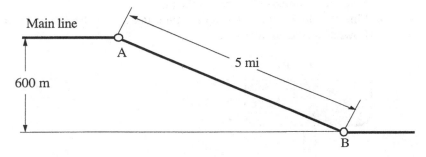

Figure P12.3 Oil pipe.

12.4   Repeat Problem 12.3 for $p_A = 1,500$ kPa.

12.5   Oil (20°C, viscosity 3 cp, specific gravity 0.95) is to flow by gravity through a 1-mile-long pipeline from a large reservoir to a lower station, the average difference in levels being 14 m. The required discharge rate is 1.6 m³/min. Calculate the diameter of steel pipe necessary for these conditions. Compare your solution and answer to those of Problem 7.10.

12.6   A pump is connected to the end of the pipe, point 2 in Fig. 12.5, and is used to recharge the reservoir. The water level in the reservoir is $h = 35$ m. The pressure at point 2, just outside the pump is 500 kPa. What is the water flow into the reservoir?

12.7   Find the water flux as in Problem 12.6 when the circular pipe is replaced by a triangular pipe with all sides equal, having the same cross-sectional area as the circular pipe.

12.8   A car travels at 80 km/h. The driver puts his hand out of the window to cool off. Assuming the hand resembles a flat strip perpendicular to the flow, 0.1 m by 0.5 m, find the additional power expanded by the car.

12.9   Repeat Problem 12.8 when the "strip" is parallel to the flow.

12.10 Figure P12.10 is a scheme of a pipe connected to a water tower open to the atmosphere. The atmospheric pressure is 100 kPa. The bend in the pipe presents a resistance to the flow equivalent to 2 m of pipe. The pipe diameter is 0.1 m, and it is a smooth pipe. The water density is 998 kg/m³, and its viscosity is $\mu = 0.001$ Pa·s. The pressure at point A, shown in the figure, is measured to be $P_1 = 160$ kPa, absolute pressure.

    a. Find the volumetric flowrate in the pipe.

    b. Find the dimensions of a square pipe that gives the same volumetric flowrate as the circular pipe.

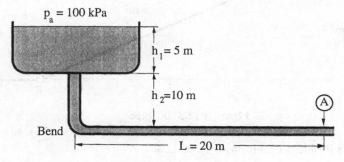

$p_a = 100$ kPa

$h_1 = 5$ m

$h_2 = 10$ m

(A)

Bend

$L = 20$ m

Figure P12.10

12.11 To simplify maintenance, it was suggested to use two 0.1-m-diameter pipes instead of one 0.1414-m-diameter pipe. For the same volume flux of water, find the ratios of the pressure drops:

    a. When the flows are laminar.

    b. When the flows are turbulent.

12.12 Measurements show that the flow between the two plates in Problem 6.7 is turbulent. The pressure between the plates is assumed to depend on the radial coordinate only; the pressure drop is approximated by using relations for pipe flow with the same hydraulic diameter as the radius-dependent circular cross section available to the flow between the plates. The plates are very smooth. With reference to Fig. P6.7, $F = 5$ N, $h = 0.002$ m, $L = 0.1$ m, find the air supply pressure and rate.

12.13 Water flows in a 100-m-long pipe with a diameter of 0.025 m. The average velocity in the pipe is 5 m/s and the pipe discharges to the atmosphere. Compute the Reynolds number.

    a. Is the flow turbulent?

    b. Calculate the pressure at the entrance to the pipe.

12.14 It is suggested to change the pipe described in Problem 12.13 and use two smaller pipes such that the average velocity of the flow remains the same, i.e., 5 m/s. The flowrate remains that of Problem 12.13.
  a. Calculate the required pressure drop in these pipes.
  b. For a pressure drop along the pipes which is only that used in Problem 12.13, find the average velocity and the flow rate in these two pipes.

12.15 The same amount of fluid as in Problem 12.13 must now be supplied using a 0.0125-m-diameter pipe.
  a. Calculate the pressure drop in this pipe and compare your result with that of Problem 12.13.
  b. Note that for a constant flow rate, the pressure drop is proportional to the diameter of the pipe raised to the power n. Find n.
  c. Also note that for a constant pressure drop the flow rate is proportional to the diameter of the pipe raised to the power s. Find s.

12.16 A fluid flows in a pipe which has the diameter D. The same fluid flows in the gap between two parallel flat plates. The size of the gap is also D. The same pressure gradient exists in both systems. Calculate the ratio between the flow rate of the fluid in the pipe and the flow rate between the plates per width of D.

12.17 The pipe and the plates of Problem 12.16 are now used such that the flow rate in the pipe is the same as that between the plates per width D. Calculate the ratio between the power needed to pump the fluid through the pipe and that needed to pump the fluid between the plates.

12.18 Water flows through a rectangular square duct with the sides $D = 0.1$ m. The average velocity is 8 m/s. The duct is 100 m long. Calculate the power needed to pump the water. Compare your results with those obtained for the same flow rate through a circular pipe.

12.19 Water flows through an annulus with an inner radius of 0.10 m and an outer radius of 0.12 m at an average velocity of 6 m/s. The length of the annulus is 100 m. Find the pressure drop along the annulus.

12.20 Repeat Problem 12.13 for a pipe of the same cross section which is:
  a. Square.
  b. Rectangular, with sides ratio of 1 : 2.
  c. Elliptic, with sides ratio of 1 : 2.

12.21 A municipal water main has the diameter of 0.1 m. The mean velocity in the main is 10 m/s. The line has a tee connection every 50 m and a bend every 100 m. Calculate the pressure drop expected per 1000 m of pipe.

12.22 A truck is designed to travel at a cruising speed of 110 km/h. A preliminary design is shown in Fig. P12.22. The width of the truck is 4 m, and to estimate the air drag force on the truck, one may use the frontal projection, i.e., the same drag as for a flat plate 4 m×4 m perpendicular to the flow. It has been suggested to streamline the geometry such that the drag coefficient becomes 0.6. Estimate the power saved at cruising speeds.

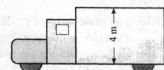

Figure P12.22 A truck.

12.23 A car has the mass of 700 kg. In normal driving the car tires are not expected to skid as long as the normal force between a tire and the road is at least 700 N. The maximum velocity technically safe for the car is 140 km/h. The projected area of the car is 6 m². Find the maximum allowable lift coefficient of the car geometry. Note that this coefficient is defined for the projected area. Find the maximum allowable driving speeds with wind velocities of 80 km/h and 30 km/h.

12.24 A plumber had to connect a 0.025-m-diameter pipe to a hose. Not having the necessary parts he inserted a connection of 0.013 m diameter in the line and then continued with the 0.025 m diameter. For a mean flow velocity of 5 m/s, estimate the extra pressure drop caused by the plumber's improvisation.

12.25 An air turbine used to measure wind velocities is shown in Fig. P12.25. The turbine is connected to an electric generator, and the energy generated is measured and wasted. For wind velocity of 30 m/s the turbine turns at 12 rpm. Estimate the energy wasted.

12.26 The airplane considered in Example in 12.9 is now using takeoff strips that permit ground speeds of 200 km/h. Find the dimensions of the airplane wings necessary under these new conditions.

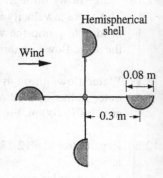

Figure P12.25

12.27 The mean velocity of a fluid having properties like water is between 1 m/s and 6 m/s. The fluid flows in a pipe with a diameter of 0.1 m. Estimate the reading of the following flowrate measuring devices at the extreme velocities:
  a. Venturi meter.
  b. Flow nozzle.
  c. Thin plate orifice.

12.28 Repeat Problem 12.27 for an airlike fluid, with the limit mean velocities 10 m/s to 60 m/s.

12.29 Consider water flowing in a smooth pipe of 0.3 m diameter at the mean velocity of 10 m/s. Calculate the pressure drop along 100 m of pipe. Now assume the flow to be a boundary layer flow along the pipe walls and calculate the pressure drop expected from the entrance to the pipe up to a point of a 100 m downstream.

12.30 Consider the turbulent velocity profile in a pipe to be approximated by a mid region of constant velocity and a boundary region, near the wall, where the velocity change is strong. Use the integral approach to the boundary layer solution and find the thickness of this region of strong velocity change such that the correct pressure drop is obtained.

12.31 Consider Fig. 12.6 and Example 12.6. We are now told that point A, where the pressure P_A = 600 kPa has been measured, is located inside a water tank, and the exit to the pipe is a perpendicular square entrance, for which Table 12.2 yields $k = 0.50$. Point B is followed by a tee connection, for which Table 12.2 suggests L_e = 60, for the direction BC and for the direction BD. Between B and C there is an additional bend with L_e = 20 and another bend of the same kind between B and D. Find the flow rates now.

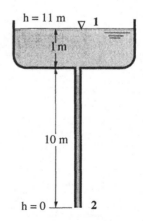

12.32 Water is being discharged from a large tank open to the atmosphere through a vertical tube, as shown in Fig. P12.32. The tube is 10 m long, 1 cm in diameter, has a roughness of 5×10^{-5} m, and its inlet is sharp edged and located 1 m below the level of the water in the tank.

Figure P12.32 Water discharge from a tank.

     a. Find the velocity and the volumetric flowrate in the pipe.

     b. Compare your answers to those of Example 7.5.

12.33 Consider Problem 12.32. A turbine is connected at the tube outlet, point 2. Find the maximum power that can be obtained from the turbine.

13. COMPRESSIBLE FLOW

Speed of Infinitesimal Disturbances – Sonic Speed

In all our considerations up to this point the density of the fluid was assumed constant. Suppose we have a fluid continuum at rest and cause a small pressure disturbance at one point in it. We may now ask how fast this disturbance travels in the continuum. To answer this question, we may use measurements or try to calculate the speed of the disturbance. Consistent with the previous assumption that

$$\rho = \text{const}, \qquad d\rho = 0,$$

we find this speed of propagation to be infinite, i.e., the disturbance is felt at once everywhere [see Eq. (13.5)]. This means, of course, that strict incompressibility does not occur, though it may be a useful approximation. This also explains why this question was not asked before, while treating incompressible flows. Once we admit compressibility, the speed of propagation of disturbances becomes finite. Then it also becomes an important dependent variable of the field of flow, which we want now to calculate.

Consider the propagation of a disturbance in a continuum, as shown in Fig. 13.1. The undisturbed part of the medium is separated from the disturbed region by a surface of discontinuity, called the *front of the disturbance*. We concentrate on a small but finite area of this front, which has the forward normal $\mathbf{n}$ and the local velocity of propagation

$$\mathbf{u} = \mathbf{n}u.$$

Consider an observer sitting in a coordinate system which moves with $\mathbf{u}$. From the observer's point of view the undisturbed medium is streaming into the considered area of the disturbance front with the velocity $(-\mathbf{u})$. As the material passes the front, its properties change:

u into $u + \Delta u$,
p into $p + \Delta p$,
ρ into $\rho + \Delta \rho$,
T into $T + \Delta T$,
etc.

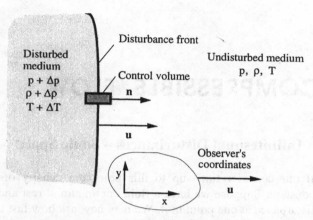

Figure 13.1 Propagation of disturbance.

Now consider a small control volume in the shape of a stream tube cutting through the disturbance front. Mass balance for the control volume requires

$$\rho u = (\rho + \Delta \rho)(u + \Delta u) \tag{13.1}$$

and the momentum theorem for the control volume states

$$p + (\rho u)u = (p + \Delta p) + \left[(\rho + \Delta \rho)(u + \Delta u)\right](u + \Delta u). \tag{13.2}$$

Substitution of Eq. (13.1) into Eq. (13.2) and rearrangement yield

$$u^2 = \frac{\Delta p}{\Delta \rho} - u \cdot \Delta u. \tag{13.3}$$

An *infinitesimal disturbance* is now defined as that which has the limits

$$\Delta u \to du, \qquad \Delta p \to dp, \qquad \Delta \rho \to d\rho, \qquad \Delta T \to dT, \dots .$$

The speed of propagation of this disturbance is denoted by c, and Eq. (13.3) becomes

$$c = \sqrt{\frac{dp}{d\rho}} . \tag{13.4}$$

The process of passing through the disturbance is taken to be adiabatic, and the undisturbed region is taken to be in thermodynamic equilibrium. Moreover, the disturbance, which is infinitesimal, causes all thermodynamic properties in the disturbed region to deviate from those in the undisturbed one only infinitesimally, and therefore the disturbed region is also assumed to be in thermodynamic equilibrium. The process of passing through the infinitesimal disturbance is thus taken to be reversible, hence isentropic. To clearly show this, Eq. (13.4) is rewritten as

$$c = \sqrt{\left(\frac{\partial p}{\partial \rho}\right)_s}, \tag{13.5}$$

where c is also called the *sonic speed,* and measurements show it to be a very good approximation to the *speed of sound,* i.e., to the speed with which sound waves propagate. Equation (13.5) is quite general and applies to solids, liquids and gases, whether ideal or not.

In what follows the considered fluid is air, assumed to obey the ideal gas law, i.e.,

$$p = \rho RT, \tag{13.6}$$

with R = 287 J/kg K for air. In an isentropic process

$$p / \rho^k = \text{const}, \tag{13.7}$$

which together with Eq. (13.6) yields

$$\left(\frac{\partial p}{\partial \rho}\right)_s = k\frac{p}{\rho} = kRT,$$

and for an ideal gas Eq. (13.5) becomes

$$c = \sqrt{k\frac{p}{\rho}} = \sqrt{kRT}. \tag{13.8}$$

Example 13.1

A large pressure vessel contains air ($k = 1.4$, $R = 287$ J/kg K) at 2 bars, 300 K. A small nozzle is fitted into the wall of the vessel through which air expands isentropically to the environment pressure of 1 bar. Find the speed of the air as it leaves the nozzle. Is this speed above or below the speed of sound, i.e., is the flow there subsonic or supersonic?

Solution

The first part of this problem is very similar to part b of Example 7.5. Since the process is isentropic,

$$T_2 = T_1 \left(\frac{p_2}{p_1}\right)^{\frac{k-1}{k}} = 300 \times \left(\frac{1}{2}\right)^{\frac{0.4}{1.4}} = 246 \text{ K.}$$

Denoting the enthalpy i and the specific heat of air $c_p = Rk/(k-1) = 1,005 \text{ J/kg}$, we obtain

$$i_1 - i_2 = c_p(T_1 - T_2) = 1,005 \times (300 - 246) = 54,270 \text{ J/kg} = \tfrac{1}{2}\left(u_2^2 - u_1^2\right),$$

$$u_1 = 0, \qquad u_2 = 329 \text{ m/s.}$$

Thus the speed of the air at the outlet of the nozzle is 329 m/s. Is this subsonic or supersonic speed?

The speed of sound of the air in the vessel is

$$c_1 = \sqrt{kRT_1} = \sqrt{1.4 \times 287 \times 300} = 347 \text{ m/s.}$$

However, the speed of sound of the air as it emerges from the nozzle is

$$c_2 = \sqrt{kRT_2} = \sqrt{1.4 \times 287 \times 246} = 314 \text{ m/s.}$$

Hence, $u_2 > c_2$ and the speed is supersonic.

Propagation of Finite Disturbances

The sonic speed, obtained for *infinitesimal disturbances,* has been uniquely determined by Eq. (13.5). Is there such a general constraint on the speed of *finite disturbances* moving in a compressible medium? An example of a finite disturbance is a solid body moving through the medium: a meteor, an airplane, a bullet, a car, a speck of dust, a feather. Such a moving body influences certain regions in the fluid by means of propagating disturbance fronts.

Consider a small body (e.g., a meteorite) moving in the x-direction at a constant velocity v, Fig. 13.2. As the body moves, it disturbs the stationary environment by pushing the air ahead of it, and the disturbance advances outward with its front shaped as a spherical shell of radius

$$r = c(t - t_o),$$

where t_o is the time of emission and r is measured from the location of the body at the moment of emission. Let the Mach number M be defined by

$$M = \frac{v}{c}.$$

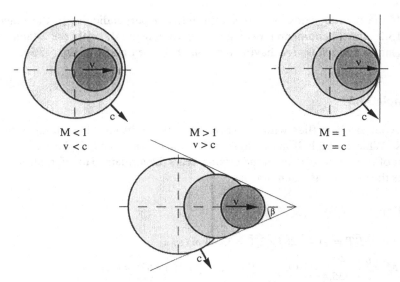

Figure 13.2 Subsonic, sonic and supersonic motions.

In *subsonic motion M* < 1, i.e., $v < c$, and the front of the disturbance, which advances with the speed c, precedes the moving body itself, Fig. 13.2. The front of the disturbance may be detected by some pressure-sensitive instrument, e.g., the human ear, and someone hearing that noise may look up and await the approach of the moving body. The front of the disturbance is called a *Mach surface*.

Now let $M = 1$. The body moves in the x-direction just as fast as the disturbance front. The front does not precede the body in the x-direction, and the region to the right of the body, Fig. 13.2, is known as the *zone of silence*. The moving body is not heard in the zone of silence, and no pressure rise precedes it. Once heard, the listener is surprised by the appearance of the body itself.

In *supersonic motion M* > 1, i.e., $v > c$, and the front of the moving body moves faster than the disturbance. It pulls the front of the disturbance forward in the direction of its own motion and the disturbance front attains the form of a cone of half apex angle

$$\beta = \arcsin\left(\frac{1}{M}\right)$$

$$(13.9)$$

which moves forward with the speed v along its axis of symmetry. The cone itself is called the *Mach cone*, and it represents a conical front which propagates with the speed c normal to itself. The region in front of the cone is a *zone of silence*, and once the moving body is heard, the listener must look rather forward, in the direction of its motion, to see the retreating body somewhat from behind.

If the moving body is a long thin cylinder perpendicular to the plane of Fig. 13.2, the phenomenon becomes two dimensional, with the Mach cone changing to a *Mach wedge,* having the same half apex angle of Eq. (13.9).

Example 13.2

A jet airplane flies with the speed of 400 m/s in air whose temperature is 280 K. What is the half apex angle of its Mach cone? The airplane flies at the height of $h = 1$ km. A man on the ground hears the airplane. How far ahead of the man is the airplane at that moment?

Solution

$$c = \sqrt{kRT} = \sqrt{1.4 \times 287 \times 280} = 335.4 \text{ m / s},$$

$$M = \frac{v}{c} = \frac{400}{335.4} = 1.19, \qquad \beta = \arcsin\left(1 / M\right) = 57^o = 1 \text{ rad}.$$

Assuming the plane flies parallel to the ground,

$$\Delta x = h \ \tan \ \beta = 1000 \tan 1 = 1550 \text{ m ahead}.$$

The Piston Analogy

The description of the motion of supersonic disturbances, as just presented, still leaves the details of what happens just in front of the body for $M \geq 1$ unclear.

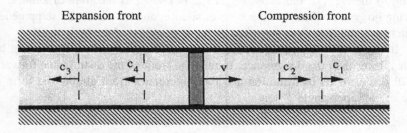

Expansion front   Compression front

Figure 13.3 Piston and disturbances.

To clarify this point let us consider a simpler, one-dimensional case. Let the moving body be an insulated frictionless piston moving in an infinitely long insulated cylinder filled with an ideal gas, Fig. 13.3. The gas is initially at rest, and at the time $t_o = 0$ the piston is suddenly set in motion at the constant speed v. The gas in the direction of the piston motion is compressed, with the disturbance running to the right as a continuous series of successive compression fronts. The gas on the reverse side of the piston expands, with the disturbances moving to the left as successive expansion fronts. All these fronts represent *Mach surfaces*.

Consider two compression fronts, 1 and 2, where the second front started later than the first one. The first moves at the speed

$$c_1 = \sqrt{kRT}$$

and leaves behind it the temperature $T + dT$, and because the compression is assumed isentropic,

$$dT > 0.$$

The second front moves into the already disturbed gas and therefore has the speed

$$c_2 = \sqrt{kR(T + dT)} > c_1.$$

The second front overtakes the first front and coalesces with it. No matter how many fronts we have on the compression side, those closer to the piston run faster and overtake those ahead of them, which eventually overtake the first one. The disturbances are concentrated into one front of finite magnitude, called a *Shock Wave*. The speed of the shock wave is no longer given by Eq. (13.5), but rather by Eq. (13.3). From the point of view of Fig. 13.1, Δu is negative, and $-u\,\Delta u > 0$. Furthermore, the process is no longer isentropic, but is still adiabatic, and therefore requires $\Delta s > 0$, which for compression results in

$$\frac{\Delta p}{\Delta \rho} > \left(\frac{\partial p}{\partial \rho}\right)_s.$$

Equation (13.3) then yields u > c. The shock wave moves at speeds higher than the sonic speed.

Now consider the expansion fronts on the left side of the piston, Fig. 13.3. The first one moves at the speed

$$c_3 = \sqrt{kRT},$$

leaving behind it the temperature $T + dT$, and because the expansion is assumed isentropic

$$dT < 0.$$

The second front moves at the speed

$$c_4 = \sqrt{kR(T + dT)} < c_3,$$

and it therefore lags behind the first front. No matter how many expansion fronts we have, no two of them coalesce, and no amplification or concentration of disturbances occur. A finite disturbance of the expansion type simply cannot be created.

Going now back to the case M > 1 in Fig. 13.2, we see that the moving body carries in front of it° a shock wave, which also moves supersonically at the speed v.

Quasi-one-dimensional Flow, Stagnation Properties

A flow which satisfies the following two requirements is called *quasi-one-dimensional:*

1. The divergence and convergence of all stream tubes in the flow are small, in the sense that the velocity component along the tube, u, and the other velocity components, v and w, which are perpendicular to u, satisfy

$$\left|\frac{v}{u}\right| \ll 1, \qquad \left|\frac{w}{u}\right| \ll 1.$$

2. The ratio of the radius of curvature of the stream tube to the linear dimension of the tube cross section, b, measured along the radius of curvature, is much larger than 1.

A quasi-one-dimensional-flow is shown in Fig. 13.4. In quasi-one-dimensional flows the x-axis points in the local direction of the flow. Conservation of mass now becomes

$$\rho u A = \dot{M} = \text{const}, \tag{13.10}$$

with A the cross-sectional area normal to u. Differentiation of Eq. (13.10) yields

$$\frac{d\rho}{\rho} + \frac{du}{u} + \frac{dA}{A} = 0, \qquad \frac{1}{\rho}\frac{d\rho}{dx} + \frac{1}{u}\frac{du}{dx} + \frac{1}{A}\frac{dA}{dx} = 0. \tag{13.11}$$

We consider here fast flows, with rather large Reynolds numbers, and we assume the conditions leading to the Euler equation (10.1) to be satisfied. Thus the momentum equation has the Euler form, Eq. (10.1), which for negligible body forces and for quasi-one-dimensional flows becomes

$$\rho u \frac{du}{dx} = -\frac{dp}{dx} \qquad \text{or} \qquad \rho u \, du = -dp. \tag{13.12}$$

Because the flow is fast, it can be assumed adiabatic. Heat transfer may still be considered, say into the wall of a nozzle. However, for the fast moving gas in that same nozzle the process is still adiabatic. The apparent contradiction here is resolved by the observation that a negligible heat interaction per unit mass of flowing gas may amount to a large heat interaction when multiplied by a very large mass flux. Thus, except for some singular locations in the field, the flow may be taken as adiabatic. Moreover, because the flow in the considered region has also been assumed frictionless, it is now both adiabatic and frictionless, i.e., isentropic.

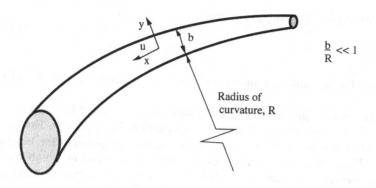

Figure 13.4 A stream tube in quasi-one-dimensional flow.

For an isentropic process the momentum equation can be integrated to yield an energy equation, and indeed Eq. (13.12) can be integrated to

$$\tfrac{1}{2}u^2 + \int_{p_o}^{p} \frac{dp}{\rho} = \text{const}$$

which is just a particular form of the Bernoulli equation, Eq. (7.7). Now, for a thermodynamic system undergoing a frictionless process the first law requires

$$\delta Q = di - \frac{1}{\rho}dp,$$

where Q is the heat per unit mass and i is the specific enthalpy. For an adiabatic process this becomes $di = 1/\rho\, dp$, and with Eq. (13.12) it takes the form

$$i + \tfrac{1}{2}u^2 = i_o = \text{const}, \tag{13.13}$$

where i_o is denoted the *stagnation enthalpy*. It is the value of the enthalpy which may be obtained by stopping the flow adiabatically. The more useful form of the energy equation is that of Eq. (13.13).

We now summarize the system of equations governing quasi-one-dimensional frictionless flows:

The energy equation

$$i + \tfrac{1}{2}u^2 = i_o = \text{const}, \tag{13.13}$$

the equation of state

$$p = \rho RT,$$
(13.6)

$$i + c_p T = \text{const}$$
(13.14)

and the equation for an isentropic process

(13.7)

$$\frac{p}{\rho^k} = \text{const}.$$

We note that the last three equations hold for an ideal gas only, while Eq. (13.13) is general.

The *stagnation properties* are those that can be obtained for a point in the flow by an *isentropic process* that ends in stagnation, i.e., in $u = 0$. The process need not be physically performed and the stagnation properties may be simply calculated. These stagnation properties "belong" to the point at which the process starts. However, as long as the flow is isentropic and contains no singular points, the same stagnation properties apply to all points in the flow.

It is convenient to render most relations for the flow dimensionless, and a natural characteristic velocity here is the sonic speed. Let Eq. (13.13) be written again using Eq. (13.14),

$$c_p T + \tfrac{1}{2} u^2 = c_p T_o.$$
(13.15)

Using Eq. (13.8) and noting $c_p = kR/(k-1)$, Eq. (13.15) becomes

$$\frac{T_o}{T} = 1 + \frac{1}{2c_p} \cdot \frac{u^2}{T} = 1 + \frac{k-1}{2} \cdot \frac{u^2}{kRT}$$

or

$$\frac{T_o}{T} = 1 + \frac{k-1}{2} M^2.$$
(13.16)

Using Eq. (13.7) and noting that $T_2/T_1 = \left(p_2/p_1 \right)^{\frac{k-1}{k}}$, Eq. (13.16) becomes

$$\frac{p_o}{p} = \left[1 + \frac{k-1}{2} M^2 \right]^{\frac{k}{k-1}},$$
(13.17)

and

$$\frac{\rho_o}{\rho} = \left(\frac{p_o}{p} \right)^{\frac{1}{k}} = \left[1 + \frac{k-1}{2} M^2 \right]^{\frac{1}{k-1}}.$$
(13.18)

Obviously, for an isentropic process

$$\frac{s_o}{s} = 1.$$

Example 13.3

Point B is in a compressible flow field of air (ideal gas, $R = 287$ J/kg·K, $k=1.4$). It is also known that

$$p_B = 2\times 10^5 \text{ Pa}, \quad T_B = 350 \text{ K}, \quad U_B = 350 \text{ m/s}.$$

Find the stagnation properties p_{Bo}, T_{Bo}, ρ_{Bo}, s_{Bo} of point B.

The streamline through B eventually passes through point D, but between B and D it passes through a shock wave (an irreversible adiabatic process). Which of the stagnation properties of B changes and in which direction?

Solution

$$\rho_B = \frac{p_B}{RT_B} = \frac{2\times 10^5}{287\times 350} = 2.0 \text{ kg} / \text{m}^3,$$

$$c_B = \sqrt{kRT_B} = 375 \text{ m} / \text{s}, \qquad M_B = \frac{u_B}{c_B} = \frac{350}{375} = 0.933.$$

For an isentropic process, $s_{Bo} = s_B$.

From Eqs. (13.17), (13.18) and (13.16), respectively,

$$p_{Bo} = p_B\left[1+\frac{k-1}{2}M_B^2\right]^{\frac{k}{k-1}} = 2\times 10^5 \times 1.754 = 3.51\times 10^5 \text{ Pa},$$

$$\rho_{Bo} = \rho_B\left[1+\frac{k-1}{2}M_B^2\right]^{\frac{l}{k-1}} 2.0\times 1.494 = 2.988 \text{ kg} / \text{m}^3,$$

$$T_{Bo} = T_B\left[1+\frac{k-1}{2}M_B^2\right] = 350 \times 1.174 = 411 \text{ K}.$$

After the shock wave:

$$i_{Do} = i_{Bo},$$

$$T_{Do} = T_{Bo},$$

$$s_{Do} > s_{Bo} = s_B,$$

$$p_{Do} < p_{Bo} \quad (\text{because } T_{Do} = T_{Bo} \text{ and } s_{Do} > s_{Bo}),$$

$$\rho_{Do} < \rho_{Bo} \quad (\text{because } T_{Do} = T_{Bo} \text{ and } s_{Do} > s_{Bo}).$$

Nozzle Flow

Consider a stream tube in quasi-one-dimensional flow, Fig. 13.4. Let du from Eq. (13.12) be substituted in Eq. (13.11) and this equation rearranged to

$$\frac{A}{\rho u^2} \cdot \frac{dp}{dx}\left[\frac{u^2}{dp/d\rho} - 1\right] + \frac{dA}{dx} = 0$$

or

$$-\frac{dA}{dx} = \frac{dp}{dx}\left(M^2 - 1\right)\frac{A}{\rho u^2} = -\frac{du}{dx}\left(M^2 - 1\right)\frac{A}{u}. \qquad (13.19)$$

Thus for subsonic flow, with $M < 1$, a converging stream tube, with $dA/dx < 0$, corresponds to expansion, with $dp/dx < 0$, and to increase in velocity, with $du/dx > 0$. For supersonic flow, however, the behavior of dp/dx and du/dx is the opposite, and the various combinations are listed in Table 13.1.

M	dA/dx	dp/dx	du/dx
$M < 1$	> 0 < 0	> 0 < 0	< 0 > 0
$M > 1$	> 0 < 0	< 0 > 0	> 0 < 0
$M = 1$	0	> 0 < 0	> 0 < 0

Table 13.1 Converging–diverging stream tube.

It is noted that $dA/dx = 0$ is a necessary condition for $M = 1$. Sonic speed can appear only where the tube has a local minimum, or a so-called *throat*. However $dA/dx = du/dx = 0$ may also occur at that location, i.e., $dA/dx = 0$ is not a sufficient condition for $M = 1$.

At this stage we consider a flow in a nozzle. A nozzle is a conduit of rigid walls satisfying the conditions for quasi-one-dimensional flow. This nozzle is connected to a vessel where a gas at the stagnation pressure p_o and the stagnation temperature T_o is assumed to be supplied indefinitely. The nozzle's exit is also assumed to be kept at a constant pressure p_e. The nozzle is assumed to have a throat, as shown in Fig. 13.5. The rigid walls of the nozzle are considered to be the stream tube envelope, and we also assume no separation of the flow from the nozzle wall. We want to obtain some relations for the design and performance of such nozzles.

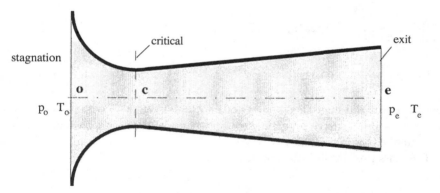

Figure 13.5 A diagram of a convergent–divergent nozzle.

Example 13.4

Show that the sonic speed can appear where the nozzle has a local minimum, i.e., a "throat," and cannot appear where the nozzle has a local maximum.

Solution

We refer to Table 13.1: For a local minimum the flow toward the location where $dA/dx = 0$ is in a convergent part of the nozzle. Thus if the approaching flow is subsonic it accelerates and may become sonic. If the approaching flow is supersonic, it decelerates and may become sonic.

On the other hand, where the nozzle has a local maximum, the approach stream is divergent. Thus a subsonic stream would slow down, and a supersonic stream would speed up, and neither can reach sonic speed.

Mass Flux through the Nozzle

Suppose the sonic speed is attained in the nozzle. This can happen only at the throat c. We then call this section "critical" and denote by asterisks the properties corresponding to it. If the sonic speed is not realized at the throat, section c, then this section is not called critical. Substitution of $M = 1$ in Eqs. (13.16), (13.17) and (13.18) yields

$$\frac{T^*}{T_o} = \frac{1}{1 + (k-1)/2} = \frac{2}{k+1} = 0.83333, \tag{13.20}$$

$$\frac{p^*}{p_o} = \left(\frac{2}{k+1}\right)^{\frac{k}{k-1}} = 0.52828, \tag{13.21}$$

$$\frac{\rho^*}{\rho_o} = \left(\frac{2}{k+1}\right)^{\frac{1}{k-1}} = 0.63394, \tag{13.22}$$

where the computed numerical values are for $k = 1.4$. We also note that because Eqs. (13.16) - (13.18) have been obtained for isentropic processes, such processes are also implied for Eqs. (13.20) - (13.22).

The mass flow through the nozzle is

$$\dot{m} = \rho u A = \rho^* u^* A^* = \rho^* c^* A = \rho_o \frac{\rho^*}{\rho_o} c_o \left(\frac{T^*}{T_o}\right)^{\frac{1}{2}} A^*.$$

And for

$$\rho_o = p_o / RT_o, \qquad c_o = \sqrt{kRT_o}, \qquad \rho_o c_o = p_o \sqrt{\frac{k}{RT_o}},$$

this becomes

$$\dot{m} = p_o A^* \sqrt{\frac{k}{RT_o}} \left(\frac{2}{k+1}\right)^{\frac{1}{k-1}+\frac{1}{2}}$$

$$= p_o A^* \sqrt{\frac{k}{RT_o}} \left(\frac{2}{k+1}\right)^{\frac{k+1}{2k-2}} = 0.6847 \frac{p_o A^*}{\sqrt{RT_o}}. \tag{13.23}$$

The mass flux is completely determined by the stagnation pressure and temperature and by the critical cross section A^*. Since the same mass flows through all sections

$$\rho u A = \rho^* u^* A^* = \rho^* c^* A^*$$

and

$$\frac{A}{A^*} = \frac{\rho^*}{\rho} \frac{c^*}{u} = \frac{\rho^*}{\rho} \cdot \frac{c^*}{c} \cdot \frac{c}{u} = \frac{\rho^*}{\rho} \left(\frac{T^*}{T}\right)^{\frac{1}{2}} \frac{1}{M}. \tag{13.24}$$

Using Eqs. (13.16) - (13.18) and (13.20) - (13.22), we obtain

$$\frac{p^*}{p} = \frac{p^*}{p_o} \cdot \frac{p_o}{p} = \left[\left(\frac{2}{k+1}\right)\left(1+\frac{k-1}{2}M^2\right)\right]^{\frac{k}{k-1}}, \tag{13.25}$$

$$\frac{\rho^*}{\rho} = \frac{\rho^*}{\rho_o} \cdot \frac{\rho_o}{\rho} = \left[\left(\frac{2}{k+1}\right)\left(1+\frac{k-1}{2}M^2\right)\right]^{\frac{1}{k-1}}, \tag{13.26}$$

$$\frac{T^*}{T}=\frac{T^*}{T_o}\cdot\frac{T_o}{T}=\left(\frac{2}{k+1}\right)\left(1+\frac{k-1}{2}M^2\right).\qquad(13.27)$$

Equations (13.26) and (13.27) are substituted into Eq. (13.24), which becomes

$$\frac{A}{A^*}=\frac{1}{M}\left[\left(\frac{2}{k+1}\right)\left(1+\frac{k-1}{2}M^2\right)\right]^{\frac{k+1}{2(k-1)}}.\qquad(13.28)$$

Using the Mach number M as a parameter, p, ρ, T and A are determined everywhere in terms of the critical properties p^*, ρ^*, T^* and A^*.

Example 13.5

Show that the mass flux per unit cross-sectional area is the highest at the critical section.

Solution

Because the same mass flux passes through all sections,

$$\dot{m}=\rho uA=\rho^*u^*A^*=\text{const.}$$

Thus $\rho u=\dot{m}/A$, and ρu has a maximum where A has a minimum. Now a minimum for A can be only where A has a local minimum. If the nozzle has only one throat and sonic speed is attained, this throat is both the critical one and the only one with a local minimum for its area; the proof is then complete.

Consider now a nozzle which has several throats and achieves sonic speed. The mass flux per unit cross-sectional area, ρu, still has a maximum where A is the smallest. Hence, of all the throats the smallest one achieves the highest value for ρu. Let the critical section, i.e., the one where sonic conditions appear and which must be one of the throats, be A^*. Let *any* other section, throat or not throat, be A. Then Eq. (13.28) gives

$$\frac{A}{A^*}=f(M)=\frac{1}{M}\left[\left(\frac{2}{k+1}\right)\left(1+\frac{k-1}{2}M^2\right)\right]^{\frac{k+1}{2(k-1)}}$$

or

$$\ln f=-\ln M+\frac{k+1}{2(k-1)}\left[\ln\left(1+\frac{k-1}{2}M^2\right)+\ln\left(\frac{2}{k+1}\right)\right].$$

Differentiation with respect to M yields

$$\frac{1}{f}\frac{df}{dM} = 0 = -\frac{1}{M} + \frac{k+1}{2(k-1)}\cdot\frac{(k-1)M}{1+[(k-1)/2]M^2} = -\frac{1}{M} + \frac{(k+1)M}{2+(k-1)M^2},$$

with the solution $M = 1$. To ensure that the function $f(M) = A/A^*$ has a true minimum at $f(1) = 1$, i.e., that the critical section is the smallest one, and the proof is complete, the reader may take the second derivative and find it positive.

The Design and Performance of a Nozzle

We return now to the nozzle in Fig. 13.5. Suppose its critical section and the mass flux are already known. Still conditions at its exit section e must be considered. Two different engineering situations may arise:

a.   A design problem: Given p_e, find A_e.
     Using p^*/p_e in Eq. (13.25), M_e is obtained and substituted in Eq. (13.28) to yield A_e/A^*.
b.   A performance problem: Given A_e, find the corresponding p_e.
     Using A_e/A^* in Eq. (13.28), two M_e values are obtained, and with them two p_e values.

This two-valued situation may be seen at once from the structure of Eq. (13.28) with its minimum (see Example 13.5). It could also, however, be inferred from Table 13.1.

The flow at c, Fig. 13.5, is assumed to be at $M = 1$. Suppose just downstream from c the flow becomes slightly subsonic, i.e., $M < 1$. Since the nozzle diverges, the flow would slow down further, ending with $M_e < 1$. Conversely, if the flow just downstream from c becomes even slightly supersonic, it would continue to accelerate because the nozzle diverges.

These two M_e values, when used in Eq. (13.25), yield two p_e values. Thus the same nozzle corresponds to two p_e values, and the flow in its diverging part, whether subsonic or supersonic, is determined by the pressure p_e actually imposed at the exit e.

The entrance section, o, which is assumed at stagnation, evades similar treatment. Because of $M_o \approx 0$, Eq. (13.28) yields $A_o \to \infty$, and because of $u \approx 0$, Eq. (13.19) yields $dA_o/dx \to \infty$. The nozzle should start, therefore, with its walls tangent to the stagnation vessel, as shown in Fig. 13.5.

Let a scheme of the convergent–divergent nozzle be drawn again, with the pressure ratio p / p_o also shown, Fig. 13.6. The two exit pressures corresponding to sonic speed at the throat are shown as point C for the subsonic exit and point D for the supersonic one. Point A corresponds to no flow at all, while point B represents a flow which is subsonic everywhere.

We find that all exit pressures between points A and C correspond to subsonic flows, while point D corresponds to a supersonic flow in the diverging part of the nozzle. Yet the nozzle can be made to discharge into a vessel where the pressure is controlled. What then happens for exit pressures below point D? Or between C and D?

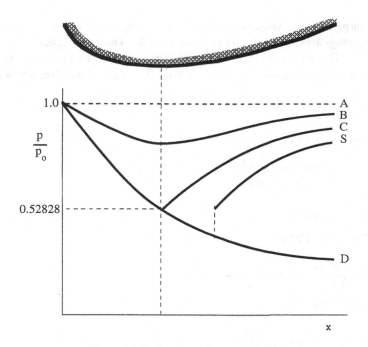

Figure 13.6 Pressure along a nozzle.

Below p_D the flow inside the nozzle is the same as for p_D The information that the pressure in the discharge vessel has dropped below p_D is expressed in the form of an expansion wave and therefore cannot propagate against the super-sonic flow into the nozzle. An expansion wave, as already shown in this chapter, propagates at the sonic speed. Hence the exit is a "zone of silence" with respect to the diverging part of the nozzle.

Coming out of the nozzle the gas encounters a zone of lower pressure and it expands further. However, this takes place outside the nozzle and the flow is free to choose its envelope for the streamlines. The phenomenon then becomes non-quasi-one-dimensional.

Between p_C and p_D the situation is more complex, because, assuming a flow down to p_D to have been established, the information that the pressure is *above* p_D is in the form of a compression wave. It can build up to a shock wave, which moves at a supersonic speed, and thus can propagate upstream into the nozzle. Indeed this is what happens, and some additional information on the behavior of shock waves is now needed in order to explain what really takes place.

Example 13.6

A large pressure vessel contains air at the stagnation pressure $p_o = 4 \times 10^5$ Pa and the stagnation temperature $T_o = 350$ K. The atmospheric pressure is $p_a = 10^5$ Pa. Design a convergent–divergent nozzle that discharges 1.5 kg/s air to the atmosphere at atmospheric pressure. Find the speed of the discharged air.

Solution

From Eq. (13.23),

$$\dot{m} = 1.5 = \frac{0.6847 \times 4 \times 10^5 A^*}{\sqrt{287 \times 350}}.$$

The critical section is

$$A^* = 1.736 \times 10^{-3} \text{ m}^2,$$

$$D^* = \sqrt{\frac{4}{\pi} A^*} = 4.70 \times 10^{-2} \text{ m} = 47 \text{ mm}.$$

From Eq. (13.17)

$$\frac{p_o}{p_e} = 4 = \left[1 + \frac{k-1}{2} M_e^2 \right]^{\frac{k}{k-1}}.$$

Hence $M_e = 1.56$, and from Eq. (13.28)

$$A_e = A^* \frac{1}{M_e} \left[\left(\frac{2}{k+1} \right) \left(1 + \frac{k-1}{2} M_e^2 \right) \right]^{\frac{k+1}{2(k-1)}}$$

$$= 1.736 \times 10^{-3} \frac{1}{1.56} \left[\left(\frac{2}{2.4} \right) \left(1 + 0.2 \times 1.56^2 \right) \right]^3 = 2.116 \times 10^{-3} \text{ m}^2,$$

$$D_e = \sqrt{\frac{4 A_e}{\pi}} = 5.19 \times 10^{-2} \text{ m} = 51.9 \text{ mm}.$$

From Eq. (13.16)

$$T_e = \frac{T_o}{\left[1 + \frac{k-1}{2}M_e^2\right]} = 235 \text{ K},$$

$$c_e = \sqrt{kRT_e} = \sqrt{1.4 \times 287 \times 235} = 307.3 \text{ m/s},$$

$$u_e = c_e M_e = 480 \text{ m/s}.$$

Half the cross section of the designed nozzle is shown in Fig. 13.7. The radius of curvature of the converging part has been chosen equal to D^*, and half the spread angle of the diverging part is chosen to be 8° to guarantee no flow separation. The reason for this is beyond the scope of this presentation.

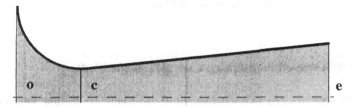

Figure 13.7 Supersonic convergent–divergent nozzle.

Example 13.7

The pressure vessel of Example 13.6 is connected to another vessel at $p_B = 3 \times 10^5 \text{Pa}$, and a nozzle is used to transfer $\dot{m} = 1.5$ kg/s air from the first vessel to the second one. Design this nozzle.

Solution

We may still utilize the concept of the critical section although in this case this section is not built but rather serves as an auxiliary computational step. Thus Eq. (13.23) yields (see Example 13.6)

$$A^* = 1.736 \times 10^{-3} \text{m}^2.$$

From Eq. (13.17) we obtain

$$\frac{p_o}{p_e} = \frac{4}{3} = \left[1 + \frac{k-1}{2}M_e^2\right]^{\frac{k}{k-1}}.$$

Hence $M_e = 0.654$. Equation (13.28) now gives

$$A_e = \frac{A^*}{M_e}\left[\frac{2}{k+1}\left(1+\frac{k-1}{2}M_e^2\right)\right]^{\frac{k+1}{2(k-1)}}$$

$$= \frac{1.736\times10^{-3}}{0.654}\left[\frac{2}{2.4}\left(1+0.2\times0.654^2\right)\right]^3 = 1.97\times10^{-3}\,\text{m}^2,$$

$$D_e = \sqrt{\frac{4A_e}{\pi}} = 5\times10^{-2}\,\text{m} = 50.0\,\text{mm}.$$

Note that this nozzle can be obtained from that designed in Example 13.6, which is shown in Fig. 13.7, by cutting it off at $D = 50.0$ mm, before the critical section.

If the nozzle of Example 13.6 is cut at $D = 50.0$ mm but after the critical section, it still satisfies the conditions of Example 13.7 .

Tables of One-dimensional Compressible Flow

Equations (13.16), (13.17), (13.18), (13.23) and (13.28) may be used for various flow conditions, and the results tabulated with the Mach Number used as a parameter. Such tables may be used to obtain a quick estimate of the principal dimensions of a nozzle. Examples of such tables are Table D-1, Table D-2 and Table D-3 for ideal gases with $k=1.4$, $k = 1.667$ and $k = 1.3$, respectively.

Example 13.8

Repeat Examples 13.6 and 13.7 for helium ($R = 2079.7$ J/kg·K, $k = 1.667$) using the compressible flow tables.

Solution

For Example 13.6, from Table D-2,

$$\dot{m} = \frac{0.7262\times\left(4\times10^5\right)A^*}{\sqrt{2079.7\times350}}$$

or

$$1.5 = 340.5\,A^*, \qquad A^* = 4.406\times10^{-3}\,\text{m}^2$$

and

$$D^* = 7.49\times10^{-2}\,\text{m} = 74.9 \text{ mm},$$

$$\frac{p_e}{p_o} = \frac{1}{4} = 0.25, \text{ hence from Table D-2}, M_e = 1.49,$$

$$\frac{\rho_e}{\rho_o} = 0.4357, \qquad \frac{T_e}{T_o} = 0.5747, \qquad \frac{A_e}{A^*} = 1.1430.$$

Hence

$$T_e = T_o \cdot \frac{T_e}{T_o} = 350 \times 0.5747 = 201 \text{ K},$$

$$c_e = \sqrt{kRT_e} = \sqrt{1.667 \times 2079.7 \times 201} = 835 \text{ m/s},$$

$$u_e = M_e c_e = 1244 \text{ m/s},$$

$$A_e = A^* \cdot \frac{A_e}{A^*} = 5.04 \times 10^{-3} \text{m}^2, \qquad D_e = 8.01 \times 10^{-2} \text{ m} = 80.1 \text{ mm}.$$

For Example 13.7

$$\frac{p_e}{p_o} = \frac{3}{4} = 0.75.$$

Hence from Table D-2, $M_e = 0.605$ and

$$\frac{A_e}{A^*} = 1.1706, \qquad A_e = 1.1706 \times \left(4.406 \times 10^{-3} \right) = 5.16 \times 10^{-3} \text{m}^2,$$

$$D_e = 8.10 \times 10^{-2} \text{ m} = 81.0 \text{ mm}.$$

Nonisentropic One-dimensional Flows

To gain some insight into the physics of shock waves as well as to extend the treatment to include additional important and frequently realized flows, we widen our scope and consider some nonisentropic flows. We do not want, how-ever, to extend our treatment too much. Therefore, having relaxed one condition, that of the flows being isentropic, we impose one condition: In this treatment the flows are not just quasi-one-dimensional, but must be really one dimensional, i.e., the continuity equations (13.10), (13.11) now become

$$\dot{m} = \rho u = \text{const}, \tag{13.29}$$

$$\frac{d\rho}{\rho} + \frac{du}{u} = 0. \tag{13.30}$$

We know of two phenomena, friction and heat transfer, which make a process nonisentropic.

Pipe Flow with Friction

The first family of nonisentropic flows we consider is *adiabatic flows with friction*. The details of the mechanism of friction are ignored, and we admit its effect by *not using the momentum equation* (13.12) in our considerations. Because the flow is adiabatic, the energy equation in its form (13.13) or (13.14) still holds, and so do the ideal gas relations.

Thus, from Eq. (13.15)

$$c_p dT + u \, du = \frac{k}{k-1} R \, dT + u^2 \frac{du}{u} = 0$$

or with Eq. (13.30)

$$\frac{R}{k-1} \cdot \frac{dT}{T} - \frac{u^2}{kRT} \cdot R \cdot \frac{d\rho}{\rho} = c_v \frac{dT}{T} - M^2 R \frac{d\rho}{\rho} = 0. \tag{13.31}$$

From basic thermodynamics for an ideal gas

$$ds = c_v \frac{dT}{T} - R \frac{d\rho}{\rho},$$

which together with Eq. (13.31) yields the forms

$$ds = \left(M^2 - 1\right) R \frac{d\rho}{\rho} = \left(1 - M^2\right) R \frac{du}{u} = \left(1 - \frac{1}{M^2}\right) c_v \frac{dT}{T}. \tag{13.32}$$

Since in an adiabatic process $ds \geq 0$, so must be $(1 - M^2) du/u \geq 0$. Thus for *supersonic* flows, $M^2 > 1$, $du < 0$, and the effect of friction is to *slow* the flow, while for *subsonic* flows, $M^2 < 1$, $du > 0$, and the flow must *accelerate*.

Repeating this same argument for the sign of

$$\left(1 - \frac{1}{M^2}\right) c_v \frac{dT}{T},$$

we find that for *supersonic* flows the gas temperature goes up along the conduit, while for *subsonic* flows the effect of friction is to *reduce* the gas temperature.

Let the x direction coincide with the direction of the flow in the pipe. The results just obtained are summarized in Table 13.2, which is somewhat analogous to Table 13.1. It is also noted that since friction, and hence *any real* adiabatic process, requires $ds > 0$, no time-independent flow with $M = 1$ may exist in a pipe, Eq. (13.32). Even an assumption of negligible friction still does not allow $M = 1$ inside the conduit, because $M = 1$ corresponds to $ds = 0$ and not to $ds \approx 0$.

M	$\dfrac{du}{dx}$	$\dfrac{dT}{dx}$	$\dfrac{d\rho}{dx}$
$M < 1$	> 0	< 0	< 0
$M > 1$	< 0	> 0	> 0
$M = 1$	no time-independent flows		

Table 13.2 Adiabatic flow with friction in a pipe.

Equation (13.32) defines a curve on the T-s diagram, having a subsonic branch and a supersonic one, Fig. 13.8, separated by the point $M = 1$ which corresponds to $ds = 0$. This curve is called the *Fanno Line* .

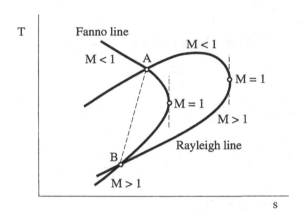

Figure 13.8 Fanno and Rayleigh lines.

A flow in a conduit of constant cross section corresponds to a particular Fanno line and a point in the flow corresponds to a point on that line. As a fluid particle proceeds on its way in the conduit, its point on the diagram moves along its Fanno line and always toward larger s. At $M = 1$ no further motion along the Fanno line toward larger s is possible. Hence $M = 1$ cannot appear in the conduit, but only at its end.

Suppose $M = 1$ does appear at the exit of a conduit. What is the velocity at that point or what is the sonic speed there? We note that Eq. (13.16) has been obtained form the energy equation (13.15) and the expression (13.8) for the sonic speed, both of which hold for adiabatic flows. Thus putting $M = 1$ in Eq. (13.16) we obtain

$$\frac{T}{T_o} = \frac{2}{k+1},$$

which is the same as Eq. (13.20) for isentropic flow! The temperature, the velocity and the sonic speed obtained at $M = 1$ by adiabatic expansion from stagnation do not depend on whether the expansion is isentropic or adiabatic with friction. The pressure and the density and, of course, the entropy come out to be different there. Still this means that in designs of convergent–divergent nozzles, where sonic speeds do appear, rough designs of the convergent part yield surprisingly good results.

Let an initial point in the conduit be chosen, e.g., at its entrance. Let the gas properties there be ρ_i, u_i, T_i, p_i, s_i and note: u_i there is not zero. Let the stagnation properties corresponding to the initial point be ρ_o, T_o, p_o, s_o, with $s_o = s_i$. Now, again, from basic thermodynamics

$$ds = c_v \frac{dT}{T} - R\frac{d\rho}{\rho} = c_v \frac{dT}{T} + R\frac{du}{u}.$$

Equation (13.15), which holds for adiabatic processes, states

$$c_p T_o = c_p T + \tfrac{1}{2}u^2, \qquad u^2 = 2c_p(T_o - T).$$

Hence

$$c_p dT + u du = 0,$$

$$\frac{du}{u} = -c_p \frac{dT}{u^2} = -\frac{dT}{2(T_o - T)}$$

and

$$ds = R\left[\frac{1}{k-1}\cdot\frac{1}{T} - \frac{1}{2(T_o - T)}\right]dT. \tag{13.33}$$

The coefficient of friction in pipe flow, as defined in Eq. (12.29), yields the head loss

$$dh_f = f\left(\frac{u^2}{2g}\right)\frac{dL}{D},$$

which affects the change of entropy

$$ds = g\frac{dh_f}{T} = f\frac{u^2}{2T}\cdot\frac{dL}{D} = f\frac{c_p}{D}\left[\frac{T_o - T}{T}\right]dL. \tag{13.34}$$

Equation (13.33) combined with Eq. (13.34) yield

$$fk\frac{dL}{D} = \left[\frac{k+1}{2}\cdot\frac{1}{1-\eta} - \frac{k-1}{2}\frac{1}{(1-\eta)^2}\right]d\eta,$$

where

$$\eta = \frac{T}{T_o} = \frac{1}{\left[1 + \frac{k-1}{2}M^2\right]} \cdot$$

Integration yields the Fanno number Fn $= f\,L/D$,

$$\text{Fn} = f\frac{L}{D} = -\frac{k+1}{2k}\ln\frac{1-\eta}{1-\eta_i} - \frac{k-1}{2k}\left(\frac{1}{1-\eta} - \frac{1}{1-\eta_i}\right), \qquad (13.35)$$

which describes how $\eta = T/T_o$ changes along the conduit. Of course, once T is known, ρ and u may also be computed. Substitution of $\eta = T^*/T_o = 2/(k+1)$ in Eq. (13.35) yields the maximum length the conduit may have, i.e., that length after which sonic speed appears. Equation (13.35) then takes the form

$$\left(\frac{L}{D}\right)_{max} = \frac{1}{2f}\left[-\frac{k+1}{2k}\ln\frac{k-1}{(k+1)(1-\eta_i)} - \frac{k-1}{2k}\left(\frac{k+1}{k-1} - \frac{1}{1-\eta_i}\right)\right]$$

with

$$\eta_i = \left[1 + \frac{k-1}{2}M_i^2\right]^{-1} \cdot$$

Some numerical values are presented in Tables D-4 , D-5 and D-6 in Appendix D arranged such that Fn $= f\,L/D$ is the dimensionless length *still remaining* before $M = 1$ is reached.

Example 13.9

A certain cooking gas, which may be considered an ideal gas with $k = 1.3$, is a mixture of CO, C_2H_6, CH_4 and N_2. It is introduced into a pipe of 0.05 m diameter and flows at this point with $M = 0.01$. The Moody diagram, Fig. 12.4, yields for this pipe the friction factor $f = 0.02$, which is assumed constant. The pipe is to be 15 km long.

Find the maximum length this pipe may have before the flow is choked. Find the Mach number at the 15-km point and the values of $T_2/T_{01}, p_2/p_{01}$ and ρ_2/ρ_{01} there.

Solution

At the point of introduction the gas has

$$M_1 = 0.01, \ \ T_1/T_{01} = 1.0, \ \ p_1/p_{01} = 1.000, \ \ \rho_1/\rho_{01} = 1.000,$$

where the various values have been read from Table D-1 which holds for isentropic relations.

Table D-6 yields, for $M = 0.01$,

$$\left(f\frac{L_{\text{max}_1}}{D}\right) = 7683.$$

Hence $L_{\text{max}_1} = 7683 \times D/f = 19182\,\text{m}.$

At $L_2 = 15,000$ m: $L_{\text{max}_2} = 19,182 - L_2 = 4,182\,\text{m}$ and $f\dfrac{L_{\text{max}_2}}{D} = 1,673.$

Table D-4 yields, by interpolation, $M_2 \approx 0.022$, and certainly $0.02 < M_2 < 0.03$. The table also yields $T_2/T_o = 0.9999$. The sonic speed at point 2 is the same as at point 1, and therefore

$$\frac{u_2}{u_1} = \frac{M_2}{M_1} = 2.2 = \frac{\rho_1}{\rho_2},$$

$$\rho_2 = \frac{\rho_1}{2.2}, \qquad \frac{\rho_2}{\rho_{01}} = \left(\frac{\rho_1}{\rho_{01}}\right)\frac{1}{2.2} = 0.4545,$$

$$\frac{p_2}{p_{01}} = \left(\frac{p_1}{p_{01}}\right)\frac{\rho_2 T_2}{\rho_1 T_1} = 0.9999 \times 0.45451 \times 1.0 = 0.4545.$$

Frictionless Pipe Flow with Heat Transfer

The second family of nonisentropic flows we consider is *frictionless flows with heat transfer*. Using the same strategy as for adiabatic flows with friction, we keep the one-dimensional continuity equations (13.29) and (13.30) and the momentum equation (13.12) and ignore the energy equation.

Substitution of $\rho\,du = -u\,d\rho$ from Eq. (13.30) into Eq. (13.12) yields

$$u^2 d\rho = dp = R(\rho\,dT + T\,d\rho).$$

Hence,

$$\frac{u^2}{RT} \cdot \frac{d\rho}{\rho} = \frac{dT}{T} + \frac{d\rho}{\rho},$$

and because $RT = c^2/k$, this becomes

$$\left(kM^2 - 1\right)\frac{d\rho}{\rho} = \frac{dT}{T}. \qquad (13.36)$$

From general thermodynamics

$$ds = c_v \frac{dT}{T} - R \frac{d\rho}{\rho},$$

and from Eq. (13.36),

$$ds = \frac{M^2 - 1}{kM^2 - 1} \cdot c_p \frac{dT}{T}. \tag{13.37}$$

Again, a relation between T and s has been obtained, which can be described as a line on the T-s diagram, the *Rayleigh Line*, Fig. 13.8.

The frictionless motion of a fluid particle along the conduit, with heat transfer, is represented by points on this line. Since heat transfer can be into and out of the moving fluid, the points on the line in the diagram move both to the left and to the right.

The Rayleigh line also breaks into subsonic and supersonic branches, with $M = 1$ at the point of maximum s. It has a point of maximum T at $M^2 = 1/k$, and for $k < M^2 < 1$ the temperature goes down with the addition of heat and increases with heat extraction. The general behavior is presented in Table 13.3, which is analogous to Table 13.2 for the Fanno line.

With heat addition

m	dT	du
$M^2 < \dfrac{1}{k}$	> 0	> 0
$M^2 > 1$	> 0	< 0

With heat extraction

M	dT	du
$M^2 < \dfrac{1}{k}$	< 0	< 0
$\dfrac{1}{k} < M^2 < 1$	> 0	
$M^2 > 1$	< 0	> 0

Table 13.3 Frictionless flow with heat transfer in a pipe.

We now endeavor to explain shock waves using the concepts just developed, i.e., those represented by the Fanno line and by the Rayleigh line.

Let a shock wave be defined as a local singularity, i.e., a local nonisentropic occurrence. Then:

- It may be considered locally one dimensional.
- It must permit local heat transfer and internal friction.

Let the gas just before the shock wave be represented by a point on the T-s diagram, say, point A in Fig. 13.8. Both Fanno and Rayleigh lines may be drawn through this point. The shock is a one-dimensional phenomenon and the state after the shock is restricted to its Fanno and Rayleigh lines; hence the shock must be represented by a jump from point A to point B or from B to A. However, $s_B > s_A$. Hence a shock can occur only at supersonic speeds.

Shock Wave Relations

Consider an observer moving with a normal shock, Fig. 13.9. The observer sees the shock stationary, with the flow going into and out of the shock. A momentum balance through a stream tube yields

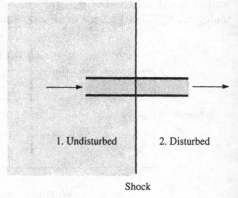

$$\rho_1 u_1^2 + p_1 = \rho_2 u_2^2 + p_2$$
$$= \rho_1 M_1^2 c_1^2 + p_1$$
$$= M_1^2 \rho_1 k \frac{p_1}{\rho_1} + p_1$$

or

$$p_1 \left[1 + kM_1^2\right] = p_2 \left[1 + kM_2^2\right]. \quad (13.38)$$

1. Undisturbed   2. Disturbed

Shock

Figure 13.9 Normal shock.

Equation (13.16) applies both before and after the shock, and because i_o and with it T_o are conserved through the shock,

$$T_1 \left[1 + \frac{k-1}{2} M_1^2\right] = T_2 \left[1 + \frac{k-1}{2} M_2^2\right]. \quad (13.39)$$

Continuity requires

$$\rho_1 u_1 = \rho_2 u_2 = \rho_1 c_1 M_1 = \rho_1 \sqrt{kR} \sqrt{T_1} M_1 = \rho_2 \sqrt{kR} \sqrt{T_2} M_2 ,$$

or

$$\rho_1 M_1 T_1^{\frac{1}{2}} = \rho_2 M_2 T_2^{\frac{1}{2}}. \quad (13.40)$$

Now $p/T = R\rho$, and p, T and ρ may be eliminated between Eqs. (13.38), (13.39) and (13.40). This done, the resulting relation between M_1 and M_2 is written as

$$M_2^2 = \frac{(k-1)M_1^2 + 2}{2kM_1^2 - (k-1)}.$$
(13.41)

It is noted that algebraically $M_1 > 1$ implies $M_2 < 1$ and $M_1 < 1$ implies $M_2 > 1$. Thus, unless $M_1 = M_2$,* the flow must change from supersonic to subsonic. A change from subsonic to supersonic flow requires decrease of entropy, which cannot be in adiabatic flows. Hence a normal shock may occur only in supersonic flows and the flow after the shock is subsonic.

The pressure ratio is given by

$$\frac{p_2}{p_1} = \frac{1 + kM_1^2}{1 + kM_2^2} = 1 + \frac{2k}{k+1}\left(M_1^2 - 1\right).$$
(13.42)

Once M_2 is obtained, Eqs. (13.38), (13.39) and (13.40) yield the temperature ratio

$$\frac{T_2}{T_1} = \frac{1 + \frac{k-1}{2}M_1^2}{1 + \frac{k-1}{2}M_2^2} = 1 + \frac{2(k-1)}{(k+1)^2} \cdot \frac{kM_1^2 + 1}{M_1^2}\left(M_1^2 - 1\right)$$
(13.43)

and the density ratio

$$\frac{\rho_2}{\rho_1} = \frac{u_1}{u_2} = \frac{M_1}{M_2}\sqrt{\frac{T_1}{T_2}} = \frac{(k+1)M_1^2}{(k-1)M_1^2 + 2}.$$
(13.44)

The relations of Eqs. (13.41) - (13.44) are also incorporated into Appendix D. Using Eqs. (13.16), (13.17) and (13.18), the stagnation properties before and after the shock can be found. Using Eqs. (13.25), (13.26) and (13.27), the critical properties can be found.

The shock is a compression wave, and p_2 is always greater than p_1. A *Shock Strength* may thus be defined as $(p_2 - p_1)/p_1 = \Delta p_1/p_1$. In general the increase of entropy may be found using properties before and after the shock. However, for shocks of small strength, i.e., weak shocks, a simpler approach is helpful. Let $m = M_1{}^2 - 1$. Then

$$s_2 - s_1 = c_v \ln\frac{p_2}{p_1} - c_p \ln\frac{\rho_2}{\rho_1} = R\ln\left[\left(1 + \frac{2km}{k+1}\right)^{\frac{1}{k-1}}(m+1)^{\frac{-k}{k-1}}\left(\frac{k-1}{k+1}m + 1\right)^{\frac{k}{k-1}}\right].$$
(13.45)

Expanding for $m \ll 1$

---

* Division by $M_1 - M_2$ has been affected. Hence, this possibility exists, which means no shock.

$$\frac{s_2 - s_1}{R} = \frac{2k}{(k+1)^2}\frac{m^3}{3} + 0(m^4) \approx \frac{2k}{(k+1)^2}\frac{\left(M_1^2 - 1\right)^3}{3}$$

or

$$\frac{s_2 - s_1}{R} = \frac{k+1}{12k^2}\left(\frac{\Delta p_1}{p_1}\right)^3.$$

(13.46)

For weak shocks the increase of entropy is proportional to the third power of the shock strength.

At this stage we can answer partially a question raised when considering nozzle flow: Exit pressures between points C and E, Fig. 13.6, are accommodated by some supersonic flow in the diverging part of the nozzle; then a normal shock occurs and the flow continues subsonically to the exit pressure. Point C corresponds to zero strength shock, at the throat, and point E to a shock at the exit.

Example 13.10

The nozzle designed in Example 13.6 is made to discharge into another vessel, where the pressure is

    a. $p_e = 3.2 \times 10^5$ Pa ;   b. 3×10^5Pa ;   c. 2×10^5 Pa.

The stagnation pressure where the flow originates is still $p_o = 4 \times 10^5$ Pa. Use values from Table D-1, without interpolation, and describe the resulting flows.

Solution

a.

$$\frac{p_e}{p_o} = \frac{3.2 \times 10^5}{4 \times 10^5} = 0.8 > 0.5283.$$

Could the flow be all subsonic?

$$\frac{A_e}{A^*} = \frac{2.12 \times 10^{-3}}{1.736 \times 10^{-3}} = 1.221.$$

From Table D-1: $M_e = 0.57$, $p_e / p_o = 0.8$.

Indeed the flow is all subsonic!

b. $\dfrac{p_e}{p_o} = \dfrac{3 \times 10^5}{4 \times 10^5} = 0.75 > 0.52,$

$A_e / A^* = 1.221.$

From Table D-1: $M_e = 0.57$, $p_e/p_o = 0.8 > 0.75$.

The flow cannot be all subsonic. Thus supersonic speed must appear and the part of the flow between the entrance and the critical section is exactly the same as in Example 13.6, with sonic conditions at A^*. Now

$p^* = (0.5283)(4 \times 10^5) = 211320$ Pa.

The air must expand further, the flow becomes supersonic and since $p_e = 3 \times 10^5$ Pa is higher than 10^5 Pa, the design exit pressure in Example 13.6, we expect a shock wave; the flow then becomes subsonic, and the compression in the divergent part of the nozzle raises the pressure up to 3×10^5 Pa.

The location of the shock wave must be found by trial and error, and we know that $1 < M_1 < 1.56$; see Example 13.6.

Try $M_1 = 1.2$. From Table D-1:
   $M_2 = 0.8442$,
$p_1 = 0.4124 p_o = 1.65 \times 10^5$ Pa,
$p_2 = 1.5133 p_1 = 2.5 \times 10^5$ Pa,
$A_1 = A^*(A_1/A^*) = 1.736 \times 10^{-3} \times 1.0304 = 1.789 \times 10^3$ m^2.

We now look again at Table D-1, at the line of $M = 0.8442$, and from $p_2 = 2.5 \times 10^5$ deduce $p_{o2} = 2.5 \times 10^5 / 0.63 = 3.97 \times 10^5$ Pa. The stagnation pressure is reduced because the flow went through an irreversible process, a shock wave. We also find there $A/A^* = 1.0237$ and deduce

$A_2^* = A_1 / 1.0237 = 1.747 \times 10^{-3}$ m^2.

The critical cross section has been increased, again because of the shock wave. Note that this critical cross section is not realized physically but is rather an answer to the question of what critical cross section is required to accommodate the flow after the shock wave. Now at the exit

$$\frac{A_e}{A_2^*} = \frac{2.12 \times 10^{-3}}{1.747 \times 10^{-3}} = 1.20,$$

and from Table D-1 we read

$M_e = 0.59$, $p_e/p_{o2} = 0.7901$ and

$p_e = 0.790 \times 3.97 \times 10^5 = 3.14 \times 10^5$ Pa $> 3 \times 10^5$ Pa.

Try $M_1 = 1.5$ and repeat:
   $M_2 = 0.7011$, $p_1 = (0.2724)p_o = 1.09 \times 10^5$ Pa,
   $p_2 = (2.4583) p_1 = 2.7 \times 10^5$ Pa,

$A_1 = 1.736 \times 10^{-3} \times 1.1762 = 2.042 \times 10^{-3}$ m^2,

$$p_{o2} = \frac{2.7 \times 10^5}{0.7209} = 3.745 \times 10^5 \text{ Pa},$$

$$A_2^* = \frac{2.042 \times 10^{-3}}{1.0944} = 1.866 \times 10^{-3}\, \mathrm{m}^2,$$

$$\frac{A_e}{A_2^*} = \frac{2.12 \times 10^{-3}}{1.866 \times 10^{-3}} = 1.14, \qquad M_e = 0.65, \quad p_e / p_{o2} = 0.7528$$

and

$$p_e = 0.7528 \times 3.745 \times 10^5 = 2.82 \times 10^5\, \mathrm{Pa} < 3 \times 10^5\, \mathrm{Pa}.$$

Try $M_1 = 1.35$ and repeat:

          $M_e = 0.6,\quad p_e = 3.03 \times 10^5\, \mathrm{Pa}.$

Try $M_1 = 1.34.$

c.   We are led in this case by the calculations in case b, and try right away

    $M_1 = 1,56, \qquad M_{2=} = 0.6809,$

    $p_1 = 0.2496 p_o = 0.9984 \times 10^5\, \mathrm{Pa},$

    $p_2 = 2.6725 p_1 = 2.668 \times 10^5\, \mathrm{Pa},$

and no solution can be obtained. The lowest exit pressure that may still correspond to a normal shock wave is $p_e = 2.668 \times 10^5$ Pa.

The flows obtained are:

a.   Complete subsonic flow.

b.   Subsonic flow up to the critical section followed by supersonic flow up to $A = 1.9 \times 10^{-3}\, \mathrm{m}^2$. Then a shock wave and isentropic compression to p_e. Solutions of the type in part b are possible for $2.668 \times 10^5\, \mathrm{Pa} < p_e < 3.2 \times 10^5\, \mathrm{Pa}$.

c.   Subsonic flow up to the critical section followed by supersonic flow up to the exit. Whatever adjusts the exit pressure to the imposed one must take place after the exit.

Oblique Shocks and Prandtl–Meyer Expansion

Consider a flow w_1 encountering an oblique shock, Fig. 13.10, i.e., the velocity component normal to the shock, u_1, undergoes a normal shock, while the component parallel to the shock, v_1, is not altered at all. For this to happen u_1 itself must be supersonic, resulting in subsonic u_2. As a result of this, the flow turns by the angle θ. A way to force an oblique shock is therefore to make a supersonic flow turn by the introduction of a wedge into the flow. Two oblique shock waves appear, Fig. 13.11.

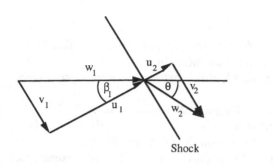

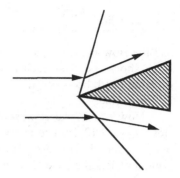

Figure 13.10 Oblique shock.   **Figure 13.11** Shocks at a wedge.

For large enough $v_2 = v_1$, w_2 may still be supersonic and $\Delta p_1/p_1$ may indeed be quite small. The flow then approaches an isentropic one. Furthermore, it can be shown that for small turning angles θ [see Eq. (13.46)],

$$\frac{\Delta p_1}{p_1} \propto \theta, \qquad \Delta s \propto \theta^3. \tag{13.47}$$

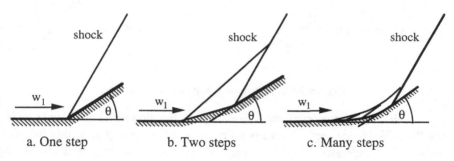

a. One step   b. Two steps   c. Many steps

Figure 13.12 Compressive turning of supersonic flow.

Consider the flow w_1 which must turn by the angle θ, Fig. 13.12. Case a is a turn as already discussed. In case b the turn is reached in two steps, say, each of $\theta/2$. The change of entropy in this case, by Eq. (13.47), is

$$\Delta s = 2\left(\frac{\theta}{2}\right)^3 = \frac{1}{4}\theta^3,$$

i.e., one quarter of case a. We note, however, that each step generates its own shock wave and that the two waves meet. In case c, the change of entropy is

$$\Delta s = \lim_{n \to \infty} n \left(\frac{\theta}{n}\right)^3 = \lim_{n \to \infty} \frac{\theta^3}{n^2} = 0$$

and the flow is isentropic. Still, each differential step generates its differentially weak shock, i.e., its Mach line, and these do meet. In the close vicinity of the rounded bend, however, these Mach lines hardly meet and the flow is isentropic, i.e., reversible.

Now consider a supersonic flow along a wall which turns *away* from the flow, Fig. 13.13a. Since every point in a supersonic flow has a right to a Mach line, we may draw some of these, and note that they diverge. We may repeat the argument leading to case c in Fig. 13.12 to show that now we have isentropic expansion. Furthermore, because the Mach lines diverge, case b in Fig. 13.13 is also an isentropic expansion.

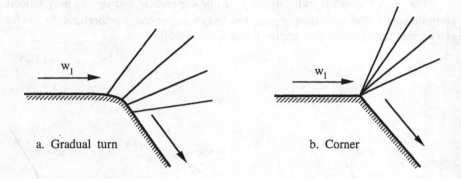

a. Gradual turn   b. Corner

Figure 13.13 Supersonic expansion by turning.

Such a supersonic expansion yields M as a function of the turn angle, called the *Prandtl–Meyer* angle. There is no compressive counterpart to this function, because in compression the Mach lines converge to generate partial shocks and these meet to form stronger shocks, in a way similar to the piston flow, Fig. 13.3. Prandtl–Meyer angle v is also given in Table D-1.

We may now complete the description of the nozzle flow, Fig. 13.6: For exit pressures between points E and D, oblique shocks occur outside the nozzle exit. For pressures below D there is a Prandtl–Meyer expansion.

Example 13.11

Complete case c in Example 13.10. Also find what flows result for the exit pressures

d. $p_e = 10^5$ Pa,
e. $p_e = 0.5 \times 10^5$ Pa.

Solution

c. There are oblique shocks outside the exit, and the phenomenon becomes three dimensional.
d. The flow in the divergent part is fully supersonic. The exit pressure is the design pressure.
e. The flow in the nozzle is at the design conditions. At the exit the flow is supersonic with $M = 1.56$. There is a Prandtl–Meyer outward expansion with the angle $\Delta v = 0.4652 - 0.2387 = 0.2265$ rad, up to $M = 2.01$. This is followed by three-dimensional phenomena.

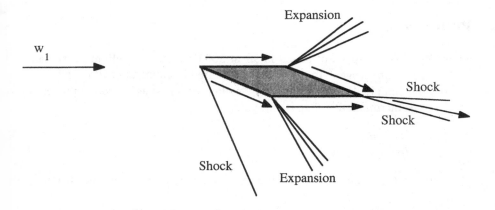

Figure 13.14 Diamond-shaped wing in supersonic flow.

Lift and Drag

Finally we note that because of the increase in entropy through shock waves, the equations for supersonic flow yield both lift and drag, as seen in Fig. 13.14. This result is different from that obtained from the equations for subsonic flows with negligible shear, which could predict only lift.

References

A.B. Cambel and B.H. Jennings, "Gas Dynamics," McGraw-Hill, New York, 1958.

A. Chapman and W.F. Walker, "Introductory Gas Dynamics," Holt, Rinehard and Winston, New York, 1971.

H.W. Liepmann and A. Roshko, "Elements of Gas Dynamics," Wiley, New York, 1957.

R. von Mises, "Mathematical Theory of Compressible Fluid Flow," Academic Press, New York, 1958.

J.A. Owczarek, "Fundamentals of Gas Dynamics," McGraw-Hill, New York, 1971.

A.H. Shapiro, "The Dynamics and Thermodynamics of Compressible Fluid Flow," 2 vols., Ronald Press, New York, 1953.

Problems

13.1a. What is the speed of sound in air at 300 K, 1 bar? In helium at the same state? In hydrogen? In a 1 : 1 molar mixture of helium and hydrogen?

b. Equation (13.4) was obtained by Newton, who first assumed the process to be isothermal rather than isentropic. Use

$$\frac{dp}{d\rho} = \left(\frac{\partial p}{\partial \rho}\right)_T$$

in Eq. (13.4) and recalculate the speed of sound c. What are the errors?

13.2a. The isentropic compressibility of water at 1 bar, 90°C, is

$$\frac{1}{\rho}\left(\frac{\partial \rho}{\partial p}\right)_s = 4.4 \times 10^{-10}\,\mathrm{m}^2\,/\,\mathrm{N}.$$

Find the speed of sound in water.

b. The isentropic compressibility of steel-like solids is

$$\frac{1}{\rho}\left(\frac{\partial \rho}{\partial p}\right)_s = \frac{3(1-2v)}{E},$$

where v is the Poisson ratio, and E is the Young modulus. Find the speed of sound in steel.

13.3 Helium at stagnation, at 2×10^5 Pa, 300 K, expands isentropically:
a. to $p_1 = 10^5$ Pa, and then continues;
b. to $p_2 = 0.5 \times 10^5$ Pa, and then continues;
c. to $p_3 \sim 0$.
Find Mach numbers, speeds, sonic speeds and stagnation pressures for a, b, c and for the pressure at which the sonic speed appears. Assume ideal gas behavior.

13.4 Assume one-dimensional adiabatic frictionless flow through the nozzle shown in Fig. P13.4. In section 1: $u = 20$ m/s, $p = 8 \times 10^6$ Pa, $T = 300$ K. Calculate the velocities in sections 2 and 3 for the fluids:

a. water,   b. molten lead,   c. air,   d. hydrogen.

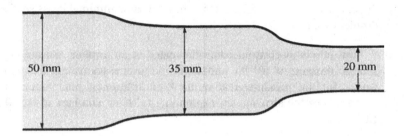

50 mm   35 mm   20 mm

Figure P13.4 Nozzle.

13.5 Design, i.e., find the dimensions at the throat and at the exit, and draw a scheme of a convergent–divergent nozzle which transfers 1 kg/s air from vessel A, where $p_0 = 10^6$ Pa, $T = 300$ K, to vessel B, where $p_e = 1.5 \times 10^5$ Pa.

13.6 In Problem 13.5 the pressure at the exit, p_e, is changed to

a. 10^5 Pa;   b. 9.5×10^5 Pa.

Find the new mass flowrate, $\dot{m}$, and the pressure at the throat for the same nozzle.

13.7. The throat cross-sectional area in Problem 13.5 is A^*, and that of the exit is A_e. Let $A_1 = (A^* + A_e)/2$. With the same stagnation conditions as in Problem 13.5, the sonic speed is reached at A^*; but then there is a shock wave at A_1. Find the Mach number and the pressure at the throat, at A_1 just before the shock and at the exit.

13.8   A compressor takes in air at the atmospheric pressure $p = 10^5$ Pa and discharges the compressed air into a high pressure settling tank. An engineer who has to measure the performance curve of the compressor, i.e., the mass flux of the compressed air as a function of the compression pressure, connects to the settling tank a thermometer and a convergent nozzle with a minimal cross-sectional area of 20×10^{-4} m². In all the measurements the temperatures were between 350 K and 355 K. The measured gauge pressures were:

   1.1×10^5 Pa, 1.6×10^5 Pa, 2.5×10^5 Pa, 4.0×10^5 Pa and 6.0×10^5 Pa.

   Find the corresponding mass flowrates.

13.9   A second engineer, who saw what the one in Problem 13.8 did, suggested that a convergent–divergent nozzle be used, such that only one equation had to be used to compute the mass fluxes in the five measurements taken there. Design such a nozzle, which has the same minimal cross section as in Problem 13.8.

13.10 A Pitot tube is used to measure the speed of an airplane. Assuming atmospheric pressure of 10^5 Pa and isentropic processes except through shock waves, find the pressures read on the Pitot differential manometer and the speeds of the airplane, corresponding to Mach numbers of 0.1, 0.3, 0.7, 1.0, 1.5, 2.0.

13.11 Helium at $p_o = 3 \times 10^5$ Pa, $T_o = 400$ K, enters a pipe of an inner diameter 0.015 m. The friction coefficient is $f = 0.023$. Find the pipe length at which sonic speed is reached.

13.12 Design a nozzle which discharges 1 kg/s air from a vessel at stagnation conditions, $p_0 = 8.0 \times 10^5$ Pa, $T_o = 400$ K, to the atmosphere at $p = 10^5$ Pa. Find A^*, A_e and the total thrust force which acts on the vessel. For reasons of material availability the nozzle designed above has been manufactured with its divergent part opening only up to the cross-sectional area of $A_1 = (A^* + A_e)/2$. This truncated nozzle is used instead of the designed one. Find the total thrust force applied to the vessel now.

13.13 Both nozzles, the designed and the truncated one, of Problem 13.12 are tested under the conditions of Problem 13.12. However, instead of using air, hydrogen is being used. Find the thrust forces acting on the vessel now.

13.14 The four tests of Problems 13.12 and 13.13 are repeated in space. Find the four new thrust forces.

13.15 A normal shock wave moves through quiescent air at $p = 10^5$ Pa, $T = 300$ K. The speed of the shock is $u_s = 694$ m/s. Find the pressure left immediately behind the shock. Is the air immediately behind the shock quiescent? If not, what is its velocity and in which direction?

13.16 The shock wave of Problem 13.15 hits a wall in a direct frontal collision. It is then reflected back. Find the highest pressure suffered by the wall. Find the speed of the reflected wave.

13.17 A large pressure vessel contains gas at the stagnation properties: $p_o = 400$ kPa, $T_o = 420$ K. The gas is approximately ideal, with $R = 287$ J/kg·K, and $k = 1.4$. The outside atmospheric pressure is $p_a = 100$ kPa. A convergent–divergent nozzle is designed to pass a mass flux of 1 kg/s from the vessel to the outside. Find:
   a. The speed of the gas at the exit from the nozzle, V_e.
   b. The exit Mach number, M_e.
   c. The critical cross-sectional area, A^*.
   d. The exit cross-sectional area, A_e.

13.18 The nozzle of Problem 13.17 is cut after the critical section at point 1, where $A_1 = (A + A_e)/2$. It still connects the pressure vessel to the outside. Find:
   e. The pressure, p_1, at section A_1.
   f. The speed of the air, V_1, at A_1.
   g. The Mach number, M_1, at A_1.
   h. The mass flux, $\dot{m}_1$, at A_1.

13.19 The nozzle of Problem 13.18 is further cut, before the critical section at point 2, where $A_2 = A_1$. It still connects the pressure vessel to the outside. Find:
   e. The pressure, p_2, at section A_2.
   f. The speed of the air, V_2, at A_2.
   g. The Mach number, M_2, at A_2.
   h. The mass flux, $\dot{m}_2$, at A_2.

13.20 The pressure vessel of Problem 13.17 is also connected to the outside by a pipe of a constant cross-sectional area of $A_3 = A_2$. At the exit of the pipe the speed of the gas is V_3, and the mass flux through the pipe is $\dot{m}$.
   i. Is V_3 smaller, equal to or greater than V_2?
   j. Is $\dot{m}_3$ smaller, equal to or greater than $\dot{m}_2$,?

13.21 A rocket, shown in Fig. P13.21, has a cylindrical exhaust pipe of the diameter $d = 0.025$ m instead of a nozzle. The rocket moves with a constant speed and exhausts gas at the rate of 0.5 kg/s. The gas is approximately ideal, with $k = 1.4$, and $R = 287$ J/kg·K. The stagnation properties of the gas in the rocket are $p_o = 400$ kPa, $T_o = 3,000$ K. The outside pressure is 100 kPa. Find the force on the rocket.

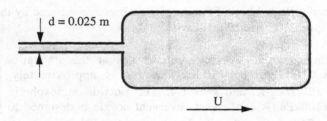

Figure P13.21 Rocket.

13.22 A jet-propelled airplane flies at 900 km/h. The air intake into the engine is designed as a diffuser, i.e., as part of a nozzle that takes in the outside air and brings it to the speed of the airplane while increasing its pressure. The outside air is at 10^5 Pa, 300 K. Find the maximum attainable pressure and the ratio between the area of the diffuser at the intake and that where the higher pressure is obtained.

13.23 The airplane of Problem 13.22, with the diffuser designed in that problem, flies now at a higher altitude, where the outside air is at 5×10^4 Pa, $T = 280$ K. Find the highest pressure which can be obtained now at the exit of the diffuser. Assuming the airplane operates at that new highest pressure, find the ratio between the mass rate of air intake under the conditions of this problem and that of Problem 13.22.

13.24 At takeoff the airplane of Problem 13.22 must have the same mass rate of air supply as in Problem 13.22. Now, however, the airplane is at zero speed. To achieve this, the engine compressor must set a lower pressure just at the exit from the diffuser. Find this pressure.

13.25 Relating to the airplane in Problem 13.22, the air coming out of the diffuser passes through a compressor where it is compressed by a compression ratio of 1 : 20. The air then flows to a combustion chamber where its temperature is raised by 1,000 K. Then the air expands in a turbine where energy is extracted just enough to run the compressor, and then the air expands in a nozzle. Assume isentropic compression and expansion. For 1 kg/s air, design the nozzle for operating under the conditions of Problem 13.22, i.e., find the critical diameter and the exit diameter. Find the thrust of the engine.

13.26   The turbine in Problem 13.25 is changed to a cheaper one, with an adiabatic efficiency of 70%. This means that the expansion in the turbine is still adiabatic but that the power extracted from the gas is only 70% of what could be extracted by isentropic expansion. To obtain the power necessary to run the compressor, the gas must exit from the turbine at a lower pressure. Find the critical diameter and the exit diameter of the jet nozzle now. Find the thrust of the engine.

13.27 The compressor in Problem 13.25 is changed to a cheaper one, with an adiabatic efficiency of 70%. This means that the compression in the compressor is still adiabatic but that the power needed to compress the gas is 1.0/0.7 times that needed in an isentropic compression. The turbine, which remains the same as in Problem 13.25, must now supply the increased power necessary to run the compressor, and the gas exits from the turbine at a lower pressure. Find the critical diameter and the exit diameter of the jet nozzle now. Find the thrust of the engine.

13.28 A continuous stream of air at 10^5 Pa, 300 K, flows with the speed of 450 m/s perpendicular to a wall. It hits the wall and a normal shock wave is reflected from the wall and propagates against the stream of the incoming air, Fig. P13.28. Now the stream of air passes through the reflected shock, and

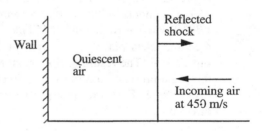

Figure P13.28

behind the shock, on the wall side, the air is stationary with respect to the wall because it must satisfy this boundary condition. Find the pressure near the wall and the speed of the reflected shock.

13.29 A diamond-shaped supersonic wing moves in air at 10^5 Pa, 300 K, at 500 m/s, Fig. P13.29.
Find the lift and the drag induced by the shocks and the Prandtl–Meyer expansion for cases a and b.

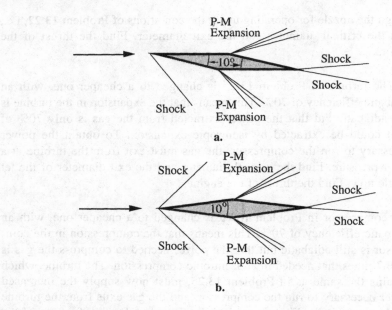

Figure P13.29 Diamond-shaped wings in supersonic flow.

13.30 A stream of air at 10^5 Pa, 300 K, moving at 450 m/s meets with another stream at 1.3×10^5 Pa, 300 K, moving at 450 m/s. The meeting takes place at an angle of 20°, Fig. P13.30. The resulting flow must have the same pressure and the same direction everywhere. Find the new pressure and direction.

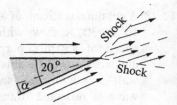

Figure P13.30

14. NON-NEWTONIAN FLUIDS

A Newtonian fluid has been defined in Chapter 5 as one which satisfies a linear relationship between its stress and its rate of strain. For a simple unidirectional flow, such as the flow between two parallel plates, this relationship is expressed as

$$\tau = \mu\left(\frac{du}{dy}\right). \tag{14.1}$$

Equation (14.1) is known as Newton's law of viscosity. This equation is a mathematical statement, and there is no reason to believe that all real fluids should obey it. Indeed, there are more fluids that do not behave according to Eq. (14.1) than those that do. However, the three fluids most abundant in nature, air, water and petroleum, follow Eq. (14.1) quite accurately. Fluids whose behavior cannot be described by Newton's law of viscosity are called *non-Newtonian* fluids.

Typical non-Newtonian fluids are paints, slurries, pastes, jellies and similar food products, foams, polymer melts, blood, etc. Some of these fluids simply exhibit nonlinear viscous effects, while others show effects of "memory" which are related to their viscoelastic behavior. A typical example of the latter is the dough climbing up the beater of a food mixer. Only nonlinear viscous effects will be dealt with in this chapter. Elastic effects which strongly influence time-dependent flows and "stretching" flows are outside the scope of this book.

Figure 14.1 shows curves of stress versus rate of strain for several classes of fluids. The Newtonian fluid of Eq. (14.1) is represented by the straight line of constant slope, μ, passing through the origin. The constant viscosity of a Newtonian fluid is defined as the ratio between a given shear stress and the resultant rate of strain. Similarly we may define an apparent viscosity of a non-Newtonian fluid by

$$\tau = \mu_{app}\left(\frac{du}{dy}\right). \tag{14.2}$$

Here, however, the apparent viscosity is not a constant but depends on the rate of strain. The relationship between the shear rate and the rate of strain is found experimentally by a series of viscometric measurements at different shear rates. Plotting the results as in Fig. 14.1, the curve obtained is called *a flow curve*. This curve is fitted by means of an equation called the *constitutive equation* of the fluid. One of the simplest constitutive equations describing the flow behavior of a non-Newtonian fluid is that of the power-law model:

$$\tau = K \left(\frac{du}{dy} \right)^n .$$

(14.3)

For cases of $n < 1$ the fluid is referred to as the *pseudoplastic* fluid. For $n > 1$ the fluid is known as *dilatant*. The Newtonian fluid is, of course, a special case of the power-law fluid with $n = 1$.

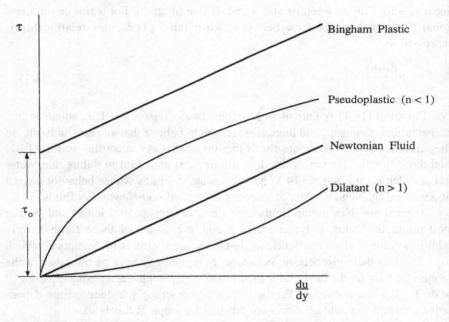

Figure 14.1 Stress vs. rate-of-strain relationships for different fluids.

Comparing Eqs. (14.2) and (14.3) it is noted that the apparent viscosity of a power-law fluid may be expressed as

$$\mu_{app} = K \left| \frac{du}{dy} \right|^{n-1} .$$

(14.4)

The absolute value sign is introduced to insure that the apparent viscosity is the same for shear rates equal in magnitude but acting in opposite directions. Equation (14.4) may be combined with (14.2) to yield

$$\tau = \left[K \left| \frac{du}{dy} \right|^{n-1} \right] \left(\frac{du}{dy} \right). \tag{14.5}$$

The power-law model as expressed by Eq. (14.5) is to be preferred over that of (14.3), because in Eq. (14.5) the term in the square brackets, i.e., the apparent viscosity, is always positive, while the sign of the shear stress is determined by the sign of du/dy. In many cases the power-law model describes the behavior of the fluid only over a limited range of shear rates. In particular, it breaks down at very low shear rates. Inspection of Eq. (14.4) shows that as the shear rate approaches zero, μ_{app} goes to infinity. On the other hand, experience shows that most non-Newtonian fluids approach Newtonian behavior at low shear rates. A slightly more complex model that takes this into account is the Ellis model:

$$\frac{du}{dy} = \left[a + b \left| \tau \right|^{s-1} \right] \tau. \tag{14.6}$$

This model includes as special cases the power-law fluid when $a = 0$ and the Newtonian fluid for $b = 0$.

Another fluid shown in Fig. 14.1 is the Bingham plastic. This fluid does not flow below a certain yield stress τ_o. Above τ_o the increase in τ is proportional to the shear rate. Thus, it can be described by

$$\tau = \tau_o + \mu \left(\frac{du}{dy} \right) \qquad \text{for} \qquad |\tau| > \tau_o,$$

$$\frac{du}{dy} = 0 \qquad \text{for} \qquad 0 \le |\tau| \le \tau_o. \tag{14.7}$$

This model may also be derived as a special case of the Ellis model.

The constitutive equations given above were written in one-dimensional form and thus could be used to solve problems of unidirectional flows only. In order to deal with higher dimensional flows, one has to extend the constitutive relationships to two- and three-dimensional cases. To do this, we start again with the Newtonian fluid and then extend our results to non-Newtonian fluids.

The stress versus rate-of-strain relationship for an incompressible Newtonian fluid is

$$\tau_{ij} = \mu \left(\frac{\partial u_i}{\partial x_j} + \frac{\partial u_j}{\partial x_i} \right) \tag{14.8}$$

or by Eq. (5.37)

$$\tau_{ij} = 2\mu\varepsilon_{ij}.$$

Similarly, one may write for a non-Newtonian fluid

$$\tau_{ij} = \mu_{app}\left(\frac{\partial u_i}{\partial x_j} + \frac{\partial u_j}{\partial x_i}\right) \tag{14.9}$$

and

$$\tau_{ij} = 2\mu_{app}\varepsilon_{ij}. \tag{14.10}$$

Again, μ_{app} is not constant but some function of the rate of deformation:

$$\mu_{app} = f\left(\varepsilon_{ij}\right). \tag{14.11}$$

Although we do not know a priori the expression for μ_{app}, we may note the following facts: Inspection of Eq. (14.10) reveals that the stress on the left-hand side stands for a second order tensor. The shear rate ε_{ij} on the right-hand side stands also for a second order tensor. In order to preserve the tensorial rank of the equation, it requires that μ_{app} be a scalar, i.e., a tensor of rank zero. The only way for a function of a tensorial quantity to be a scalar is by being a function of the scalar *invariants* of the tensor. The scalar invariants of a tensor are those combinations of the components that do not vary under rotation of the coordinate system.

The three scalar invariants of the rate-of-strain tensor are

$$I_1 = \varepsilon_{ii} = \varepsilon_{11} + \varepsilon_{22} + \varepsilon_{33},$$
$$I_2 = \varepsilon_{ij}\varepsilon_{ij} = \varepsilon_{11}^2 + \varepsilon_{22}^2 + \varepsilon_{33}^2 + 2\varepsilon_{12}^2 + 2\varepsilon_{23}^2 + 2\varepsilon_{13}^2,$$
$$I_3 = \det|\varepsilon_{ij}|,$$

where the third invariant is the determinant of the matrix of ε_{ij}. The first invariant, I_1, equals simply $\nabla \cdot \mathbf{q}$, which vanishes for incompressible fluids. Hence

$$\mu_{app} = f(I_2, I_3).$$

For two-dimensional flows $I_3 = 0$. Hence for this case

$$\mu_{app} = f(I_2). \tag{14.12}$$

Now whatever form this $f(I_2)$ attains, once the flow is unidirectional, it must reduce to a simple one-dimensional form. For example, the apparent viscosity of the power-law model for a two-dimensional flow may be put in the following form which satisfies this condition:

$$\mu_{\text{app}} = K \left| \sqrt{2\varepsilon_{ij}\varepsilon_{ij}} \right|^{n-1}. \tag{14.13}$$

Hence, the stress versus rate-of-strain relationship becomes

$$\tau_{ij} = \left[K \left| \sqrt{2\left(\frac{\partial u}{\partial x}\right)^2 + 2\left(\frac{\partial v}{\partial y}\right)^2 + \left(\frac{\partial u}{\partial y} + \frac{\partial v}{\partial x}\right)^2} \right|^{n-1} \right] \left(\frac{\partial u_i}{\partial x_j} + \frac{\partial u_j}{\partial x_i}\right). \tag{14.14}$$

The scalar expression in the square brackets is, of course, the apparent viscosity. For unidirectional flows where $u = u(y)$, Eq. (14.14) reduces to

$$T_{xy} = \left[K \left| \frac{\partial u}{\partial y} \right|^{n-1} \right] \frac{\partial u}{\partial y}, \tag{14.15}$$

which corresponds to the one-dimensional form of Eq. (14.5).

The constitutive equation for a power-law fluid in polar coordinates is derived similarly. It reads

$$\tau_{r\theta} = \left[K \left| 2\left(\frac{\partial v_r}{\partial r}\right)^2 + 2\left(\frac{1}{r}\frac{\partial v_\theta}{\partial \theta} + \frac{v_r}{r}\right)^2 + \left(r\frac{\partial}{\partial r}\left(\frac{v_\theta}{r}\right) + \frac{1}{r}\frac{\partial v_r}{\partial \theta}\right)^2 \right|^{\frac{n-1}{2}} \right] \left(r\frac{\partial}{\partial r}\left(\frac{v_\theta}{r}\right) + \frac{1}{r}\frac{\partial v_r}{\partial \theta}\right), \tag{14.16}$$

which for a flow with $v_\theta = v_\theta(r)$ simplifies to

$$\tau_{r\theta} = \left[K \left| r\frac{\partial}{\partial r}\left(\frac{v_\theta}{r}\right) \right|^{n-1} \right] \left(r\frac{\partial}{\partial r}\left(\frac{v_\theta}{r}\right)\right). \tag{14.17}$$

Similarly for axisymmetrical flows

$$\tau_{rz} = \left[K \left| 2\left(\frac{\partial v_r}{\partial r}\right)^2 + 2\left(\frac{\partial v_z}{\partial z}\right)^2 + \left(\frac{\partial v_z}{\partial r} + \frac{\partial v_r}{\partial z}\right)^2 \right|^{\frac{n-1}{2}} \right] \left(\frac{\partial v_z}{\partial r} + \frac{\partial v_r}{\partial z}\right), \tag{14.18}$$

and for unidirectional flows, such as the flow in a circular pipe, with $v_z = v_z(r)$,

$$\tau_{rz} = \left[K \left| \frac{\partial v_z}{\partial r} \right|^{n-1} \right] \frac{\partial v_z}{\partial r}. \tag{14.19}$$

The stress versus rate-of-strain relationships given by the constitutive equations may now be substituted into the momentum equations, Eqs. (5.38) - (5.40) or Eq. (5.41), which can be solved for particular cases. The following examples illustrate this point.

Flow of a Power-law Fluid in a Circular Tube

This problem, solved in Chapter 6 for a Newtonian fluid, is considered again using a power-law model. The momentum equations for steady flow in a circular pipe, Eqs. (5.48) - (5.50), simplify to

$$0 = \frac{\partial P}{\partial r}, \tag{6.45}$$

$$0 = \frac{1}{r}\frac{\partial P}{\partial \theta}, \tag{6.46}$$

$$0 = -\frac{\partial P}{\partial z} + \frac{1}{r}\frac{\partial}{\partial r}(r\,\tau_{rz}), \tag{14.20}$$

where $P = p + \rho g h$ is the modified pressure. From Eq. (6.47) we conclude that P is a function of z only and $\partial P/\partial z$ in Eq. (14.20) may be replaced by $\Delta P/\Delta z$. Substitution of τ_{rz} from (14.19) results in

$$\frac{\partial}{\partial r}\left[Kr\left|\frac{\partial v_z}{\partial r}\right|^{n-1}\frac{\partial v_z}{\partial r}\right] = \frac{\Delta P}{\Delta z}r \tag{14.21}$$

with the boundary conditions

$$v_z(R) = 0 \tag{14.22}$$

and

$$\frac{\partial v_z(0)}{\partial r} = 0. \tag{14.23}$$

Equation (14.21) is rewritten in a slightly more convenient form by making use of a sign coefficient ε:

$$\varepsilon = \frac{\partial u/\partial y}{|\partial u/\partial y|}. \tag{14.24}$$

The value of ε is -1 or $+1$, depending on the sign of $\partial v_z/\partial r$. Thus

$$\varepsilon K \frac{\partial}{\partial r}\left[r\left|\frac{\partial v_z}{\partial r}\right|^n\right] = \frac{\Delta P}{\Delta z}r.$$

Integration and the use of the boundary condition Eq. (14.23) result in

$$\left|\frac{\partial v_z}{\partial r}\right|^n = \frac{r}{2K\varepsilon}\frac{\Delta P}{\Delta z}.$$

The absolute value sign may now be dropped because $\partial v_z/\partial r$ is always negative in tube flow, i.e., $\varepsilon = -1$; thus

$$-\frac{\partial v_z}{\partial r} = \left[\frac{1}{2K}\left(-\frac{\Delta P}{\Delta z}\right)\right]^{1/n} r^{1/n}.$$

A second integration yields

$$v_z = \frac{n}{n+1}\left[\frac{R^{n+1}}{2K}\left(-\frac{\Delta P}{\Delta z}\right)\right]^{1/n}\left[1-\left(\frac{r}{R}\right)^{n+1}\right]. \qquad (14.25)$$

The volumetric flow rate is obtained as

$$Q = \int_0^R 2\pi v_z r\, dr = \pi \frac{n}{3n+1}\left[\frac{1}{2K}\left(-\frac{\Delta P}{\Delta z}\right)\right]^{1/n} R^{\frac{3n+1}{n}}. \qquad (14.26)$$

The centerline velocity, v_o, is obtained from Eq. (14.25) by setting $r = 0$:

$$v_o = \frac{n}{n+1}\left[\frac{R^{n+1}}{2K}\left(-\frac{\Delta P}{\Delta z}\right)\right]^{1/n}, \qquad (14.27)$$

which combined with (14.25) yields

$$v_z = v_o\left[1-\left(\frac{r}{R}\right)^{n+1}\right]. \qquad (14.28)$$

The volumetric flow rate is given in terms of the centerline velocity as

$$Q = \frac{n+1}{3n+1}\pi R^2 v_o, \qquad (14.29)$$

and the average velocity is

$$\bar{v} = \frac{Q}{A} = \frac{n+1}{3n+1}v_o. \qquad (14.30)$$

The appropriate expressions in Chapter 6 for Newtonian flow in a tube may be obtained by setting $n = 1$.

Example 14.1

Find the change in the volumetric flowrate of a fluid flowing in a pipe when the pressure drop along the pipe is doubled.
a.   For a Newtonian fluid.
b.   For a power-law fluid with $n = 0.2$.

Solution

The volumetric flowrate given by Eq. (14.27) for a power-law fluid also holds for a Newtonian fluid with $n = 1$. Thus:

a.   For a Newtonian fluid

$$\frac{Q_2}{Q_1} = \frac{\Delta P_2}{\Delta P_1} = 2.$$

b.   For a power-law fluid

$$\frac{Q_2}{Q_1} = \left(\frac{\Delta P_2}{\Delta P_1}\right)^{1/n} = 2^{\frac{1}{0.2}} = 2^5 = 32.$$

Hence, doubling the pressure drop will increase the flowrate 32-fold.

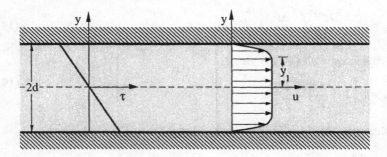

Figure 14.2 Flow of a Bingham plastic between two parallel plates.

Flow of a Bingham Plastic Material between Parallel Plates

Consider the flow of a Bingham plastic material between two stationary parallel plates separated by a distance $2d$, Fig. 14.2. The momentum equation for this case is derived similarly to Eq. (6.21):

$$0 = -\frac{\Delta P}{\Delta x} + \frac{\partial \tau_{xy}}{\partial y} \tag{14.31}$$

with the boundary conditions

$$u(d) = 0, \tag{14.32}$$

$$\frac{\partial u(0)}{\partial y} = 0, \quad \text{or} \quad \tau_{xy}(0) = 0. \tag{14.33}$$

Integration of Eq. (14.31) with the boundary condition Eq. (14.33) results in

$$\tau_{xy} = \left(\frac{\Delta P}{\Delta x}\right)y. \tag{14.34}$$

The shear stress given in Eq. (14.34) is a linear function of y as shown in Fig. 14.2. It has a maximum at $y = -d$, goes through zero at the centerline and has a minimum at $y = +d$. The constitutive relationship for a Bingham plastic is given by Eq. (14.7). Substitution of τ_o for τ_{xy} in Eq. (14.34) yields the distance $\pm y_1$, within which $du/dy = 0$, i.e.,

$$u = u_1 = \text{const}, \quad -y_1 \le y \le y_1, \tag{14.35}$$

with

$$\tau_o = \left(\frac{\Delta P}{\Delta x}\right)y_1, \quad y_1 = \tau_o \Big/\left(\frac{\Delta P}{\Delta x}\right). \tag{14.36}$$

Equation (14.7) is combined with (14.34) resulting in

$$\frac{\Delta P}{\Delta x}y = \tau_o + \mu\frac{du}{dy},$$

which upon substitution of τ_o from Eq. (14.36) yields

$$\frac{\Delta P}{\Delta x}(y - y_1) = \mu\frac{du}{dy}. \tag{14.37}$$

Integration between the limits y_1 and y gives

$$u - u_1 = \frac{1}{2\mu}\frac{\Delta P}{\Delta x}(y - y_1)^2. \tag{14.38}$$

Equation (14.38) holds for the regions $y_1 < y < d$ and $-d < y < y_1$. The constant velocity u_1 may now be evaluated from the boundary condition at $y = d$, Eq. (14.32):

$$u_1 = \frac{1}{2\mu}\left(-\frac{\Delta P}{\Delta x}\right)(d - y_1)^2, \quad -y_1 \le y \le +y_1 \tag{14.39}$$

and finally

$$u = \frac{1}{2\mu}\left(-\frac{\Delta P}{\Delta x}\right)\left[(d - y_1)^2 - (y - y_1)^2\right], \quad \begin{matrix} y > y_1, \\ -y < -y_1. \end{matrix} \tag{14.40}$$

Equation (14.40) reduces to the Newtonian profile for the special case of $y_1 = 0$.

Example 14.2

Many paints behave like a Bingham plastic. For such a paint, derive an expression for the maximum film thickness which just will not flow down a vertical wall.

Solution

Assume that creeping flow conditions apply in this case. With the coordinates of Fig. 14.3, Eqs. (5.42) and (5.45) yield

$$0 = -\frac{\partial p}{\partial x} + g + \frac{1}{\rho}\frac{\partial \tau_{yx}}{\partial y},$$

$$0 = \frac{\partial p}{\partial y}.$$

From the second equation, $p \neq p(y)$, and therefore

$$\frac{\partial p}{\partial x} = \frac{dp}{dx} \neq f(y).$$

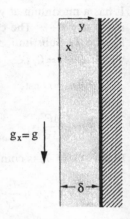

$g_x = g$

Figure 14.3

This equation holds in the bulk as well as on the air–liquid interface, where $dp/dx = 0$. Hence $dp/dx = 0$ holds everywhere. The momentum equation for this film flow simplifies to

$$0 = g + \frac{1}{\rho}\frac{\partial \tau_{yx}}{\partial y}.$$

With the boundary conditions of zero shear at the interface and zero velocity on the solid wall, respectively,

$$\tau_{yx}\big|_{y=0} = 0 \quad \text{and} \quad u(\delta) = 0.$$

Integration of this momentum equation with the first boundary condition yields

$$\tau_{yx} = -\rho g \delta$$

and combining this with the constitutive equation for a Bingham plastic fluid, Eq. (14.7), yields

$$\rho g \delta = -\tau_o + \mu\left(\frac{du}{dy}\right) \qquad \text{for} \quad |\tau| > \tau_o$$

and

$$\frac{du}{dy} = 0 \qquad \text{for} \quad 0 \leq \tau \leq \tau_o.$$

The maximum thickness, δ_{max}, is obtained for $\tau = \tau_o$ and $du/dy = 0$. Hence

$$\rho g \delta_{max} = \tau_o$$

and

$$\delta_{max} = \frac{\tau_o}{\rho g}.$$

A film thicker than δ_{max} results in the sagging of the paint.

The Boundary Layer Equations for a Power-law Fluid

Consider the flow of a power-law fluid in a two-dimensional boundary layer. The momentum and continuity equations for the two-dimensional flow are

$$\rho\left(u\frac{\partial u}{\partial x} + v\frac{\partial u}{\partial y}\right) = -\frac{\partial P}{\partial x} + \frac{\partial \tau_{xx}}{\partial x} + \frac{\partial \tau_{xy}}{\partial y}, \tag{14.41}$$

$$\rho\left(u\frac{\partial v}{\partial x} + v\frac{\partial v}{\partial y}\right) = -\frac{\partial P}{\partial y} + \frac{\partial \tau_{xy}}{\partial x} + \frac{\partial \tau_{yy}}{\partial y}, \tag{14.42}$$

$$\frac{\partial u}{\partial x} + \frac{\partial v}{\partial y} = 0, \tag{14.43}$$

where the components of the stress tensor are given by

$$\tau_{ij} = \left[K\left|\sqrt{2\left(\frac{\partial u}{\partial x}\right)^2 + 2\left(\frac{\partial v}{\partial y}\right)^2 + \left(\frac{\partial u}{\partial y} + \frac{\partial v}{\partial x}\right)^2}\right|^{n-1}\right]\left(\frac{\partial u_i}{\partial x_j} + \frac{\partial u_j}{\partial x_i}\right). \tag{14.14}$$

The considerations of orders of magnitudes that led to the boundary layer equations in Chapter 8 may be repeated here. The order of magnitude analysis simplifies the expression for the apparent viscosity given in the square brackets in Eq. (14.14), and Eq. (14.14) is modified to

$$\tau_{ij} \approx \left[K\left|\frac{\partial u}{\partial y}\right|^{n-1}\right]\left(\frac{\partial u_i}{\partial x_j} + \frac{\partial u_j}{\partial x_i}\right). \tag{14.44}$$

This expression may now be substituted in Eqs. (14.41) and (14.42):

$$\rho\left(u\frac{\partial u}{\partial x} + v\frac{\partial u}{\partial y}\right) = -\frac{\partial P}{\partial x} + \frac{\partial}{\partial x}\left[\varepsilon K\left|\frac{\partial u}{\partial y}\right|^n \frac{2\frac{\partial u}{\partial x}}{\frac{\partial u}{\partial y}}\right] + \frac{\partial}{\partial y}\left[\varepsilon K\left|\frac{\partial u}{\partial y}\right|^n\left(1 + \frac{\frac{\partial v}{\partial x}}{\frac{\partial u}{\partial y}}\right)\right], \tag{14.45}$$

$$\rho\left(u\frac{\partial v}{\partial x}+v\frac{\partial v}{\partial y}\right)=-\frac{\partial P}{\partial y}+\frac{\partial}{\partial x}\left[\varepsilon K\left|\frac{\partial u}{\partial y}\right|^{n}\left(1+\frac{\frac{\partial v}{\partial x}}{\frac{\partial u}{\partial y}}\right)\right]+\frac{\partial}{\partial y}\left[\varepsilon K\left|\frac{\partial u}{\partial y}\right|^{n}\frac{2\frac{\partial v}{\partial x}}{\frac{\partial u}{\partial y}}\right],$$

$$(14.46)$$

where again

$$\varepsilon=\frac{\partial u/\partial y}{|\partial u/\partial y|}.$$

Repeated application of order of magnitude analysis reduces Eq. (14.46) to

$$0=\frac{\partial P}{\partial y},\qquad(14.47)$$

leading to $\partial P/\partial x=dP/dx$ in Eq. (14.45); and Eq. (14.45) reduces to

$$\rho\left(u\frac{\partial u}{\partial x}+v\frac{\partial u}{\partial y}\right)=-\frac{dP}{dx}+K\frac{\partial}{\partial y}\left|\frac{\partial u}{\partial y}\right|^{n}.\qquad(14.48)$$

Equation (14.48) together with the continuity equation, Eq. (14.43), constitute the boundary layer equations for a power-law fluid.

Equation (14.48) may be put in dimensionless form to yield the definition of a Reynolds number for a power-law fluid,

$$\mathrm{Re}=\frac{L^{n}U^{2-n}r}{K}.\qquad(14.49)$$

Again, for $n=1$, this reduces to the Reynolds number for a Newtonian fluid.

References

R. B. Bird, R.C. Armstrong and O. Hassager, "Dynamics of Polymeric Liquids," Vol. 1, "Fluid Mechanics," Wiley, New York, 1977.

J. A. Brydson, "Flow Properties of Polymer Melts," Van Nostrand, London, 1970.

B.D. Coleman, H. Markowitz and W. Noll, "Viscometric Flows of Non-Newtonian Fluids," Springer-Verlag, New York, 1966.

S. Middleman, "The Flow of High Polymers," Wiley-Interscience, New York, 1968.

M. Reiner, "Deformation, Strain and Flow," Wiley-Interscience, New York, 1960.

A.V. Tobolsky, "Properties and Structure of Polymers," Wiley, New York, 1960.

Problems

14.1 Show that for the laminar flow of a Bingham plastic fluid the velocity distribution in a circular pipe is given by

$$u = \frac{1}{\mu}\left[\frac{\Delta P}{4L}\left(R^2 - r^2\right) - \tau_o(R - r)\right], \qquad r_o \leq r < R,$$

and

$$u = \frac{1}{\mu}\left[\frac{\Delta P\,R^2}{4L} + \frac{\tau_o^2}{\Delta P / L} - \tau_o R\right] = \text{const}, \qquad 0 \leq r < r_o,$$

where r_o is the pipe radius corresponding to τ_o, $\tau_o \geq 0$. Further, show that this may be integrated to give

$$Q = \frac{\pi R^4 \Delta P}{8\mu L}\left[1 - \frac{8}{3}\frac{L\tau_o}{\Delta P\,R} + \frac{1}{3}\left(\frac{2L\tau_o}{PR}\right)^4\right].$$

14.2 Consider film flow of a Bingham plastic material on the vertical wall described in Example 14.2. Derive expressions for the velocity profile, average velocity and volumetric flowrate for this case.

14.3 The following data relate the flow rate Q to the pressure drop ΔP for the flow of a non-Newtonian fluid through a capillary tube of diameter 2 mm and of length 25 cm.

ΔP (N/cm^2)	12	16	20	24	27	29	30
Q (cm^3/s)	0.02	0.055	0.107	0.193	0.306	0.490	0.70

Use the data to predict the pressure drop required to cause the same fluid to flow at $u_{ave} = 0.1$ m/s through a pipe with a diameter of 3 cm and a length of 100 m.

14.4 It is conventional to represent the momentum associated with fluid passing through a given cross section of a tube as $\dot{m}V$, where $\dot{m}$ is the mass flowrate and V is the average fluid velocity. In reality a velocity distribution exists, and a factor β should be introduced to take this into account, i.e., $M = \beta \dot{m} V$. Determine the numerical value of β for the laminar flow of a power-law fluid in a circular pipe.

14.5 A power-law fluid flows in a pipe which has the diameter D. The same fluid flows in the gap between two parallel flat plates. The size of the gap is also D. The same pressure gradient exists in both systems. Calculate the ratio between the flowrate of the fluid in the pipe and the flowrate between the plates per width of D. Perform your calculations for $n = 0.2$, 0.4, 0.8. Compare your results with those of Problem 6.15.

14.6 The pipe and the plates of Problem 14.5 are now used such that the flowrate in the pipe is the same as that between the plates per width D. Calculate the ratio between the power needed to pump the fluid through the pipe and that needed to pump between the plates.

14.7 A power-law fluid with $n = 0.5$ and $K = 0.001$ (in SI units) flows at the rate of $0.05\,\text{m}^3/\text{s}$ through the annular gap with an inner radius of $0.10\,\text{m}$ and an outer radius of $0.12\,\text{m}$. The length of the annulus is $100\,\text{m}$. Assume the flow laminar and fully developed.
   a. Find the pressure drop along the annulus.
   b. Find the magnitude and location of the maximum velocity in the annular gap.
   c. Compare your results with those of Problem 6.25.

14.8 Measurements of pressure drop along a fully developed flow in a pipe of diameter 0.05 m gave the following readings:

Q (m³/s)	0.001	0.002
$\Delta P/L$ (Pa/m)	10^4	1.5×10^4

Assuming a power-law fluid, find K and n.

14.9 For the power-law fluid of Problem 14.8 find the volumetric flowrates in a pipe of 0.1 m diameter for pressure drops of 0.5×10^4 Pa/m and of 2.0×10^4 Pa/m.

14.10 In a plane shear flow the upper plate is 0.1 m above the lower plate and moves at 1.0 m/s. The fluid is that of Problem 14.8. Find the shear stress on the plates.

14.11 The layer of a power-law fluid ($K = 10$, $n = 1.2$) in a plane shear flow between two parallel plates is 0.01 m thick. The shear stress on the plates is 100 N/m². Find how fast the upper plate moves.

14.12 The lower plate in the shear flow system of Problem 14.11 is cooled while the upper plate is insulated. As a result the power-law fluid changes its n value from 0.8 at the upper plate to 1.0 at the lower plate, linearly.

    a. Find the velocity profile between the plates.

    b. Find the speed of the upper plate.

APPENDIXES

APPENDIX A: UNIT CONVERSION FACTORS

Each of the following conversion factors is dimensionless and equals unity. Hence, any expression may be multiplied or divided by any of these factors without change in. its physical magnitude.

Length

2.540 cm/in.	3.281 ft/m	1.609 km/mi	12 in./ft
30.480 cm/ft	5,280 ft/mi	10^{10} Å/m	3 ft/yd

Area

6.452 $cm^2/in.^2$	10.764 ft^2/m^2	144 $in.^2/ft^2$	929.0 cm^2/ft^2

Volume

16.387 $cm^3/in.^3$	28.317 L/ft^3	7.48 gal/ft^3
3.7845 L/gal	1,728 $in.^3/ft^3$	4 qt/gal

Mass

453.59 g/lbm	14.594 kg/slug	2,000 lbm/ton
2.2046 lbm/kg	32.174 lbm/slug	28.349 g/oz

Force

9.807 N/kgf	7.233 poundal/N	10^5 dyn/N
4.448 N/lbf	32.174 poundal/lbf	2.205 lbf/kgf

Density

$$62.428 \frac{\text{lbm} / \text{ft}^3}{\text{g} / \text{cm}^3} \qquad 1,000 \frac{\text{kg} / \text{m}^3}{\text{g} / \text{cm}^3} \qquad 32.174 \frac{\text{lbm} / \text{ft}^3}{\text{slug} / \text{ft}^3}$$

$$1.9403 \frac{\text{slug} / \text{ft}^3}{\text{g} / \text{cm}^3} \qquad 16.018 \frac{\text{kg} / \text{m}^3}{\text{lbm} / \text{ft}^3} \qquad 8.345 \frac{\text{lbm} / \text{gal}}{\text{g} / \text{cm}^3}$$

Viscosity

$$100 \frac{\text{cp}}{\text{g} / (\text{cm} \cdot \text{s})} \qquad 1,000 \frac{\text{cp}}{\text{kg} / (\text{m} \cdot \text{s})} \qquad 0.6723 \frac{\text{lbm} / (\text{ft} \cdot \text{s})}{\text{kg} / (\text{m} \cdot \text{s})}$$

$$2.42 \frac{\text{lbm} / (\text{ft} \cdot \text{h})}{\text{cp}} \qquad 0.06723 \frac{\text{lbm} / (\text{ft} \cdot \text{s})}{\text{g} / (\text{cm} \cdot \text{s})}$$

Thermal Conductivity

$$1.730278 \frac{\text{W/m} \cdot {}^\circ\text{C}}{\text{Btu/h} \cdot \text{ft} \cdot {}^\circ\text{F}} \qquad 241.9 \frac{\text{Btu/h} \cdot \text{ft} \cdot {}^\circ\text{F}}{\text{cal/s} \cdot \text{cm} \cdot {}^\circ\text{C}}$$

Pressure

$$1.01325 \frac{\text{bars}}{\text{atm}} \qquad 1.0332 \frac{\text{kgf} / \text{cm}^2}{\text{atm}} \qquad 144 \frac{\text{lbf} / \text{ft}^2}{\text{psi}} \qquad 27.71 \frac{\text{in.} \cdot \text{H}_2\text{O}}{\text{psi}}$$

$$10^5 \frac{\text{N} / \text{m}^2}{\text{bar}} \qquad 14.696 \frac{\text{psi}}{\text{atm}} \qquad 0.1 \frac{\text{dyn} / \text{cm}^2}{\text{Pa}} \qquad 760 \frac{\text{mm Hg}}{\text{atm}}$$

$$100 \frac{\text{kPa}}{\text{bar}} \qquad 2,116.2 \frac{\text{lbf} / \text{ft}^2}{\text{atm}} \qquad 2.036 \frac{\text{in.} \cdot \text{Hg}}{\text{psi}} \qquad 6.8949 \frac{\text{kPa}}{\text{psi}}$$

Energy

$$1.8 \frac{\text{Btu} / \text{lbm}}{\text{kcal} / \text{kg}} \qquad 550.0 \frac{\text{ft} \cdot \text{lbf} / \text{s}}{\text{hp}} \qquad 737.56 \frac{\text{ft} \cdot \text{lbf} / \text{s}}{\text{kW}}$$

4.18676 kJ / kcal   3,412.8 Btu / kWh   1.35582 J / (ft · lbf)

2.6552×10^6 ft · lbf / kWh   3,600.0 kJ / kWh   0.25199 kcal / Btu

1.3410 hp / kW   778.16 ft · lbf / Btu   860.0 kcal / kWh

0.252 kcal / Btu   2,544.46 Btu / (hp · h)   101.92 kgf · m / kJ

1.05505 kJ / Btu 1.98×10^6 ft · lbf / (hp · h)   0.746 kW / hp

APPENDIX B: PHYSICAL PROPERTIES OF WATER AND AIR

TABLE B-1: PROPERTIES OF WATER

Temperature T [°C]	Specific weight γ [N/m^3]	Density ρ [kg/m^3]	Viscosity μ [kg/m·s]	Kinematic viscosity ν [m^2/s]	Surface tension σ [N/m]
0	9,805	999.9	1.792×10^{-3}	1.792×10^{-6}	7.62×10^{-2}
5	9,806	1000.0	1.519	1.519	7.54
10	9,803	999.7	1.308	1.308	7.48
15	9,798	999.1	1.140	1.141	7.41
20	9,789	998.2	1.005	1.007	7.36
25	9,779	997.1	0.894	0.897	7.26
30	9,767	995.7	0.801	0.804	7.18
35	9,752	994.1	0.723	0.727	7.10
40	9,737	992.2	0.656	0.660	7.01
45	9,720	990.2	0.599	0.605	6.92
50	9,697	988.1	0.549	0.556	6.82
60	9,658	983.2	0.469	0.477	6.68
70	9,600	977.8	0.406	0.415	6.50
80	9,557	971.8	0.357	0.367	6.30
90	9,499	965.3	0.317	0.328	6.12
100	9,438	958.4	0.284	0.296	5.94

TABLE B-2: PROPERTIES OF AIR AT ATMOSPHERIC PRESSURE

Temperature		Density	Viscosity	Kinematic viscosity
T		ρ	μ	ν
[°C]	[K]	[kg/m³]	[kg/m·s]	[m²/s]
-50	223	1.582	1.46×10^{-5}	0.921×10^{-5}
-40	233	1.514	1.51	0.998
-30	243	1.452	1.56	1.08
-20	253	1.394	1.61	1.16
-10	263	1.342	1.67	1.24
0	273	1.292	1.72	1.33
10	283	1.247	1.76	1.42
20	293	1.204	1.81	1.51
30	303	1.164	1.86	1.60
40	313	1.127	1.91	1.69
50	323	1.092	1.95	1.79
60	333	1.060	2.00	1.89
70	343	1.030	2.05	1.99
80	353	1.000	2.09	2.09
90	363	0.973	2.13	2.19
100	373	0.946	2.17	2.30
150	423	0.834	2.38	2.85
200	473	0.746	2.57	3.45
250	523	0.675	2.75	4.08
300	573	0.616	2.93	4.75

APPENDIX C: PROPERTIES OF SOME COMMON FLUIDS

(At 300 K and atmospheric pressure)

Fluid	Density ρ [kg/m^3]	Viscosity μ [kg/m·s]	Kinematic viscosity ν [m^2/s]	Thermal conductivity k [W/m·K]	Thermal diffusivity α [m^2/s]	Prandtl number Pr
Air	1.1774	1.846×10^{-5}	1.568×10^{-5}	0.0262	0.222×10^{-4}	0.708
Hydrogen	0.0819	8.963×10^{-6}	1.095×10^{-4}	0.182	1.554×10^{-4}	0.706
Carbon dioxide	1.797	1.496×10^{-5}	8.321×10^{-6}	0.0166	0.106×10^{-4}	0.770
Water	995.7	8.60×10^{-4}	8.64×10^{-7}	0.614	1.475×10^{-4}	5.85
Mercury	13,580	1.55×10^{-3}	1.14×10^{-7}	8.690	4.606×10^{-6}	0.025
Glycerin	1,264	1.4915	1.18×10^{-3}	0.286	0.947×10^{-7}	21.5
Engine oil *	840	1.752×10^{-2}	2.03×10^{-5}	0.137	0.738×10^{-4}	276

---

* At 373 K.

APPENDIX D: COMPRESSIBLE FLOW DATA FOR IDEAL GASES

TABLE D-1:   COMPRESSIBLE NOZZLE FLOW, $k = c_p/c_v = 1.4000$

$$\text{Mass Flux} = 0.6847\, p_o A^*/\sqrt{RT_o}$$

$$p^*/p_o = 0.5283 \quad T^*/T_o = 0.8333 \quad \rho^*/\rho_o = 0.6339$$

SUBSONIC FLOW, $k = c_p/c_v = 1.4000$

M	T/T_o	p/p_o	ρ/ρ_o	A/A^*	M	T/T_o	p/p_o	ρ/ρ_o	A/A
0.010	1.000	1.000	1.000	57.874	0.020	1.000	1.000	1.000	28.9
0.030	1.000	0.999	1.000	19.301	0.040	1.000	0.999	0.999	14.4
0.050	1.000	0.998	0.999	11.591	0.060	0.999	0.997	0.998	9.6
0.070	0.999	0.997	0.998	8.292	0.080	0.999	0.996	0.997	7.2
0.090	0.998	0.994	0.996	6.461	0.100	0.998	0.993	0.995	5.8
0.110	0.998	0.992	0.994	5.299	0.120	0.997	0.990	0.993	4.8
0.130	0.997	0.988	0.992	4.497	0.140	0.996	0.986	0.990	4.1
0.150	0.996	0.984	0.989	3.910	0.160	0.995	0.982	0.987	3.6
0.170	0.994	0.980	0.986	3.464	0.180	0.994	0.978	0.984	3.2
0.190	0.993	0.975	0.982	3.112	0.200	0.992	0.972	0.980	2.9
0.210	0.991	0.970	0.978	2.829	0.220	0.990	0.967	0.976	2.
0.230	0.990	0.964	0.974	2.597	0.240	0.989	0.961	0.972	2.
0.250	0.988	0.957	0.969	2.403	0.260	0.987	0.954	0.967	2.
0.270	0.986	0.951	0.964	2.238	0.280	0.985	0.947	0.962	2.
0.290	0.983	0.943	0.959	2.098	0.300	0.982	0.939	0.956	2.
0.310	0.981	0.936	0.954	1.977	0.320	0.980	0.932	0.951	1.
0.330	0.979	0.927	0.948	1.871	0.340	0.977	0.923	0.944	1.
0.350	0.976	0.919	0.941	1.778	0.360	0.975	0.914	0.938	1.
0.370	0.973	0.910	0.935	1.696	0.380	0.972	0.905	0.931	1.
0.390	0.970	0.900	0.928	1.623	0.400	0.969	0.896	0.924	1.
0.410	0.967	0.891	0.921	1.559	0.420	0.966	0.886	0.917	1.
0.430	0.964	0.881	0.913	1.501	0.440	0.963	0.875	0.909	1.
0.450	0.961	0.870	0.906	1.449	0.460	0.959	0.865	0.902	1
0.470	0.958	0.860	0.898	1.402	0.480	0.956	0.854	0.893	1
0.490	0.954	0.849	0.889	1.359	0.500	0.952	0.843	0.885	1
0.510	0.951	0.837	0.881	1.321	0.520	0.949	0.832	0.877	1
0.530	0.947	0.826	0.872	1.286	0.540	0.945	0.820	0.868	1
0.550	0.943	0.814	0.863	1.255	0.560	0.941	0.808	0.859	1
0.570	0.939	0.802	0.854	1.226	0.580	0.937	0.796	0.850	1
0.590	0.935	0.790	0.845	1.200	0.600	0.933	0.784	0.840	1
0.610	0.931	0.778	0.836	1.177	0.620	0.929	0.772	0.831	1
0.630	0.926	0.765	0.826	1.155	0.640	0.924	0.759	0.821	
0.650	0.922	0.753	0.816	1.136	0.660	0.920	0.746	0.812	
0.670	0.918	0.740	0.807	1.118	0.680	0.915	0.734	0.802	
0.690	0.913	0.727	0.797	1.102	0.700	0.911	0.721	0.792	
0.710	0.908	0.714	0.787	1.087	0.720	0.906	0.708	0.781	
0.730	0.904	0.702	0.776	1.074	0.740	0.901	0.695	0.771	

SUBSONIC FLOW, $k = c_p/c_v = 1.4000$ (cont.)

M	T/T_o	p/p_o	ρ/ρ_o	A/A^*	M	T/T_o	p/p_o	ρ/ρ_o	A/A^*
0.750	0.899	0.689	0.766	1.062	0.760	0.896	0.682	0.761	1.057
0.770	0.894	0.676	0.756	1.052	0.780	0.892	0.669	0.750	1.047
0.790	0.889	0.663	0.745	1.043	0.800	0.887	0.656	0.740	1.038
0.810	0.884	0.650	0.735	1.034	0.820	0.881	0.643	0.729	1.030
0.830	0.879	0.636	0.724	1.027	0.840	0.876	0.630	0.719	1.024
0.850	0.874	0.624	0.714	1.021	0.860	0.871	0.617	0.708	1.018
0.870	0.869	0.611	0.703	1.015	0.880	0.866	0.604	0.698	1.013
0.890	0.863	0.598	0.692	1.011	0.900	0.861	0.591	0.687	1.009
0.910	0.858	0.585	0.682	1.007	0.920	0.855	0.578	0.676	1.006
0.930	0.853	0.572	0.671	1.004	0.940	0.850	0.566	0.666	1.003
0.950	0.847	0.559	0.660	1.002	0.960	0.844	0.553	0.655	1.001
0.970	0.842	0.547	0.650	1.001	0.980	0.839	0.541	0.645	1.000
0.990	0.836	0.534	0.639	1.000	1.000	0.833	0.528	0.634	1.000

SUPERSONIC FLOW, $k = c_p/c_v = 1.4000$

M	T/T_o	p/p_o	ρ/ρ_o	A/A^*	ν	M_2	p_2/p_1	T_2/T_1	$\rho_2/$
1.10	0.8052	0.4684	0.5817	1.0079	0.02	0.9118	1.2450	1.0649	1.16
1.20	0.7764	0.4124	0.5311	1.0304	0.06	0.8422	1.5133	1.1280	1.34
1.30	0.7474	0.3609	0.4829	1.0663	0.11	0.7860	1.8050	1.1909	1.51
1.40	0.7184	0.3142	0.4374	1.1149	0.16	0.7397	2.1200	1.2547	1.68
1.50	0.6897	0.2724	0.3950	1.1762	0.21	0.7011	2.4583	1.3202	1.86
1.60	0.6614	0.2353	0.3557	1.2502	0.26	0.6684	2.8200	1.3880	2.03
1.70	0.6337	0.2026	0.3197	1.3376	0.31	0.6405	3.2050	1.4583	2.19
1.80	0.6068	0.1740	0.2868	1.4390	0.36	0.6165	3.6133	1.5316	2.35
1.90	0.5807	0.1492	0.2570	1.5553	0.41	0.5956	4.0450	1.6079	2.51
2.00	0.5556	0.1278	0.2300	1.6875	0.46	0.5774	4.5000	1.6875	2.66
2.10	0.5313	0.1094	0.2058	1.8369	0.51	0.5613	4.9783	1.7704	2.81
2.20	0.5081	0.0935	0.1841	2.0050	0.55	0.5471	5.4800	1.8569	2.95
2.30	0.4859	0.0800	0.1646	2.1931	0.60	0.5344	6.0050	1.9468	3.08
2.40	0.4647	0.0684	0.1472	2.4031	0.64	0.5231	6.5533	2.0403	3.21
2.50	0.4444	0.0585	0.1317	2.6367	0.68	0.5130	7.1250	2.1375	3.33
2.60	0.4252	0.0501	0.1179	2.8960	0.72	0.5039	7.7200	2.2383	3.44
2.70	0.4068	0.0430	0.1056	3.1830	0.76	0.4956	8.3383	2.3429	3.5
2.80	0.3894	0.0368	0.0946	3.5001	0.80	0.4882	8.9800	2.4512	3.6
2.90	0.3729	0.0317	0.0849	3.8498	0.83	0.4814	9.6450	2.5632	3.7
3.00	0.3571	0.0272	0.0762	4.2346	0.87	0.4752	10.3333	2.6790	3.8
3.10	0.3422	0.0234	0.0685	4.6573	0.90	0.4695	11.0450	2.7986	3.9
3.20	0.3281	0.0202	0.0617	5.1209	0.93	0.4643	11.7800	2.9220	4.0
3.30	0.3147	0.0175	0.0555	5.6286	0.96	0.4596	12.5383	3.0492	4.1
3.40	0.3019	0.0151	0.0501	6.1837	0.99	0.4552	13.3200	3.1802	4.1
3.50	0.2899	0.0131	0.0452	6.7896	1.02	0.4512	14.1250	3.3150	4.2
3.60	0.2784	0.0114	0.0409	7.4501	1.05	0.4474	14.9533	3.4537	4.3
3.70	0.2675	0.0099	0.0370	8.1691	1.08	0.4439	15.8050	3.5962	4.3
3.80	0.2572	0.0086	0.0335	8.9506	1.10	0.4407	16.6800	3.7426	4.4
3.90	0.2474	0.0075	0.0304	9.7990	1.12	0.4377	17.5783	3.8928	4.5
4.00	0.2381	0.0066	0.0277	10.7187	1.15	0.4350	18.5000	4.0469	4.5
4.10	0.2293	0.0058	0.0252	11.7146	1.17	0.4324	19.4450	4.2048	4.6
4.20	0.2208	0.0051	0.0229	12.7916	1.19	0.4299	20.4133	4.3666	4.6
4.30	0.2129	0.0044	0.0209	13.9549	1.21	0.4277	21.4050	4.5322	4.7
4.40	0.2053	0.0039	0.0191	15.2098	1.23	0.4255	22.4200	4.7017	4.7
4.50	0.1980	0.0035	0.0174	16.5622	1.25	0.4236	23.4583	4.8751	4.8
4.60	0.1911	0.0031	0.0160	18.0178	1.27	0.4217	24.5200	5.0523	4.8
4.70	0.1846	0.0027	0.0146	19.5828	1.29	0.4199	25.6050	5.2334	4.8
4.80	0.1783	0.0024	0.0134	21.2637	1.31	0.4183	26.7133	5.4184	4.9
4.90	0.1724	0.0021	0.0123	23.0671	1.33	0.4167	27.8450	5.6073	4.9
5.00	0.1667	0.0019	0.0113	24.9999	1.34	0.4152	29.0000	5.8000	5.
5.10	0.1612	0.0017	0.0104	27.0695	1.36	0.4138	30.1783	5.9966	5.

TABLE   D-2:   COMPRESSIBLE NOZZLE FLOW, $k = c_p/c_v = 1.6667$

Mass Flux $= 0.7262\, p_o A^*/\sqrt{RT_o}$

$p^*/p_o = 0.4871$ $T^*/T_o = 0.7500$ $\rho^*/\rho_o = 0.6495$

SUBSONIC FLOW, $k = c_p/c_v = 1.6667$

M	T/T_o	p/p_o	ρ/ρ_o	A/A^*	M	T/T_o	p/p_o	ρ/ρ_o	A/A^*
0.010	1.000	1.000	1.000	56.254	0.020	1.000	1.000	1.000	28.132
0.030	1.000	0.999	1.000	18.761	0.040	0.999	0.999	0.999	14.077
0.050	0.999	0.998	0.999	11.269	0.060	0.999	0.997	0.998	9.397
0.070	0.998	0.996	0.998	8.062	0.080	0.998	0.995	0.997	7.061
0.090	0.997	0.993	0.996	6.284	0.100	0.997	0.992	0.995	5.663
0.110	0.996	0.990	0.994	5.155	0.120	0.995	0.988	0.993	4.733
0.130	0.994	0.986	0.992	4.376	0.140	0.994	0.984	0.990	4.071
0.150	0.993	0.981	0.989	3.806	0.160	0.992	0.979	0.987	3.576
0.170	0.990	0.976	0.986	3.373	0.180	0.989	0.974	0.984	3.193
0.190	0.988	0.971	0.982	3.032	0.200	0.987	0.967	0.980	2.888
0.210	0.986	0.964	0.978	2.758	0.220	0.984	0.961	0.976	2.640
0.230	0.983	0.957	0.974	2.533	0.240	0.981	0.954	0.972	2.435
0.250	0.980	0.950	0.970	2.345	0.260	0.978	0.946	0.967	2.262
0.270	0.976	0.942	0.965	2.186	0.280	0.975	0.938	0.962	2.115
0.290	0.973	0.933	0.959	2.050	0.300	0.971	0.929	0.957	1.989
0.310	0.969	0.924	0.954	1.933	0.320	0.967	0.920	0.951	1.880
0.330	0.965	0.915	0.948	1.831	0.340	0.963	0.910	0.945	1.784
0.350	0.961	0.905	0.942	1.741	0.360	0.959	0.900	0.939	1.700
0.370	0.956	0.894	0.935	1.662	0.380	0.954	0.889	0.932	1.626
0.390	0.952	0.884	0.929	1.592	0.400	0.949	0.878	0.925	1.560
0.410	0.947	0.873	0.921	1.530	0.420	0.944	0.867	0.918	1.501
0.430	0.942	0.861	0.914	1.474	0.440	0.939	0.855	0.910	1.449
0.450	0.937	0.849	0.907	1.424	0.460	0.934	0.843	0.903	1.401
0.470	0.931	0.837	0.899	1.380	0.480	0.929	0.831	0.895	1.359
0.490	0.926	0.825	0.891	1.339	0.500	0.923	0.819	0.887	1.320
0.510	0.920	0.812	0.883	1.302	0.520	0.917	0.806	0.879	1.286
0.530	0.914	0.800	0.874	1.269	0.540	0.911	0.793	0.870	1.254
0.550	0.908	0.786	0.866	1.239	0.560	0.905	0.780	0.861	1.225
0.570	0.902	0.773	0.857	1.212	0.580	0.899	0.767	0.853	1.200
0.590	0.896	0.760	0.848	1.187	0.600	0.893	0.753	0.844	1.176
0.610	0.890	0.747	0.839	1.165	0.620	0.886	0.740	0.835	1.155
0.630	0.883	0.733	0.830	1.145	0.640	0.880	0.726	0.825	1.135
0.650	0.877	0.719	0.821	1.126	0.660	0.873	0.713	0.816	1.118
0.670	0.870	0.706	0.811	1.110	0.680	0.866	0.699	0.807	1.102
0.690	0.863	0.692	0.802	1.094	0.700	0.860	0.685	0.797	1.088
0.710	0.856	0.678	0.792	1.081	0.720	0.853	0.671	0.787	1.075
0.730	0.849	0.664	0.783	1.069	0.740	0.846	0.658	0.778	1.063
0.750	0.842	0.651	0.773	1.058	0.760	0.839	0.644	0.768	1.053
0.770	0.835	0.637	0.763	1.048	0.780	0.831	0.630	0.758	1.043
0.790	0.828	0.623	0.753	1.039	0.800	0.824	0.617	0.748	1.035

SUBSONIC FLOW, $k = c_p/c_v = 1.6667$ (cont.)

M	T/T_o	p/p_o	ρ/ρ_o	A/A^*	M	T/T_o	p/p_o	ρ/ρ_o	A/A^*
0.810	0.821	0.610	0.743	1.031	0.820	0.817	0.603	0.738	1.02
0.830	0.813	0.596	0.733	1.025	0.840	0.810	0.590	0.728	1.02
0.850	0.806	0.583	0.723	1.019	0.860	0.802	0.576	0.719	1.01
0.870	0.799	0.570	0.714	1.014	0.880	0.795	0.563	0.709	1.01
0.890	0.791	0.557	0.704	1.010	0.900	0.787	0.550	0.699	1.00
0.910	0.784	0.544	0.694	1.006	0.920	0.780	0.537	0.689	1.00
0.930	0.776	0.531	0.684	1.004	0.940	0.772	0.524	0.679	1.00
0.950	0.769	0.518	0.674	1.002	0.960	0.765	0.512	0.669	1.00
0.970	0.761	0.506	0.664	1.001	0.980	0.757	0.499	0.659	1.00
0.990	0.754	0.493	0.654	1.000	1.000	0.750	0.487	0.650	1.00

SUPERSONIC FLOW, $k = c_p/c_v = 1.6667$

M	T/T_o	p/p_o	ρ/ρ_o	A/A^*	v	M_2	p_2/p_1	T_2/T_1	ρ_2/ρ_1
1.10	0.7126	0.4286	0.6015	1.0071	0.02	0.9131	1.2625	1.0982	1.1496
1.20	0.6757	0.3753	0.5554	1.0267	0.05	0.8462	1.5500	1.1948	1.2973
1.30	0.6396	0.3272	0.5116	1.0575	0.09	0.7934	1.8625	1.2922	1.4414
1.40	0.6048	0.2845	0.4704	1.0983	0.14	0.7508	2.2000	1.3919	1.5806
1.50	0.5714	0.2468	0.4320	1.1484	0.18	0.7157	2.5625	1.4948	1.7143
1.60	0.5396	0.2139	0.3963	1.2076	0.22	0.6864	2.9500	1.6018	1.8417
1.70	0.5093	0.1851	0.3635	1.2754	0.26	0.6618	3.3625	1.7133	1.9626
1.80	0.4808	0.1603	0.3334	1.3520	0.30	0.6407	3.8000	1.8297	2.0769
1.90	0.4538	0.1388	0.3058	1.4372	0.34	0.6226	4.2625	1.9512	2.1845
2.00	0.4286	0.1202	0.2806	1.5312	0.38	0.6070	4.7500	2.0782	2.2857
2.10	0.4048	0.1043	0.2576	1.6341	0.42	0.5933	5.2625	2.2107	2.3805
2.20	0.3826	0.0906	0.2367	1.7462	0.45	0.5813	5.8000	2.3488	2.4693
2.30	0.3619	0.0788	0.2177	1.8675	0.48	0.5707	6.3625	2.4927	2.5524
2.40	0.3425	0.0686	0.2004	1.9983	0.52	0.5613	6.9500	2.6425	2.6301
2.50	0.3243	0.0599	0.1847	2.1390	0.55	0.5530	7.5625	2.7982	2.7026
2.60	0.3074	0.0524	0.1704	2.2898	0.58	0.5455	8.2000	2.9598	2.7704
2.70	0.2915	0.0459	0.1574	2.4510	0.60	0.5388	8.8626	3.1275	2.8338
2.80	0.2767	0.0403	0.1456	2.6228	0.63	0.5327	9.5501	3.3012	2.8929
2.90	0.2629	0.0354	0.1348	2.8057	0.66	0.5272	10.2626	3.4809	2.9482
3.00	0.2500	0.0313	0.1250	2.9999	0.68	0.5222	11.0001	3.6668	2.9999
3.10	0.2379	0.0276	0.1160	3.2058	0.70	0.5177	11.7626	3.8587	3.0483
3.20	0.2266	0.0244	0.1079	3.4236	0.72	0.5136	12.5501	4.0568	3.0936
3.30	0.2160	0.0217	0.1004	3.6539	0.75	0.5098	13.3626	4.2610	3.1360
3.40	0.2060	0.0193	0.0935	3.8968	0.77	0.5063	14.2001	4.4714	3.1757
3.50	0.1967	0.0172	0.0873	4.1527	0.79	0.5031	15.0626	4.6880	3.2130
3.60	0.1880	0.0153	0.0815	4.4220	0.80	0.5002	15.9501	4.9107	3.2480
3.70	0.1797	0.0137	0.0762	4.7051	0.82	0.4974	16.8626	5.1396	3.2809
3.80	0.1720	0.0123	0.0713	5.0023	0.84	0.4949	17.8001	5.3747	3.3118
3.90	0.1647	0.0110	0.0669	5.3139	0.85	0.4926	18.7626	5.6160	3.3409
4.00	0.1579	0.0099	0.0627	5.6403	0.87	0.4904	19.7501	5.8635	3.3683
4.10	0.1514	0.0089	0.0589	5.9819	0.89	0.4884	20.7626	6.1172	3.3941
4.20	0.1453	0.0081	0.0554	6.3390	0.90	0.4865	21.8001	6.3771	3.4185
4.30	0.1396	0.0073	0.0522	6.7121	0.91	0.4848	22.8626	6.6432	3.4415
4.40	0.1342	0.0066	0.0491	7.1014	0.93	0.4831	23.9501	6.9156	3.4632
4.50	0.1290	0.0060	0.0464	7.5073	0.94	0.4816	25.0626	7.1941	3.4838
4.60	0.1242	0.0054	0.0438	7.9302	0.95	0.4801	26.2002	7.4789	3.5032
4.70	0.1196	0.0049	0.0413	8.3705	0.96	0.4788	27.3627	7.7699	3.5216
4.80	0.1152	0.0045	0.0391	8.8285	0.98	0.4775	28.5502	8.0672	3.5391
4.90	0.1111	0.0041	0.0370	9.3046	0.99	0.4763	29.7627	8.3706	3.5556
5.00	0.1071	0.0038	0.0351	9.7992	1.00	0.4752	31.0002	8.6803	3.5713
5.10	0.1034	0.0034	0.0333	10.3126	1.01	0.4741	32.2627	8.9963	3.5862

TABLE D-3: COMPRESSIBLE NOZZLE FLOW, $k = c_p/c_v = 1.3000$

Mass Flux $= 0.6673\, p_o A*/\sqrt{RT_o}$

$p*/p_o = 0.5457 \qquad T*/T_o = 0.8696 \qquad \rho*/\rho_o = 0.6276$

SUBSONIC FLOW, $\quad k = c_p/c_v = 1.3000$

M	T/T_o	p/p_o	ρ/ρ_o	$A/A*$	M	T/T_o	p/p_o	ρ/ρ_o	$A/A*$
0.010	1.000	1.000	1.000	58.526	0.020	1.000	1.000	1.000	29.26
0.030	1.000	0.999	1.000	19.518	0.040	1.000	0.999	0.999	14.64
0.050	1.000	0.998	0.999	11.721	0.060	0.999	0.998	0.998	9.77
0.070	0.999	0.997	0.998	8.384	0.080	0.999	0.996	0.997	7.34
0.090	0.999	0.995	0.996	6.533	0.100	0.999	0.994	0.995	5.88
0.110	0.998	0.992	0.994	5.357	0.120	0.998	0.991	0.993	4.91
0.130	0.997	0.989	0.992	4.546	0.140	0.997	0.987	0.990	4.22
0.150	0.997	0.986	0.989	3.952	0.160	0.996	0.984	0.987	3.71
0.170	0.996	0.981	0.986	3.500	0.180	0.995	0.979	0.984	3.31
0.190	0.995	0.977	0.982	3.145	0.200	0.994	0.974	0.980	2.99
0.210	0.993	0.972	0.978	2.858	0.220	0.993	0.969	0.976	2.7
0.230	0.992	0.966	0.974	2.623	0.240	0.991	0.963	0.972	2.5
0.250	0.991	0.960	0.969	2.426	0.260	0.990	0.957	0.967	2.3
0.270	0.989	0.954	0.964	2.260	0.280	0.988	0.951	0.962	2.1
0.290	0.988	0.947	0.959	2.117	0.300	0.987	0.944	0.956	2.0
0.310	0.986	0.940	0.953	1.994	0.320	0.985	0.936	0.950	1.9
0.330	0.984	0.932	0.947	1.887	0.340	0.983	0.928	0.944	1.8
0.350	0.982	0.924	0.941	1.793	0.360	0.981	0.920	0.938	1.7
0.370	0.980	0.916	0.934	1.710	0.380	0.979	0.911	0.931	1.6
0.390	0.978	0.907	0.928	1.636	0.400	0.977	0.902	0.924	1.6
0.410	0.975	0.898	0.920	1.570	0.420	0.974	0.893	0.917	1.5
0.430	0.973	0.888	0.913	1.511	0.440	0.972	0.883	0.909	1.4
0.450	0.971	0.878	0.905	1.459	0.460	0.969	0.873	0.901	1.4
0.470	0.968	0.868	0.897	1.411	0.480	0.967	0.863	0.893	1.3
0.490	0.965	0.858	0.889	1.368	0.500	0.964	0.853	0.885	1.3
0.510	0.962	0.847	0.880	1.329	0.520	0.961	0.842	0.876	1.3
0.530	0.960	0.836	0.871	1.293	0.540	0.958	0.831	0.867	1.2
0.550	0.957	0.825	0.863	1.261	0.560	0.955	0.819	0.858	1.2
0.570	0.954	0.814	0.853	1.232	0.580	0.952	0.808	0.849	1.2
0.590	0.950	0.802	0.844	1.206	0.600	0.949	0.796	0.839	1.
0.610	0.947	0.790	0.834	1.181	0.620	0.945	0.784	0.830	1.
0.630	0.944	0.778	0.825	1.159	0.640	0.942	0.772	0.820	1.
0.650	0.940	0.766	0.815	1.139	0.660	0.939	0.760	0.810	1.
0.670	0.937	0.754	0.805	1.121	0.680	0.935	0.748	0.800	1.
0.690	0.933	0.742	0.795	1.105	0.700	0.932	0.735	0.789	1.
0.710	0.930	0.729	0.784	1.090	0.720	0.928	0.723	0.779	1.
0.730	0.926	0.717	0.774	1.077	0.740	0.924	0.710	0.769	1.
0.750	0.922	0.704	0.763	1.064	0.760	0.920	0.698	0.758	1.
0.770	0.918	0.691	0.753	1.054	0.780	0.916	0.685	0.747	1.
0.790	0.914	0.679	0.742	1.044	0.800	0.912	0.672	0.737	1.
0.810	0.910	0.666	0.731	1.035	0.820	0.908	0.659	0.726	1.

SUBSONIC FLOW, $k = c_p/c_v = 1.3000$ (cont.)

M	T/T_o	p/p_o	ρ/ρ_o	A/A*	M	T/T_o	p/p_o	ρ/ρ_o	A/A*
0.830	0.906	0.653	0.721	1.028	0.840	0.904	0.647	0.715	1.025
0.850	0.902	0.640	0.710	1.021	0.860	0.900	0.634	0.704	1.019
0.870	0.898	0.628	0.699	1.016	0.880	0.896	0.621	0.693	1.013
0.890	0.894	0.615	0.688	1.011	0.900	0.892	0.608	0.682	1.009
0.910	0.890	0.602	0.677	1.007	0.920	0.887	0.596	0.671	1.006
0.930	0.885	0.589	0.666	1.004	0.940	0.883	0.583	0.660	1.003
0.950	0.881	0.577	0.655	1.002	0.960	0.879	0.571	0.649	1.001
0.970	0.876	0.564	0.644	1.001	0.980	0.874	0.558	0.639	1.000
0.990	0.872	0.552	0.633	1.000	1.000	0.870	0.546	0.628	1.000

SUPERSONIC FLOW, $k = c_p/c_v = 1.3000$

M	T/T_o	p/p_o	ρ/ρ_o	A/A^*	ν	M_2	p_2/p_1	T_2/T_1	$\rho_2/$
1.10	0.8464	0.4854	0.5735	1.0083	0.02	0.9112	1.2374	1.0506	1.1:
1.20	0.8224	0.4285	0.5211	1.0321	0.07	0.8403	1.4974	1.0995	1.3(
1.30	0.7978	0.3757	0.4709	1.0703	0.11	0.7825	1.7800	1.1480	1.5:
1.40	0.7728	0.3273	0.4235	1.1227	0.17	0.7346	2.0852	1.1971	1.7
1.50	0.7477	0.2836	0.3793	1.1895	0.22	0.6942	2.4130	1.2473	1.9
1.60	0.7225	0.2446	0.3385	1.2712	0.28	0.6599	2.7635	1.2991	2.1:
1.70	0.6976	0.2100	0.3011	1.3690	0.33	0.6304	3.1365	1.3529	2.3
1.80	0.6729	0.1797	0.2671	1.4841	0.39	0.6048	3.5322	1.4087	2.5-
1.90	0.6487	0.1533	0.2363	1.6182	0.45	0.5825	3.9504	1.4668	2.6'
2.00	0.6250	0.1305	0.2087	1.7732	0.50	0.5629	4.3913	1.5274	2.8
2.10	0.6019	0.1108	0.1841	1.9514	0.55	0.5455	4.8548	1.5905	3.0
2.20	0.5794	0.0939	0.1621	2.1556	0.61	0.5301	5.3409	1.6562	3.2
2.30	0.5576	0.0795	0.1427	2.3885	0.66	0.5163	5.8496	1.7245	3.3
2.40	0.5365	0.0673	0.1255	2.6535	0.71	0.5040	6.3809	1.7956	3.5
2.50	0.5161	0.0569	0.1103	2.9545	0.75	0.4929	6.9348	1.8694	3.7
2.60	0.4965	0.0481	0.0969	3.2954	0.80	0.4829	7.5113	1.9459	3.8
2.70	0.4777	0.0407	0.0852	3.6811	0.85	0.4738	8.1104	2.0253	4.C
2.80	0.4596	0.0344	0.0749	4.1165	0.89	0.4655	8.7322	2.1075	4.1
2.90	0.4422	0.0291	0.0659	4.6073	0.93	0.4580	9.3765	2.1925	4.2
3.00	0.4255	0.0247	0.0580	5.1598	0.97	0.4511	10.0435	2.2804	4.4
3.10	0.4096	0.0209	0.0510	5.7807	1.01	0.4448	10.7330	2.3711	4.!
3.20	0.3943	0.0177	0.0450	6.4776	1.05	0.4389	11.4452	2.4648	4.(
3.30	0.3797	0.0151	0.0396	7.2586	1.09	0.4336	12.1800	2.5613	4.:
3.40	0.3658	0.0128	0.0350	8.1328	1.12	0.4287	12.9374	2.6607	4.!
3.50	0.3524	0.0109	0.0309	9.1098	1.16	0.4241	13.7174	2.7630	4.!
3.60	0.3397	0.0093	0.0273	10.2004	1.19	0.4199	14.5200	2.8681	5.
3.70	0.3275	0.0079	0.0242	11.4160	1.22	0.4160	15.3452	2.9762	5.
3.80	0.3159	0.0068	0.0215	12.7693	1.25	0.4123	16.1930	3.0873	5.
3.90	0.3047	0.0058	0.0190	14.2737	1.28	0.4089	17.0635	3.2012	5.
4.00	0.2941	0.0050	0.0169	15.9441	1.31	0.4058	17.9565	3.3180	5.
4.10	0.2840	0.0043	0.0151	17.7963	1.34	0.4028	18.8722	3.4378	5.
4.20	0.2743	0.0037	0.0134	19.8475	1.37	0.4000	19.8104	3.5605	5.
4.30	0.2650	0.0032	0.0120	22.1161	1.39	0.3975	20.7713	3.6861	5.
4.40	0.2561	0.0027	0.0107	24.6221	1.42	0.3950	21.7548	3.8147	5.
4.50	0.2477	0.0024	0.0095	27.3869	1.44	0.3927	22.7608	3.9462	5
4.60	0.2396	0.0020	0.0085	30.4334	1.47	0.3906	23.7895	4.0806	5
4.70	0.2318	0.0018	0.0077	33.7863	1.49	0.3886	24.8408	4.2180	5
4.80	0.2244	0.0015	0.0069	37.4719	1.51	0.3867	25.9147	4.3582	5
4.90	0.2173	0.0013	0.0062	41.5185	1.53	0.3849	27.0113	4.5015	6
5.00	0.2105	0.0012	0.0056	45.9564	1.56	0.3832	28.1304	4.6476	6
5.10	0.2040	0.0010	0.0050	50.8177	1.58	0.3816	29.2721	4.7967	6

TABLE D-4: FANNO LINE, COMPRESSIBLE PIPE FLOW

$$k = c_p/c_v = 1.400$$

$f(L/D)$ = Dimensionless distance to point of $M = 1$

M	T/T_o	p/p_o	R/R_o	$f(L/D)$
0.01	1.000	1.000	1.000	7134.405
0.02	1.000	1.000	1.000	1778.450
0.04	1.000	1.000	0.999	440.352
0.05	1.000	0.999	0.999	280.020
0.10	0.998	0.997	0.995	66.922
0.15	0.996	0.994	0.989	27.932
0.20	0.992	0.989	0.980	14.533
0.25	0.988	0.983	0.969	8.483
0.30	0.982	0.975	0.956	5.299
0.35	0.976	0.967	0.941	3.452
0.40	0.969	0.957	0.924	2.308
0.45	0.961	0.946	0.906	1.566
0.50	0.952	0.934	0.885	1.069
0.55	0.943	0.921	0.863	0.728
0.60	0.933	0.907	0.840	0.491
0.65	0.922	0.893	0.816	0.325
0.70	0.911	0.877	0.792	0.208
0.75	0.899	0.861	0.766	0.127
0.80	0.887	0.845	0.740	0.072
0.85	0.874	0.828	0.714	0.036
0.90	0.861	0.810	0.687	0.015
0.95	0.847	0.793	0.660	0.003
1.00	0.833	0.775	0.634	0.000
1.25	0.762	0.683	0.507	0.049
1.50	0.690	0.594	0.395	0.136
1.75	0.620	0.512	0.303	0.225
2.00	0.556	0.439	0.230	0.305
2.25	0.497	0.376	0.174	0.374
2.50	0.444	0.321	0.132	0.432
2.75	0.398	0.275	0.100	0.481
3.00	0.357	0.237	0.076	0.522
3.25	0.321	0.204	0.059	0.557
3.50	0.290	0.177	0.045	0.586
3.75	0.262	0.154	0.035	0.612
4.00	0.238	0.134	0.028	0.633
4.25	0.217	0.118	0.022	0.652
4.50	0.198	0.104	0.017	0.668
4.75	0.181	0.092	0.014	0.682
5.00	0.167	0.081	0.011	0.694
5.25	0.154	0.073	0.009	0.705
5.50	0.142	0.065	0.008	0.714
5.75	0.131	0.058	0.006	0.722
6.00	0.122	0.053	0.005	0.730
6.25	0.113	0.048	0.004	0.737
6.50	0.106	0.043	0.004	0.743
7.00	0.093	0.036	0.003	0.753

TABLE D-5: FANNO LINE, COMPRESSIBLE PIPE FLOW
$$k = c_p/c_v = 1.667$$
$f(L/D)$ = Dimensionless distance to point of $M = 1$

M	T/T_o	p/p_o	R/R_o	$f(L/D)$
0.01	1.000	1.000	1.000	5992.262
0.02	1.000	1.000	1.000	1493.371
0.03	1.000	1.000	1.000	660.686
0.04	0.999	0.999	0.999	369.480
0.05	0.999	0.999	0.999	234.836
0.10	0.997	0.994	0.995	55.943
0.15	0.993	0.988	0.989	23.255
0.20	0.987	0.978	0.980	12.044
0.25	0.980	0.966	0.970	6.996
0.30	0.971	0.952	0.957	4.347
0.35	0.961	0.935	0.942	2.816
0.40	0.949	0.917	0.925	1.873
0.45	0.937	0.897	0.907	1.263
0.50	0.923	0.875	0.887	0.857
0.55	0.908	0.852	0.866	0.580
0.60	0.893	0.828	0.844	0.389
0.65	0.877	0.803	0.821	0.256
0.70	0.860	0.777	0.797	0.163
0.75	0.842	0.751	0.773	0.099
0.80	0.824	0.724	0.748	0.056
0.85	0.806	0.698	0.723	0.028
0.90	0.787	0.671	0.699	0.011
0.95	0.769	0.645	0.674	0.002
1.00	0.750	0.619	0.650	0.000
1.25	0.658	0.497	0.533	0.036
1.50	0.571	0.393	0.432	0.098
1.75	0.495	0.310	0.348	0.159
2.00	0.429	0.244	0.281	0.211
2.25	0.372	0.192	0.227	0.255
2.50	0.324	0.153	0.185	0.291
2.75	0.284	0.123	0.151	0.321
3.00	0.250	0.099	0.125	0.346
3.25	0.221	0.081	0.104	0.366
3.50	0.197	0.067	0.087	0.383
3.75	0.176	0.055	0.074	0.397
4.00	0.158	0.046	0.063	0.409
4.25	0.142	0.039	0.054	0.419
4.50	0.129	0.033	0.046	0.428
4.75	0.117	0.028	0.040	0.436
5.00	0.107	0.024	0.035	0.442
5.25	0.098	0.021	0.031	0.448
5.50	0.090	0.018	0.027	0.453
5.75	0.083	0.016	0.024	0.458
6.00	0.077	0.014	0.021	0.462
6.25	0.071	0.012	0.019	0.465
6.50	0.066	0.011	0.017	0.468
6.75	0.062	0.010	0.015	0.471
7.00	0.058	0.009	0.014	0.474

TABLE D-6: FANNO LINE, COMPRESSIBLE PIPE FLOW
$$k = c_p/c_v = 1.300$$
$f(L/D)$ = Dimensionless distance to point of $M = 1$

M	T/T_o	p/p_o	R/R_o	$f(L/D)$
0.01	1.000	1.000	1.000	7683.515
0.02	1.000	1.000	1.000	1915.510
0.03	1.000	1.000	1.000	847.851
0.04	1.000	1.000	0.999	474.429
0.05	1.000	1.000	0.999	301.746
0.10	0.999	0.998	0.995	72.202
0.15	0.997	0.996	0.989	30.183
0.20	0.994	0.992	0.980	15.732
0.25	0.991	0.988	0.969	9.201
0.30	0.987	0.983	0.956	5.759
0.35	0.982	0.977	0.941	3.760
0.40	0.977	0.970	0.924	2.520
0.45	0.971	0.962	0.905	1.714
0.50	0.964	0.953	0.885	1.172
0.55	0.957	0.944	0.863	0.800
0.60	0.949	0.934	0.839	0.541
0.65	0.940	0.923	0.815	0.359
0.70	0.932	0.912	0.789	0.230
0.75	0.922	0.900	0.763	0.141
0.80	0.912	0.888	0.737	0.080
0.85	0.902	0.875	0.710	0.041
0.90	0.892	0.862	0.682	0.016
0.95	0.881	0.848	0.655	0.004
1.00	0.870	0.834	0.628	0.000
1.25	0.810	0.761	0.496	0.055
1.50	0.748	0.685	0.379	0.156
1.75	0.685	0.612	0.284	0.261
2.00	0.625	0.543	0.209	0.357
2.25	0.568	0.480	0.152	0.441
2.50	0.516	0.423	0.110	0.514
2.75	0.469	0.373	0.080	0.575
3.25	0.387	0.291	0.042	0.673
3.50	0.352	0.258	0.031	0.711
3.75	0.322	0.229	0.023	0.744
4.00	0.294	0.204	0.017	0.773
4.25	0.270	0.182	0.013	0.797
4.50	0.248	0.163	0.010	0.819
4.75	0.228	0.146	0.007	0.838
5.00	0.211	0.132	0.006	0.854
5.25	0.195	0.119	0.004	0.869
5.50	0.181	0.108	0.003	0.882
5.75	0.168	0.098	0.003	0.893
6.00	0.156	0.090	0.002	0.904
6.25	0.146	0.082	0.002	0.913
6.50	0.136	0.075	0.001	0.921
6.75	0.128	0.069	0.001	0.929
7.00	0.120	0.063	0.001	0.935

INDEX

Printed in the United States
By Bookmasters